Sex and
Scientific
Inquiry

Sex and Scientific Inquiry

Edited by
Sandra
Harding
and
Jean F.
O'Barr

The University of Chicago Press
Chicago and London

The material in this volume originally appeared in various issues of SIGNS: JOURNAL OF WOMEN IN CULTURE AND SOCIETY. Acknowledgments of the original publication date can be found on the first page of each article.

The University of Chicago Press, Chicago 60637
The University of Chicago Press, Ltd., London
© 1975, 1978, 1980, 1982, 1983, 1986, 1987 by The University of Chicago
All rights reserved. Published 1987
Printed in the United States of America
96 95 94 93 92 91 90 89   87 5 4 3 2

Library of Congress Cataloging-in-Publication Data
Sex and scientific inquiry.
    Bibliography: p.
    Includes index.
    1. Women in science. 2. Feminism. 3. Science—
Social aspects. I. Harding, Sandra G. II. O'Barr,
Jean F.
HQ1397.S49 1987   305.4'2   86-27213
ISBN 0-226-31626-2 (alk. paper)
ISBN 0-226-31627-0 (pbk.:alk. paper)

The paper used in this publication meets the minimum requirements of American National Standard for Information Sciences—Permanence of Paper for Printed Library Materials. ANSI Z39.48-1984. ∞™

Cover design by Christine Leonard Raquepaw.

Contents

Acknowledgments

The staff of *Signs,* Mary Wyer and Anne Vilen—as well as Marguerite Rogers, Sharon Klavins, Amy Thomas, and Robin Skaist—have made important contributions to this collection. We are deeply grateful to all of them for the time and care they have put into the project.

As editors we dedicate the book to the women and men who have struggled on behalf of women in science that they might find a place and make their contributions.

Introduction

Science it would seem is not sexless; she is a man, a father and infected too. [VIRGINIA WOOLF, *Three Guineas*, 1938]

Though Virginia Woolf was not the first to remark on sex bias in the social institution we call science, her comment fifty years ago does foreshadow the range and the depth of the critique that has only recently emerged from the combined efforts of biologists, sociologists, historians, philosophers, political theorists, and psychologists, as well as scholars and researchers in other disciplines.

This critique asks all of us to reconsider science—its pervasive rationality and its widely disseminated applications and technologies—in light of what we have learned about gender symbols and images and the characteristic sex-segregated practices of our culture. Are the masculinity and the infection to which Woolf referred intrinsic to the scientific enterprise, as some have claimed? Or can they be eliminated, leaving a residue of claims, practices, and meanings that can be used for the kinds of emancipatory ends envisioned by many of science's most stalwart defenders?

From its very first year, *Signs* has played a significant role in generating and publishing a broad array of material on gender and science. The essays reprinted here are selected from nearly fifty that *Signs* has published on the natural sciences. The issues addressed in these essays reflect five major focuses of the more general feminist science critiques. We have organized the essays reprinted here to reflect this diversity.

In the first section, the social structure of science is critically examined by two historians and complemented by photographs of students at the Women's Medical College of Pennsylvania around 1900. We open this collection with Londa Schiebinger's very recent essay because it begins by reflecting on several streams in the study of the social structure of science. It helps the reader to map the political terrain within each critique, as well as build links between these critiques and the others that follow. She points out that we can detect three distinct political perspectives within each of the conceptual approaches to science she examines: the view that women cannot be great scientists; the view that women should get their fair share of power within the sciences as they now exist; and the view that women "have at this historical moment a unique opportunity to make a difference" in the practices and goals of science. She raises a central issue for many of the authors in this volume: "The question is: who will accommodate whom? Will women mold their values and methods to accommodate science, or will science mold its methods and practices to accommodate women?" (9).

Margaret W. Rossiter's 1978 paper on occupational segregation in the sciences, which we reprint next, is based on work that she subsequently published in her influential historical study, *Women Scientists in America: Struggles and Strategies to 1940* (Baltimore: Johns Hopkins University Press, 1982)—one of the social studies of science Schiebinger discusses. In the paper, Rossiter discusses some of the social forces that have succeeded in channeling women scientists into those fields that are most open or receptive to them. Sandra L. Chaff's four photographs of women medical students give us glimpses of the courageous women who first entered medicine and hints of the areas of health service in which they might have gone on to practice. We see the medical student who is reported to have become the first Black physician in South Carolina.

Another focus of feminist concern has been the applications of science, its technologies, and their social meanings. The essays reprinted in this section represent three different approaches to these topics. "Was the female experience of technological change significantly different from the male experience?" Judith A. McGaw's essay reviews research on the history of American technology that suggests women have experienced technological changes in ways specific to gender-based assumptions. In identifying many of the questions earlier scholars left unasked or unanswered, McGaw also provides a rich list of projects for future research.

A different kind of history of technology is provided by Sally G. Allen and Joanna Hubbs's analysis of an important seventeenth-century alchemical treatise. They argue that the "particularly vehement absorption and denial of the feminine by the masculine" that we can see in the images of this historic text may be related to the masculine meanings of the successes of the young sciences of the day. Harnessing the energies of nature "for the benefit of humanity" appears to require the mastery of the feminine—a theme we will see other scholars explore in later sections of this volume. Finally, Inez Smith Reid's searing indictment of the racist uses of the results of genetic and psychological research challenges feminists to insist on "a new scientific responsibility, one which sheds the shield of indifference for a more humane armor which protects not only the scientist in his investigation but also humanity from the capricious uses of scientific inquiry" (124).

The next section properly contains the most essays. Challenges to the claimed "value neutrality" of science have emerged most forcefully from critical scrutiny of particular theories in biology and the social sciences that draw on biological models. Critics of these theories have identified androcentric and/or sexist biases in the selection of problems and the definition of what is problematic, the construction of concepts and theories, the design of research, the collection of evidence, and the interpretation of results of research. Many of the feminist criticisms of the social sciences proper—of sociology, psychology, economics, history, and anthropology—also have important implications for the *natural* sciences. After all, the history, practices, and meanings of the natural sciences are properly the subject matter for social inquiry. If history's focus on only public actors and dramatic situations distorts our understandings of the effect of women's activities on history, then we probably have a distorted understanding of the *causes* of many events in the history of science. [If sociology's assumption that "the community" in any culture is structured primarily by men's relationships with each other distorts our understandings of how communities function, then we probably have a distorted vision of how scientific communities function.] In many other ways, feminist inquiry challenges the prevailing historical, economic, sociological, and psychological understandings of even those sciences that at first appear the most immune from feminist critique.

We begin this section on bias in the sciences with Helen H. Lambert's reminder that even though the literature of sex differences is indeed sexist in its elaborations of various forms of biological determinism, it is a mistake for feminists to ground our demands for social justice on the claim that sex differences are due entirely to culture rather than partially to biology. She warns us that this terrain is itself a sexist one: "If a particular sex difference is incompatible with important aspects of social equality, we should argue for compensatory measures, independent of biological causation" (145). In the next paper, Patricia Y. Miller and Martha R. Fowlkes critically survey the various behavioral and social models of sexuality that have developed in opposition to the Freudian model. While the research studies by Kinsey, Masters and Johnson, Gagnon and Simon, and others reveal that "the modernization of sex has advanced under the standard of democracy," the authors find in these studies "scant attention [paid] to the sexuality of adult women in the context of their normative social roles and relationships" (148). The authors argue that "there is every reason to believe that a woman's sexuality is emergent and takes on new meanings and qualities in the context of the emergent marriage relationship itself" (156). They also critically examine the limited and sometimes distorting images these studies provide of lesbian sexuality.

Helen Longino and Ruth Doell carefully examine the various moments in inquiry where sexism and androcentrism can enter the research process. Their examples are evolutionary studies and endocrinological research on behavioral sex differences. They argue that because there are so many different ways in which masculine bias expresses itself in science, feminists can call on a variety of tactical responses. "It is not necessary for us to turn our backs on science as a whole or to condemn it as an enterprise. In a number of ways, the logical structure of science itself provides opportunities for the expression of the creative and self-conscious sensibility that has characterized recent feminist attempts to transform the sciences" (186).

The next two essays examine the histories of particular scientific theories. Stephanie A. Shields reports on the rise and, finally, demise of the thesis that men's natural variability in mental powers made them more likely candidates than women for scholarly work. Not surprisingly, this thesis was enthusiastically endorsed exactly during the period in the nineteenth century when women began to enter universities in significant numbers. In her conclusion, she raises the issue Lambert addressed: "If one acknowledges the biological bases of behavior, must one also assume their inevitability and immutability? Is a feminist/humanist scientific model nonbiological by definition?" (215). Donna Haraway is certainly not the first observer of science to point out how science reads into nature the social relations supported by scientists and then insists we all find "natural" those social relations. But her essays on the history of primatology have made vivid some precise ways in which this has occurred during the twentieth century. In her widely cited paper reprinted here, she scrutinizes the work of Clarence Ray Carpenter, who linked functionalism in sociology and biology in his studies of dominant/submissive behavior in rhesus monkeys and thereby promoted "scientific management of every phase of society" (224). Carpenter's studies had the consequence that "the political principle of domination has been transformed here into the legitimating scientific principle of dominance as a natural property with a physical-chemical base" (231). Haraway argues that we cannot transform our sciences without also transforming the cultures in which they exist. "If our experience is of domination, we will theorize our lives according to principles of dominance. As we transform the foundations of our lives, we will know how to build natural sciences to underpin new relations with the world" (232).

The next section turns to the sexual meanings of science. Using techniques borrowed from literary criticism, psychoanalysis, and interpretive social science, a number of feminist observers of science have begun to "read the text" of Western science's descriptions, explanations, and institutional practices to discover the moral and political resources patriarchal science and society have provided for each other. We have already seen examples of this third focus of feminist critique in earlier essays—particularly in those by Londa Schiebinger and by Sally G. Allen and Joanna Hubbs. In the first paper included in this section, Evelyn Fox Keller draws on psychoanalytic insights to argue that it is a specifically male consciousness that projects an emphasis on power and control in the rhetoric of Western science. Nevertheless, she thinks that there need not be a conflict between our commitment to feminism and our commitment to science. In order to envision a different science— one "less restrained by the impulse to dominate—we need only look to the thematic pluralism in the history of our own science" (245). There we can find alternative nonhierarchical images of the transformative science feminism seeks. Susan Bordo identifies the specific gender contours of "a certain instability, a dark underside, to the bold rationalist vision" of the seventeenth-century French philosopher, Descartes (247). This "great Cartesian anxiety [is] manifestly expressed in epistemological terms, as anxiety over separation from the organic female universe of the Middle Ages and the Renaissance. Cartesian objectivism, correspondingly, will be explored as a defensive response to that separation anxiety, an aggressive intellectual 'flight from the feminine' rather than (simply) the confident articulation of a positive new epistemological ideal" (248). Both Bordo and Keller are showing us the mutually supportive relationship between men's understandings of gender relations and the sexual meanings of science: each reinforces the other.

In the final section on the epistemology and metatheory of science, Hilary Rose traces the history of the "prefeminist" critiques of science emerging from orthodox Marxism and then from the radical social movements of the 1970s. "Indifferent to the second system of production—reproduction—[this] analysis excluded the relationship of science to patriarchy" (265). It is left to feminism to develop an alternative epistemology for the natural sciences, one that arises from a feminist understanding of women's social experience and that heals the damaging divisions in nonfeminist theory and practice between body, mind, and emotions—between "hand, brain, and heart." Rose points to the emergence of such an epistemology in the theories and practices of the women's health movement. Finally, Sandra Harding argues that we should try to preserve a number of the tensions and contradictions that exist within feminist theory since these represent ways in which feminist thought is attempting to break out of and transform the categories of the patriarchal theories we inherit. Her examples come from feminist science discussions, where she focuses on five such oppositions.

Many insightful analyses could not be included in this collection. A list of all the feminist critiques of the natural sciences published in *Signs* appears on page 305. We have attempted to demonstrate the wide range of the feminist critique of the natural sciences through reprinting analyses that represent still current assumptions and interests and that present the most recent data. We have also tried to select papers that would be especially interesting to readers new to these issues.

To a certain extent the differing approaches reflect the differing concerns of groups of feminists. Moreover, the criticisms emerging from one group are not always appreciated by feminists in another group. For instance, the many women scientists who have struggled persistently, fiercely, and at great personal cost to create opportunities for more women to enter the higher echelons of science are often distressed by charges that the whole social hierarchy of science must be reorganized before science will be free of androcentrism. Others protest that the "misuses and abuses" of science and its technologies have nothing to do with the inherent characteristics of the "pure science" they practice:

these are matters of individual pathology and public policy. Still other scientists are skeptical of the value of the critiques of science's sexual meanings and of science's epistemology because such critiques appear to them to have little to do with the practice of science. It should not be surprising if, as has been the case for other institutions charged with androcentrism, feminists inside and outside this institution have different but equally important contributions to make to the improvement of science.

There are other tensions between the critiques collected here. We think they are valuable, for they reveal the different ways in which this influential institution affects the condition of women both within science and within the larger culture that science helps to generate, maintain, and justify. We hope that the juxtaposition of all five kinds of critique in this anthology contributes to a stimulating and fruitful dialogue across these critical boundaries.

The History and Philosophy of Women in Science: A Review Essay

Londa Schiebinger

> Until now there has been a longstanding gap in the
> history of the natural sciences. . . . There has been no
> historical and evaluative survey of all the women, who
> from the earliest times until our own have distinguished
> themselves in the various sciences.[1] [CHRISTIAN
> FRIEDRICH HARLESS, 1830]

We often congratulate ourselves on the gains women are making in the
sciences. Every day, it seems, more female faces are appearing in the
laboratories of official science. Yet when we consider that the question of
women's place in science is an old one, dating back to Christian Harless's
Germany of 1830 and beyond, our congratulations may turn to concern
or even dismay. If women have been working in science over a number of
centuries, why, we might ask ourselves, does women's participation in
science remain limited? Why so few?[2]

A number of colleagues and friends read and commented on this essay; my thanks to
Robert Proctor, Autumn Stanley, Carolyn Merchant, Leora Auslander, members of the
Values, Technology, Science, and Society Forum, Stanford University, and the participants
in the "Women, Technology, and Education" conference, Rockefeller Foundation Conference Center, Bellagio, Italy, Autumn, 1985.

1. Christian Friedrich Harless, *Die Verdienste der Frauen um Naturwissenschaft, Gesund-
sheits- und Heilkunde, so wie auch um Länder- Völker- und Menschenkunde von der ältesten Zeit bis
auf die neueste* (Göttingen: Van den Hoeck-Ruprecht, 1830), ix.

2. In 1982, women represented about one-quarter of all scientists (National Science
Foundation, *Women and Minorities in Science and Engineering* [January 1984], 1). This figure

This essay originally appeared in *Signs*, vol. 12, no. 2, Winter 1987.

From these dual moods of optimism and dismay there has emerged a new field of study: the history and philosophy of women in science. Those working in this field—a small but growing number—write from different perspectives, reflecting varied expertise, interests, strategies, and political orientations. Though the study of gender and science has historical roots, it has become an important field only within the last ten years, for several reasons. There is, first, the growing number of women scientists. Since the Sputnik years, more women than ever before have gone into science. Second, the field of gender and science arose out of the contemporary women's movement and a general concern for women's position in the professions.[3] Third, the study of gender-based distortions within science emerged along with contemporary critiques of science. Since the 1960s, many have been concerned with the dangers of modern science, in particular its military connections and its impact on the environment.

A survey of the field of gender and science requires examination of very diverse literatures. What interests me here is not only what we can learn by studying the strengths and weaknesses of each type of literature considered in isolation but also what we can learn by considering the relationships among the different approaches.

To map the field of gender and science,[4] I identify four conceptual

---

should be set in perspective; women constitute 52 percent of the population and 45 percent of the U.S. labor force. See Alice Rossi, "Women in Science: Why So Few?" *Science* 148, no. 3,674 (1965): 1196–1202.

3. In the mid-1960s, the women's movement fostered a greater concern for the participation of women in science. The women's movement encouraged women to pursue "nontraditional" careers, and, at the same time, new antidiscrimination legislation made it possible for women to pursue those careers. Donna Haraway, however, attributes the growing numbers of women in science not to the women's movement of the 1960s and 1970s but, rather, to the growing needs of high-tech industries, fostered by the Cold War and the nuclear arms race. As Haraway puts it, the need for more science and more science personnel required the enlistment of women into the scientific enterprise. See Donna Haraway, "Class, Race, Sex, Scientific Objects of Knowledge: A Socialist-Feminist Perspective on the Social Construction of Productive Nature and Some Political Consequences," in *Women in Scientific and Engineering Professions*, ed. Violet Haas and Carolyn Perrucci (Ann Arbor: University of Michigan Press, 1984), 212–29, esp. 212–13.

4. For bibliographies on women and science, see Patricia Siegel and Kay Finley, *Women in the Scientific Search: An American Bio-bibliography, 1724–1979* (Metuchen, N.J.: Scarecrow Press, 1985); Audrey B. Davis, *Bibliography on Women: With Special Emphasis on Their Roles in Science and Society* (New York: Science History Publications, 1974); Phyllis Zweig Chinn, *Women in Science and Mathematics: Bibliography* (Arcata, Calif.: Humboldt State University, 1980); Ruth Hubbard, Mary Sue Henifin, and Barbara Fried, eds., *Biological Woman—the Convenient Myth: A Collection of Feminist Essays and a Comprehensive Bibliography* (Cambridge, Mass.: Schenkman Publishing Co., 1982), esp. 289–376. See also Michele Aldrich, "Review Essay: Women in Science," *Signs: Journal of Women in Culture and Society* 4, no. 1 (1978): 126–35. While I do not intend to review the vast literature on women and medicine, there is an excellent bibliography on that topic by Sandra Chaff et al., *Women in Medicine: A Bibliography of the Literature on Women Physicians* (Metuchen, N.J.: Scarecrow Press, 1977).

approaches. The first, involving literature from Christine de Pizan to Margaret Rossiter, seeks to recover the achievements of Hypatia's sisters, to brush off the dust of obscurity from those women whose scientific contributions have been neglected by mainstream (or in Mary O'Brien's phrase, "malestream") historians of science.[5] The second approach complements the first by analyzing the history of women's participation in the institutions of science, focusing on the history of women's (limited) access to the means of scientific production and on the current status of women within the scientific profession. The third approach looks at how the sciences—most often the biological and medical sciences—have defined (and misdefined) the nature of women. The fourth approach analyzes the masculinist nature of science and seeks to unveil distortions in the very norms and methods of science that have resulted from the historic absence of women from any significant role in the making of modern science.[6]

In addition to looking at these four approaches, it is also my purpose to contrast three fundamental points of view within the four approaches. The first is that women simply cannot do science as well as men—that something in the physical, psychological, and intellectual nature of women prohibits them from producing great science. The second standpoint sees the absence of women from science as an access-to-education-and-employment issue and advocates a straightforward integration of women into science. The third standpoint maintains that it is not enough for women to be scientists if science is to proceed along its present course; the task of opening science to women must be combined with the task of making science more responsible. From their position as outsiders, women (like other "outsiders," ethnic minorities and nonelites) have at this historical moment an opportunity to make a difference. Women have a choice: they can enter science and carry on "science as usual," or they can use their culturally defined difference as a platform for critique. The question is, Who will accommodate whom? Will women mold their values and methods to accommodate science, or will science mold its methods and practices to accommodate women?

Not So Few: In Search of Lost Women

The first of my four approaches to gender and science, the history of women's contributions to the sciences, fits best under the rubric of

5. Mary O'Brien, *The Politics of Reproduction* (London: Routledge & Kegan Paul, 1981), 5.

6. Carolyn Merchant has suggested a similar typology for the study of the history and philosophy of women in science. The first approach is to consider the role of gender in the modern scientific worldview—in the methods and ideals of science. The second is to focus on the role played in science by language and image, in particular the image of nature as feminine. The third is to analyze the role of women in science vis-à-vis the role of women in society at large ("Isis' Consciousness Raised," *Isis* 73, no. 268 [1982]: 398–409).

women's history and shares with this field the goal of recovering the history of women's achievements. Authors of recent monographs and biographies argue that women's contributions have been greater than generally recognized, whether this be the early discovery of the process of distillation (by Maria the Jewess) or more recent discoveries of the mechanisms of genetic transposition (by Barbara McClintock).

Although many consider the field of the history of women scientists a phenomenon of the 1970s, the question of women's presence in science is not a new one. Already in 1405, Christine de Pizan asked if women had made original contributions in the arts and sciences:

> I realize that you are able to cite numerous and frequent cases of women learned in the sciences and the arts. But I would then ask you whether you know of any women who, through strength of emotion and of subtlety of mind and comprehension, have themselves discovered any new arts and sciences which are necessary, good, and profitable, and which have hitherto not been discovered or known. For it is not such a great feat of mastery to study and learn some field of knowledge already discovered by someone else, as it is to discover by oneself some new and unknown thing.[7]

Christine's "Reason" gave the answer of the modern historian of women: "Rest assured, dear friend, that many noteworthy and great sciences and arts have been discovered through the understanding and subtlety of women, both in cognitive speculation, demonstrated in writing, and in the arts, manifested in manual works of labor."[8]

Christine de Pizan's work of the fifteenth century was preceded and followed by a number of encyclopedias of famous women. The first was Giovanni Boccaccio's 1355–59 De claris mulieribus, containing short biographies of 104 women, mostly queens (real and mythical) of the ancient world.[9] This encyclopedia format—the most common type of history of

---

7. Christine de Pizan, The Book of the City of Ladies (1405), trans. Earl Jeffrey Richards (New York: Persea Books, 1982), 70. On Christine, see Susan Groag Bell, "Christine de Pizan (1364–1430): Humanism and the Problem of the Studious Woman," Feminist Studies 3, nos. 3, 4 (1976): 173–84.

8. Christine, 71.

9. Giovanni Boccaccio, De claris mulieribus (1355–59) (Concerning famous women), trans. Guido A. Guarino (New Brunswick, N.J.: Rutgers University Press, 1963). Among the encyclopedias of famous women, see Augustin della Chiesa, Theatrum literatar feminarum (1620); Johann Frauenlob, Die Lobwürdige Gesellschaft der gelehrten Weiber (1631); Marguerite Buffet, Eloges des illustres sçavantes anciennes et modernes (Paris, 1668); J. C. Eberti, Eröffnetes Cabinet des gelehrten Frauenzimmers (1706); C. F. Paulini, Hoch- und Wohlgelahrtes teutsches Frauenzimmer (1712); G. Ballard, Memoirs of Several Ladies of Great Britain: Who Have Been Celebrated for Their Writings or Skill in the Learned Languages, Arts, and Sciences (1752), ed. Ruth Perry (Detroit: Wayne State University Press, 1985).

women in science from the fourteenth through the nineteenth centuries—was developed as a strategy by those who wished to argue for women's greater participation in science. Encyclopedists collected names of distinguished women in order to prove that women were, indeed, capable of great achievements and should be admitted to institutions of science. In 1690, for example, Gilles Ménage prepared an encyclopedia of women accomplished in ancient and modern philosophy as part of his proposal to admit women to the Académie française.[10]

It was not until the late eighteenth century, however, that the first encyclopedia appeared that was devoted exclusively to the history of women's achievements in the natural sciences and medicine. In 1786, French astronomer Jérome de Lalande included in his *Astronomie des dames* (Astronomy for Ladies) the first short history of women astronomers.[11] In the 1830s, German medical doctor Christian Friedrich Harless intended his *Die Verdienste der Frauen um Naturwissenschaft, Gesundsheits- und Heilkunde* (The contribution of women to natural science, health, and healing) to fill the gap that he believed to exist in the history of science of his time; Harless thus proposed an evaluative history of the contributions of women to all fields of the natural sciences as well as to geology, anthropology, and medicine.[12]

The European women's movement of the 1880s to the 1920s precipitated a flurry of interest in the question of women's ability to participate in

10. Gilles Ménage, *Historia mulierum philosopharum* (1690). This work has recently been translated by Beatrice Zedler as *The History of Women Philosophers* (Lanham, Md.: University Press of America, 1984). Ménage is reported to have stated, "My treatise, *Mulierum philosopharum*, could have furnished more ancient examples of marks of distinction granted erudite women. Nevertheless, the proposal made to the Académie produced no results" (P.-L. Joly, *Remarques critiques sur le dictionnaire de Bayle* [Paris, 1752], 2:605). I thank Joan DeJean for calling this passage to my attention.

11. Joseph-Jérôme Le Français de Lalande, *Astronomie des dames* (Paris: Ménard et Desenne fils, 1820), 5–6.

12. Harless (n. 1 above), ix. Although Harless emphasized that men and women were equally capable of doing science, he noted differences between men's and women's relationships to nature and between their methods of science. The man, he wrote, as soon as he is moved by the spirit, searches to uncover the causes underlying appearances and strives to discover laws in life and nature. In contrast, the woman searches nature over for expressions of love, and, Harless concluded, this is the most natural as well as the most beautiful and highest conception of the external world. Women, he affirmed, have made their distinctive contribution through their sympathetic care for the well-being of their fellow creatures (p. 2). The body of Harless's book consists of a list in chronological order of women active in the sciences from mythical times to the present. The first half of the book is devoted to an illustrated history of the myths of Isis and Minerva, among others—figures that he, like Christine, took to represent real women. The second half of the book lists the names and accomplishments of women who contributed to science as patrons, natural philosophers, or practicing scientists. Much of Harless's information was gathered from the earlier encyclopedias of famous women. There seems to have been a general interest in women's contributions to science in early nineteenth-century Germany.

science.[13] In 1894, the Saint-Simonians held the first conference in modern times on women and science, in Paris, from which grew Alphonse Rebière's book *Les femmes dans la science*, and in the same year, Elise Oelsner published her *Die Leistungen der deutschen Frau* (The achievements of German women), in which she paid close attention to women's intellectual achievements. Both books followed the encyclopedia format, listing the women's names alphabetically, giving dates of birth, social conditions, contributions, and publications.[14] Rebière included professional scientists, as he called them, as well as amateurs who had made contributions, and also included those patrons whose contribution had been to aid the progress of science. Appended to his work is a section of diverse opinions of famous people on the question of whether or not woman is capable of scientific pursuits as well as a short collection of facts about women and science.

By the late nineteenth century, the encyclopedia approach seemed doomed. Antifeminists such as Gino Loria in Italy pointed out that, even if there were enough distinguished intellectual women to fill three hundred pages, an equivalent project for men would run to three thousand pages. Loria argued that the accomplishments of the most distinguished woman had not rivaled those of a Pythagoras or Newton, Archimedes or Leibniz.[15] In response to this kind of argument, European and American feminists turned from the strategy of emphasizing the achievements of a few exceptional women to the strategy of emphasizing the barriers to women's participation in science. The first detailed work on women in science was published in America in 1913 under the pseudonym H. J.

---

13. In 1885, there appeared a minor history of women in astronomy by E. Lagrange, "Les femmes-astronomes," *Ciel et terre* 5 (1885): 513–27. Lagrange attacked those who denied women's right to do science and argued that France was far behind the United States in allowing women to enter the liberal professions. His history of outstanding women astronomers—Hypatia, Maria Kirch, Caroline Herschel—was intended to prove his conviction that women could indeed make contributions to the sciences. G. Weyer of the University of Kiel published a similar history of women in astronomy (*Die Akademische Frau: Gutachten hervorragender Universitätsprofessoren, Frauenlehrer und Schriftsteller über die Befähigung der Frau zum wissenschaftlichen Studium und Berufe*, ed. Arthur Kirchhoff [Berlin: H. Steinitz, 1897], 243–55). See also Mary Whitney, "Scientific Study and Work for Women," *Education* 3 (1882): 58–69; Williamina P. Fleming, "A Field for Woman's Work in Astronomy," *Astronomy and Astrophysics* 12 (1893): 683–89; Herman S. Davies, "Women Astronomers, 400 A.D.–1750," *Popular Astronomy* 6 (1898): 128–229; and Caroline Hunt, *The Life of Ellen H. Richards* (Boston: Whitcomb & Barrows, 1912).

14. Alphonse Rebière, *Les femmes dans la science*, 2d ed. (Paris: Noni & Cie, 1897); Elise Oelsner, *Die Leistungen der deutschen Frau in der letzten vierhundert Jahren auf wissenschaftlichen Gebiete* (Guhrau: M. Lemke, 1894). Interestingly, Rebière and Oelsner were the first such historians to drop the mythical characters, such as Athena and Isis, that were featured so prominently in both Christine's and Harless's work.

15. Gino Loria, "Les femmes mathématiciennes," *Revue scientifique* 20, no. 13 (1903): 386.

Mozans.[16] This work represents an impassioned attempt to show that whatever women have achieved in science has been accomplished through "defiance of that conventional code which compelled them to confine their activities to the ordinary duties of the household."[17] Mozans provided a summary of current discussions concerning women's capacity to do science, focusing largely on attempts by nineteenth-century craniologists to prove that the female brain was too small for scientific reasoning. Drawing much of his information from the earlier works of Rebière and Harless, Mozans also discussed women's achievements in the fields of mathematics, astronomy, physics, chemistry, medicine, and archaeology.[18]

The works of Christine de Pizan, Harless, Oelsner, Rebière, and Mozans represent major landmarks in the field of the history of women in science. Yet it should be noted that these authors, who wrote about outsiders, were themselves by and large outsiders—the study of women in science was no more welcome than women scientists themselves within European or American universities. Thus, despite scattered interest since the time of Christine such studies did not become part of historical canon.[19]

Nor was this picture to change with the emergence of the history of science as a discipline in the 1920s and 1930s.[20] This new field, purporting

16. H. J. Mozans, *Woman in Science: With an Introductory Chapter on Women's Long Struggle for Things of the Mind* (1913; reprint, Cambridge, Mass.: MIT Press, 1974). Mozans was the pseudonym for John Augustine Zahm, a Roman Catholic priest who headed a small congregation in Indiana. He was a prolific author and wrote on a wide variety of subjects. For a review of this book, see Alison Kelly, "Upon an Even Pedestal with Man," *Studies in Science Education* 3 (1976): 131–37. See also Ida Welt, "The Jewish Woman in Science," *Hebrew Standard* 50, no. 11 (1907): 4; Simon Flexner, "The Scientific Career for Women," *Scientific Monthly* 13 (1921): 97–105; Marie Farnsworth, "Women in Chemistry: A Statistical Study," *Industrial and Engineering Chemistry* 3 (1925): 4; Florence Wall, "The Status of Women Chemists," *Chemists* 15 (1938): 175–92.

17. Mozans, 391.

18. Mozans was careful to point out that their work in science did not prevent women from being devoted wives and mothers. Mozans called for women to become knowledgeable in the sciences in order to become mothers better able to educate their children. Moreover, Mozans argued that women joining the scientific enterprise would unleash half the energies of humanity. Each woman, he wrote, will act as a Beatrice inspiring a Dante to achieve his full potential; man and woman would thus for the first time fully complement each other in a perfect androgyne. Only then, according to Mozans, would the world enter a "Golden Age of Science—the Golden Age of cultured, noble, and perfect womanhood" (ibid., 415–16).

19. Dale Spender has pointed out that this means that women's history must be rediscovered and rewritten by each new generation (*Women of Ideas and What Men Have Done to Them* [London: Routledge & Kegan Paul, 1982], 9–11).

20. On the development of the field "history of science," see the following collection of short articles in *Isis* 75, no. 276 (1984): John T. Edsall, "Lawrence J. Henderson and George Sarton"; I. B. Cohen, "A Harvard Education"; A. Rupert Hall, "Beginnings in Cambridge"; A. C. Crombie, "Beginnings at Oxford"; and Thomas Kuhn, "Professionalization Recollected in Tranquility."

to study the relation between science and society, did not consider the role of women in science. Even the women working in the field—Marie Boas, Martha Ornstein, and Dorothy Stimson—took little note of women's participation. Nor did theorists exploring the social origins of modern science—Robert Merton, Edgar Zilsel, or Boris Hessen—make any mention of women. Social historians studied participation in science from many important vantage points, including religious affiliation, class, age, and vocation, but they entirely ignored questions of gender. Merton, for example, pointed out that 62 percent of the initial membership of the Royal Society was Puritan.[21] Merton did not, however, explore the implications of the even more striking fact that the early membership in the Royal Society, and indeed in all seventeenth-century academies of science, was 100 percent male.[22]

In the 1940s and 1950s, those who did work on the history of women in science did so largely from outside the history profession.[23] In the 1970s, however, with increasing numbers of women entering science on one hand and the discipline of history on the other, the study of the history of women in science took off. Women scientists contributed thoughtful autobiographies providing firsthand accounts of their struggle to make a mark on science.[24] Historians provided biographies of

21. Robert Merton, *The Sociology of Science: Theoretical and Empirical Investigations* (Chicago: University of Chicago Press, 1973), 208; Dorothy Stimson, "Puritanism and the New Philosophy in 17th Century England," *Bulletin of the Institute of the History of Medicine* 3 (1935): 321–34.

22. Among leading early sociologists of science, only Dorothy Stimson pointed to this fact in a discussion of about two pages in her book *Scientists and Amateurs: A History of the Royal Society* (New York: H. Schurman, 1948), 82–83.

23. Eve Curie, *Madame Curie: A Biography*, trans. Vincent Sheean (Garden City, N.Y.: Doubleday, Doran, 1938); A. W. Richeson, "Hypatia of Alexandria," *Natural Mathematics Magazine* 15 (November 1940): 74–82; Edna Yost, *American Women of Science* (Philadelphia: J. B. Lippincott Co., 1943); Alice Goff, *Women CAN Be Engineers* (Youngstown, Ohio, 1946); P. S. Codellas, "Ancient Greek Women Leaders in Science," *Actes de Ve Congres International d'histoire des sciences* (Lausanne, 1947); Marie-Louise Dubreil-Jacotin, "Figures de mathématiciennes," in *Les grands courants de la pensée mathématique*, ed. F. Le Lionnais (Marseilles: Cahiers du Sud, 1948); Julian Coolidge, "Six Female Mathematicians," *Scripta Mathematica* 17 (1951): 21–31; Denis Duveen, "Madame Lavoisier: 1758–1836," *Chymia Annual: Studies in the History of Chemistry* 4 (1953): 13–29; Margaret Ingels, "Petticoats and Slide Rules," *Midwest Engineer* 5 (1952): 2–4, 10–16; P. V. Rizzo, "Early Daughters of Urania," *Sky and Telescope* 14 (1954): 7–10.

24. Marie Curie, "Autobiographical Notes," in *Pierre Curie*, trans. Charlotte Kellogg and Vernon Kellogg (New York, 1923); Ida Hyde, "Before Women Were Human Beings: Adventures of an American Fellow in German Universities of the '90s," *Journal of the American Association of University Women* 31 (1938): 226–36; Lise Meitner, "The Status of Women in the Professions," *Physics Today* 13, no. 8 (1960): 16–21; Kathleen Lonsdale, "Women in Science: Reminiscences and Reflections," *Impact of Science on Society* 20, no. 1 (1970): 45–59; Sara Ruddick and Pamela Daniels, *Working It Out: 23 Women Writers, Artists, Scientists, and Scholars Talk about Their Lives and Work* (New York: Pantheon Books, 1977);

women scientists that deepened and broadened the work inherited from the nineteenth century.[25] In the past few years, several important intellectual biographies have appeared on individual women scientists such as Sophie Germaine, Mary Somerville, Sofia Kovalevskia, and Clemence Royer.[26] These books are valuable both for documenting the lives of these women and for assessing their contributions to science. These biographies address what has become a standard set of questions about the lives of women scientists: What sparked their initial interest in science? How did they obtain access to the scientific world? How did they make their scientific discoveries? What recognition did these achievements receive in the broader community of scholars?

Most of the work on women scientists fits the "history of great men" mold, with women simply substituted for men. Thus we have many biographies of great women scientists. These biographical studies of women scientists, for the most part, place the achievements of Marie

---

Vivian Gornick, *Women in Science: Portraits from a World in Transition* (New York: Simon & Schuster, 1983). For personal accounts by women scientists in India, Japan, France, Italy, Iran, Sweden, USSR, Kenya, and the United Kingdom, see the anthology edited by Derek Richter, *Women Scientists: The Road to Liberation* (London: Macmillan, 1982). See also Naomi Weisstein, "Adventures of a Woman in Science," in Hubbard, Henifin, and Fried, eds. (n. 4 above), 265–82; Cecilia Payne-Gaposchkin, *An Autobiography and Other Recollections*, ed. Katherine Haramundanis (Cambridge: Cambridge University Press, 1984).

25. Work in this genre includes Robert Reid, *Marie Curie* (New York: Saturday Review Press, 1974); Lois Arnold, *Four Lives in Science: Women's Education in the Nineteenth Century* (New York: Schocken Books, 1983); Lynn M. Osen, *Women in Mathematics* (Cambridge, Mass.: MIT Press, 1974); Anne Sayre, *Rosalind Franklin and DNA: A Vivid View of What It Is Like to Be a Gifted Woman in an Especially Male Profession* (New York: W. W. Norton & Co., 1975); June Goodfield, *An Imagined World: A Story of Scientific Discovery* (New York: Penguin Books, 1982); Olga Opfell, *The Lady Laureates: Women Who Have Won the Nobel Prize* (Metuchen, N.J.: Scarecrow Press, 1978); Margaret Alic, *Hypatia's Heritage* (Boston: Beacon Press, 1986). Much of this work remains buried in journal articles. Eva Armstrong, "Jane Marcet and Her Conversation on Chemistry," *Journal of Chemical Education* 15 (February 1983): 53–57; George Basalla, "Mary Somerville: A Neglected Popularizer of Science," *New Scientist* 16 (1963): 531–33; Sherida Houlihan and John Wotiz, "Women in Chemistry before 1900," *Journal of Chemical Education* 52 (1975): 362–36; Carolyn Merchant, "The Vitalism of Anne Conway: Its Impact on Leibniz's Concept of the Monad," *Journal of the History of Philosophy* 17 (1979): 255–69, and "Madame du Châtelet's Metaphysics and Mechanics," *Studies in History and Philosophy of Science* 8 (1977): 29–48; Sally Kohlstedt, "In from the Periphery: American Women in Science, 1830–1880," *Signs* 4, no. 1 (1978): 81–96, and "Maria Mitchell: The Advancement of Women in Science," *New England Quarterly* 51 (1978): 39–63; Deborah Warner, "Women Astronomers," *Natural History* 88, no. 5 (1979): 12–16; Peggy Kidwell, "Women Astronomers in Britain, 1780–1930," *Isis* 75 (1984): 534–46.

26. Louis Bucciarelli and Nancy Dworsky, *Sophie Germaine: An Essay in the History of the Theory of Elasticity* (Dordrecht: D. Reidel Publishing Co., 1980); Elizabeth Patterson, *Mary Somerville and the Cultivation of Science, 1815–1840* (The Hague: Nijhoff, 1983); Ann Hibner Koblitz, *A Convergence of Lives, Sofia Kovalevskia: Scientist, Writer, Revolutionary* (Cambridge, Mass.: Birkhauser Boston, Inc., 1983); Geneviève Fraisse, *Clemence Royer: Philosophe et femme de science* (Paris: Editions la Decouverte, 1985).

Curie or Rosalind Franklin within the male world, demonstrating that women have, in fact, made important contributions to what has been defined as mainstream science. Yet the focus remains on the woman as exceptional—the woman who defied convention to claim a prominent position in an essentially male world.

One of the problems with this approach to history is that it retains the male norm as the measure of excellence. Some biographical work, however, breaks out of this mold. Margaret Rossiter's analysis of women scientists in America (1790–1940) has shifted the focus from the exceptional woman to the more usual patterns of women working in science.[27] Rossiter analyzes the structure of the scientific community and women's place within it. Rossiter's story documents two forms of discrimination that have shaped women's experience within American science. The first of these is territorial discrimination—the channeling of women into what Rossiter calls women's work or sex-typed work, such as the tedious computing of astronomical data or the classifying and cataloging of natural history collections. Women also suffered from hierarchical discrimination that contained bright and capable women within the ranks of invisible and poorly paid assistants. A second part of Rossiter's story explores the dialectic between the social barriers to women's advancement and the strategies women employed to secure a place in science. These strategies ranged from attempts to break down barriers to graduate education, scientific societies, and networks to attempts to create unique fields for women, such as home economics.

Evelyn Fox Keller is another biographer whose work does not fit the traditional historical mold. In *A Feeling for the Organism: The Life and Work of Barbara McClintock*, though she retains the focus on an exceptional woman, Keller does not simply measure McClintock against traditional male standards. In addition to telling McClintock's story (largely in McClintock's own words), Keller uses this story as a vehicle for evaluating current methods of experimental science. Keller emphasizes that McClintock's relation to her material in her genetic research was characterized not by the conventional practice of distinguishing sharply between subject and object but, rather, by a merging of self with the material—a "feeling for the organism." McClintock's unconventional style, arising in part from individual idiosyncrasies and in part from the isolation she experienced as a woman in a man's world, reflects a unity with and a deep reverence for nature—a kind of reverence that McClintock believes could forestall environmental disasters.[28]

27. Margaret Rossiter, *Women Scientists in America: Struggles and Strategies to 1940* (Baltimore: Johns Hopkins University Press, 1982).

28. Evelyn Fox Keller, *A Feeling for the Organism: The Life and Work of Barbara McClintock* (San Francisco: W. H. Freeman & Co., 1983). See also S. J. Gould's review "Triumph of a Naturalist," *New York Review of Books* (March 1984).

Others have gone even further to argue that we must reevaluate science in order to appreciate fully women's contributions. An interesting example of this approach is provided in the work of Christine de Pizan, who celebrated women's contributions to the development of fields such as law, writing, and numbers, fields considered central to the development of Western civilization. Yet Christine also emphasized women's contributions to what we consider the domestic arts. It was Isis, she wrote, who discovered the art of constructing gardens and of planting. Ceres taught humankind how to grind grain and make bread. Arachne invented the art of dyeing wool, and of making tapestries, flax, and linen. Here Christine made a point that we have only recently begun to appreciate. In order to evaluate women's contributions to science, our definitions of science may need to be broadened. By ascribing equal significance to arts now commonly devalued as women's arts, Christine could claim that the achievements of women were both ancient and significant.[29]

Why So Few? Identifying Structural Barriers

Anyone who reads about the lives of women in science becomes keenly aware of the battle that women have had to wage in order to gain recognition and support within the scientific community. When looking at women's opportunities in science, one central fact emerges: women have never fared well in official institutions of science—past or present. Until the twentieth century, all but a few privileged women were officially barred from universities and scientific academies. Even now the vast majority of women scientists work only at the periphery. Women simply do not hold senior positions in science from which they can guide the future course of science. To what extent has this been true in the past, and what is the present status of women in science?

There has been little study of women's role in scientific institutions in ancient and early modern times. We know that in the ancient world

29. Christine (n. 7 above), 70–83. As was common in this period, Christine presented her history of Minerva, Isis, and others as the accomplishments of real women. Francis Bacon also considered Ceres a real woman and credited her with the discovery of the art of baking bread. In the *Novum Organum* he wrote, "In things mechanical [take notice of] the works of Bacchus and Ceres—that is, of the arts of preparing wine and beer, and of making bread" (Francis Bacon, *The New Organon and Related Writings of Francis Bacon* [1620], ed. Fulton Anderson [Indianapolis: Liberal Arts Press, 1960], bk. 1, aphorism 85). In 1982, Autumn Stanley argued that, in order to appreciate women's contributions to technology, women's activities such as methods of food processing, detoxification, cooking, preserving, and herbal preparations for childbirth must be discussed as technology ("Women Hold Up Two-Thirds of the Sky: Notes for a Revised History of Technology," *Machina ex Dea: Feminist Perspectives on Technology*, ed. Joan Rothschild [New York: Pergamon Press, 1983], 5).

Pythagoras's wife ran his academy after his death,[30] though we do not have a full sense of how many women attended this or other Greek academies. Over the ages, women's opportunities to participate in science have changed with the changing institutions of knowledge. In the Middle Ages, convents provided an important place where women could pursue learning.[31] The rise of European universities in the twelfth through fifteenth centuries actually reduced educational opportunities for women, for from their beginnings, these universities were, in principle, closed to women. Nonetheless, a few exceptional women—principally in Italy but also in Germany—studied and taught at universities beginning in the thirteenth century. Most notable perhaps was Laura Bassi (1711–78), who held a chair in physics at the University of Bologna. As a group, however, women were not formally admitted to European universities until the 1860s in Switzerland, the 1870s in England, the 1880s in France, and the 1900s in Germany.

Universities were not the only centers of scientific life, however. Guilds of the Middle Ages and the courts of the Renaissance also contributed to the development of modern science. Women often participated in science through these and other institutions where membership embraced broader groups.[32] With the founding of the scientific academies of the seventeenth century, obstacles were again thrown in the path of women's participation in science. These new academies, like the older universities, were closed to women from the outset, even to qualified women such as the astronomer Maria Winckelmann, the mathematician Sophie Germaine, or the Nobel prize winner Marie Curie. The Royal Society of London, the oldest permanent scientific academy, though founded in the 1660s, did not admit its first women (Marjory Stephenson and Kathleen Lonsdale) until 1945. The prestigious Académie des Sciences in Paris, founded in 1666, did not admit a woman (Yvonne Choquet-Bruhat) until 1979.

Despite this resistance to women's participation in science, by the late nineteenth century women in Europe and America had leveled mighty barriers. By the 1870s, women gained admittance to universities; and by 1920, American women had gained admittance to Ph.D. programs—the

30. Mozans (n. 16 above), 199.

31. Perhaps the most notable female author on medicine, natural history, and cosmology was the twelfth-century abbess Hildegard von Bingen (Hildegard von Bingen, *Naturkunde*, trans. Peter Riethe [Salzburg: Müller, 1959]). See also Charles Singer, "Hildegard of Bingen," in *From Magic to Science: Essays on the Scientific Twilight* (New York: Boni & Liveright, 1928), 199–239. See also Mozans; Kate Campbell Hurd-Mead, *A History of Women in Medicine: From the Earliest Times to the Beginning of the Nineteenth Century* (New York: AMS Press, 1977).

32. Londa Schiebinger, "Women and the Origins of Modern Science" (Ph.D. diss., Harvard University, 1984), chaps. 1, 2.

usual prerequisite for serious work in science.[33] The story has not, however, been one of a slow and steady progress toward equality for women.
Between 1930 and 1960, for example, the proportion of women Ph.D.'s
and faculty in the sciences took a downward turn; 1920s levels were not
regained until the 1970s.[34] Women did, however, take a step forward in
1964 when discrimination on the basis of sex in education and employment was outlawed by the Civil Rights Act. The time had passed when the
chairman of a biochemistry department could tell a scientist like Anne
Briscoe that Cornell University simply did not want a woman.[35]

Yet women have not advanced in the scientific professions as fast or
as far as men even after discrimination became illegal. As Margaret
Rossiter points out, women scientists working in government, industry,
and academia have faced double standards and insufficient recognition.[36]
A large body of work documenting the status of women in modern
scientific professions differs from the traditional historical literature in
drawing heavily on statistical analyses to clarify women's current status.[37]
Even though overt restrictions on women's participation in science have
been removed, it is clear that other less obvious barriers remain.

Women have made some gains in the past decade. The employment
of women scientists and engineers in U.S. government, industry, and

33. Though some institutions admitted women to graduate work as early as 1877, the
more prestigious institutions were slow in admitting women. See Rossiter (n. 27 above); Roy
MacLeod and Russell Moseley, "Fathers and Daughters: Reflections on Women, Science and
Victorian Cambridge," *History of Education* 8, no. 4 (1979): 321–33.

34. Jessie Bernard, *Academic Women* (New York: Meridian Books, 1964), 73.

35. For a summary of affirmative action legislation, see National Research Council
Committee on the Education and Employment of Women in Science and Engineering (NRC
Committee), *Climbing the Academic Ladder: Doctoral Women Scientists in Academe* (Washington,
D.C.: National Academy of Sciences, 1979), 135–36; Anne Briscoe, "Scientific Sexism: The
World of Chemistry," in Haas and Perrucci, eds. (n. 3 above), 147–59, esp. 153.

36. Rossiter (n. 27 above); Betty Vetter, "Changing Patterns of Recruitment and Employment," in Haas and Perrucci, eds. (n. 3 above), 59–74.

37. Jacquelyn Mattfeld and Carol Van Aken, eds., *Women and the Scientific Professions: The
M.I.T. Symposium on American Women in Science and Engineering* (Cambridge, Mass.: MIT
Press, 1965); Anna Sexton, *A Chronicle of the Division of Laboratories and Research: New York
State Department of Health: The First Fifty Years* (Lunenburg, Vt.: Stinehour Press, 1967); A. E.
Bayer and H. S. Astin, "Sex Differentials in the Academic Reward System," *Science* 188
(1975): 796–802; H. Zuckerman and J. Cole, "Women in American Science," *Minerva* 13
(1975): 82–102; Anne Briscoe and Sheila Pfafflin, eds., *Expanding the Role of Women in the
Sciences* (New York: New York Academy of Sciences, 1979); NRC Committee, *Climbing the
Academic Ladder* (n. 35 above); National Research Council, *Women Scientists in Industry and
Government: How Much Progress in the 1970s?* (Washington, D.C.: National Academy of
Sciences, 1980); Sue Berryman, *Who Will Do Science? Minority and Female Attainment of Science
and Mathematics Degrees: Trends and Causes* (New York: Rockefeller Foundation, 1983); NRC
Committee, *Climbing the Ladder: An Update on the Status of Doctoral Women Scientists and
Engineers* (Washington, D.C.: National Academy Press, 1983); Haas and Perrucci, eds. (n. 3
above); National Science Foundation (n. 2 above).

academe increased by over 200 percent between 1972 and 1982, compared with about 40 percent for men.[38] Yet the numbers *still* remain small. Physics, a prestigious field, employed less than 4 percent women in 1980—significantly fewer than the number of women trained and ready to work in the field.[39] As Betty Vetter has documented, all women scientists are two to five times as likely to be unemployed or underemployed as men scientists.[40] Similarly, women more often work in less prestigious jobs than men with identical education and work experience.[41]

Though the number of women working in science is up, pay for women is not. The National Science Foundation report of 1984 showed that annual salaries for women scientists and engineers averaged not even 80 percent of those for men ($27,000 for women compared to $35,000 for men).[42] This gap in pay has remained constant over the past decade. Moreover, and more shocking, women's income in any particular field is generally inversely proportional to the prestige of that field. In 1984, women physicists and astronomers earned only one-fourth of what men earned. According to U.S. Department of Labor statistics, this repre-

38. National Science Foundation (n. 2 above), vii. All statistics are for the United States. Statistical analyses of women in science in other parts of the world are just becoming available. See Nancie Gonzalez, "Professional Women in Developing Nations: The United States and the Third World Compared," in Haas and Perrucci, eds. (n. 3 above), 19–42.

39. The number of women working in physics is fewer (in terms of proportional relation to representation in the population) than the percentage of male and female blacks (1.3), American Indians (0.1), or Hispanics (2.1) working in physics. Yolanda Scott George, "Affirmative Action Programs that Work," in *Women and Minorities in Science: Strategies for Increasing Participation*, ed. Sheila Humphreys (Boulder, Colo.: American Association for the Advancement of Science, 1982), 88; S. M. Malcom et al., *The Double Bind: The Price of Being a Minority Woman in Science* (Washington, D.C.: American Association for the Advancement of Science, 1976). See also Vera Kistiakowsky, "Women in Physics and Astronomy," in Briscoe and Pfafflin, eds. (n. 37 above), 35–47.

40. Vetter (n. 36 above), 61; M. G. Finn, "Understanding the Higher Unemployment Rate of Women Scientists," *American Economic Review* 73 (1983): 1137–40.

41. National Research Council (n. 37 above), 39. In 1977, only 7.9 percent of women scientists and engineers employed by the government held senior positions. Men in the same age group were twice as likely to hold the top positions (ibid., 34). A similar story can be told for industry. In 1977, 13 percent of women employed in industry held managerial positions, compared to 28 percent of the men (ibid., 12). By 1981, this had improved only slightly: 16.4 percent of women employed in industry held managerial positions, compared to 28.7 percent of men (NRC Committee, *Climbing the Ladder: An Update* [n. 37 above], table 5.4). Women's position within industry also varies by field. In the prestigious field of physics, e.g., women find it difficult to advance. In 1983, of the 469 physicists employed at ten top corporations—Bell Telephone Laboratories, Exxon, General Electric, GTE, IBM, Westinghouse, Xerox, Ford Motor Co., RCA Laboratories—only twenty-four (or 5 percent) were women. There were no women among the thirty-five research managers in the survey (*Bulletin of American Physical Societies* [May/June 1982], 836). See also S. Traweek, "High Energy Physics: A Male Preserve," *Technology Review* 87 (November/December 1984): 42–43.

42. National Science Foundation (n. 2 above), 6.

sented the widest pay gap for any profession in the United States.[43] Nor has increased education increased women's pay. In 1981, the average salary of women Ph.D. scientists and engineers was only 75 percent of that which men of equivalent background earned working in the same field. Salaries of women scientists and engineers are lower in all fields and at every level of experience than those of men. The 1984 report by the National Science Foundation concluded that, though some salary differences can be explained by experience, almost half of the salary differential between men and women in science fields remains unexplained by factors other than sexual discrimination.[44]

Turning to academia, where 67 percent of women scientists work, we find that the pay is generally lower than in industry, and, as in government and industry, women appear at lower ranks and at less prestigious institutions.[45] At the undergraduate level women appear to be making gains: in 1981, women received 37 percent of all science and engineering bachelor's degrees. This contrasts dramatically with women's presence only ten years ago, when they earned only 26 percent of these degrees. Inequalities increase at higher levels of education, however. Few women go on to graduate school, and of those, fewer women than men will finish their Ph.D. degrees. In 1982, women received only 23 percent of Ph.D. degrees in science and engineering.[46] Although women receive fewer degrees, they receive degrees from prestigious departments in the same proportion as men;[47] but graduation from a prestigious department does not translate automatically into a good job. Women are two to three times more likely than men to be in marginal, non-tenure-track positions. Furthermore, as rank increases, the proportion of women drops dramatically. In 1981 the National Research Council reported that, at major research universities, women constitute only 3 percent of the full professors in science and engineering, 10 percent of associate professors, and 24 percent of assistant professors.[48] The statistics for women in science vary

43. Men in the fields of physics and astronomy earn $674 per week on the average. Women in these fields earn only $166 per week (Carole Bodger, "Salary Survey: Who Does What and for How Much?" *Working Woman* [January 1985], 72).

44. National Science Foundation, 6. See also Vetter (n. 36 above), 63.

45. Only 55 percent of men scientists work in academic jobs (Lilli Hornig, "Professional Women in Transition," in Haas and Perrucci, eds. [n. 3 above], 43–58, esp. 51).

46. National Science Foundation, vii–viii. Again, women's participation in science varies by field. Few women, e.g., receive undergraduate degrees in physics. At Stanford University, only one or two women out of a class of twenty-five graduate each year with a major in physics. This figure has remained unchanged over the past ten years. See Londa Schiebinger, "Women Left Out of the Physics Club," *Stanford Daily* (April 3, 1985).

47. Rachel Rosenfeld, "Academic Career Mobility for Women and Men Psychologists," in Haas and Perrucci, eds. (n. 3 above), 89–127, esp. 92–93.

48. These figures are for faculty at the top fifty U.S. institutions (NRC Committee, *Climbing the Ladder: An Update* [n. 37 above], xvi). In 1976, in cell biology women constituted

according to field, but the pattern is always the same: many female faces at the bottom of the pyramid, few at the top. Between 1977 and 1981 at the top fifty U.S. universities, three-quarters of male assistant professors were promoted to a higher rank, but only half of the women at this rank were promoted.[49] Only at the level of assistant professor did women approximate their representation in the Ph.D. pool. As Lilli Hornig has pointed out, equal opportunity has been reserved for entry-level positions.[50]

These statistics have consequences that go beyond the size of a paycheck. Rank in the scientific community deeply influences individual scientific productivity. Lilli Hornig has found that after graduate school men more often receive research postdoctoral fellowships, while women move into low-level teaching. Consequently, Hornig argues, women often do not establish the publishing record required to move to the next stage in professional advancement.[51] Furthermore, as women take up positions at lower ranks and at lower-prestige institutions, they lose access to funding or the latest scientific equipment. It is in the top-rated universities— places like Harvard, MIT, Cal Tech, and Stanford—that the biggest and best linear accelerators, electron microscopes, and computers are found, and these are the institutions in which women have found it most difficult to make inroads.[52]

Sociologists and psychologists have tried to explain the divergent career patterns of men and women.[53] Alice Rossi's classic paper "Women

---

11 percent of full professors, 20 percent of associate professors, 20.9 percent of assistant professors, and 44 percent of all research associates. These numbers mean something different for women than they would for men. Whereas 39 percent of all men in academic cell biology are full professors, only 18 percent of women are. Whereas only 7 percent of the men work as research associates, 20 percent of the women do (Elizabeth O'Hern, "Women in the Biological Sciences," in Briscoe and Pfafflin, eds. [n. 37 above], 116). The situation for women in physics is the poorest. In 1977 and 1978, women physicists held only 2 percent of tenured or tenure-track faculty positions in the ten top physics departments in the United States. Claire Max has shown that this poor showing does not result from an inadequate supply of qualified candidates. She reports that the overall percentage of women Ph.D.'s of the appropriate age was roughly 50 percent higher than their employment record (Claire Ellen Max, "Career Paths for Women in Physics," in Humphreys, ed. [n. 39 above], 111–12).

49. NRC Committee, *Climbing the Ladder: An Update* (n. 37 above), xvi.

50. Hornig (n. 45 above), 53.

51. Lilli Hornig, "Demographics and Women and Science" (Massachusetts Institute of Technology, Women's Studies, 1984, typescript).

52. Karen Messing, "The Scientific Mystique: Can a White Lab Coat Guarantee Purity in the Search for Knowledge about the Nature of Women?" in *Woman's Nature: Rationalizations of Inequality*, ed. Ruth Hubbard and Marian Lowe (New York: Pergamon Press, 1983), 77–78.

53. Janice Trecker, "Sex, Science and Education," *American Quarterly* 26 (1974): 353–66; L. H. Fox, L. Brody, and D. Tobin, eds., *Women and the Mathematical Mystique: Proceedings of the Eighth Annual Hyman Blumberg Symposium on Research in Early Childhood Education* (Baltimore: Johns Hopkins University Press, 1976); Barbara Reskin, "Academic

in Science: Why So Few?" given at the 1964 Massachusetts Institute of Technology conference on women and science, posed a question that has become the focus of subsequent research. What, Rossi asked, are the "social and psychological influences [that] restrict women's choice and pursuit of careers in science?"[54] Her research shows that girls' unwillingness to major in science in college has deep cultural roots—ranging from the kinds of toys they play with to the kinds of education they receive. Studies show that erector sets and chemistry sets help children develop different skills and aspirations than do Barbie dolls. Boys and girls show equal skills in mathematics until about the age of thirteen—about the time (at least in the past) that boys go into shop and girls go into home economics. At this age, boys begin to move ahead in mathematics, and the gap continues to widen with additional years of schooling. A 1972 survey of University of California, Berkeley, freshmen showed that 93 percent of the boys had taken three or more years of high school math, compared with only 84 percent of the girls.[55] Other studies have shown that, even when girls take the same number of math courses as boys, their math educations may not be equivalent. Teachers may not expect (and thus do not encourage) the girls to perform at the same level as boys. These high school experiences limit young women's opportunities in college. No one forces girls out of the study of math and science, yet a process of self-selection occurs as a result of the differing career aspirations of boys and girls, as well as the differing expectations of their parents, peers, and teachers.

Some have correlated the low numbers of women in science with the effects on academic productivity and job advancement of such factors as marriage, geographic mobility, and having children. Although marriage and family responsibilities are often blamed for women's lower scientific productivity, it is not clear that single women are advanced as readily as men (married or single). Even women who publish regularly are not as likely to hold high-ranking positions as their equally prolific male colleagues.[56]

Sponsorship and Scientists' Careers," *Sociology of Education* 52 (1979): 129–46; J. R. Becker, "Differential Treatment of Females and Males in Mathematics Classes," *Journal for Research in Mathematics Education* 12 (1981): 40–53; Patricia Campbell and Susan Geller, "Early Socialization: Causes and Cures of Mathematics Anxiety," in Haas and Perrucci, eds. (n. 3 above), 173–80; Lucy Sells, "Leverage for Equal Opportunity through Mastery of Mathematics," in Humphreys, ed. (n. 39 above), 7–26.

54. It should be pointed out that the initiative for this conference came from students--a characteristic feature of many innovations in women's studies (Rossi [n. 2 above]).

55. Lucy Sells, "The Mathematics Filter and the Education of Women and Minorities," in Fox, Brody, and Tobin, eds., 70.

56. Rita Simon, Shirley Clark, and Kathleen Galway, "The Women Ph.D.'s: A Recent Profile," *Social Problems* 15, no. 2 (1967–68): 221–37; G. Marwell, R. Rosenfeld, and

Then, too, other more subtle factors keep women on the periphery of science. Barbara Reskin has shown that, even when women are admitted to the laboratory, they are often excluded from the networks of informal communication crucial to the development of scientific ideas. Because women are such a minority in science, men and women scientists often fail to establish the collegial relationships that have characterized relationships among men scientists and fall instead into traditional male and female relationships—those of romance, father and daughter, brother and sister. Reskin argues that these kinds of relationships are not beneficial to a scientific career.[57]

As Elizabeth Fee has pointed out, this approach often locates the problem in women (their socialization, aspirations, and values) and does not ask the broader questions of whether or in what way the institutions of science could be reshaped to accommodate women. Institutions of science are presently structured with the assumption that a scientist is equipped with a wife to rear children. No real arrangements have been made for men or women who rear children. Maternity and paternity leave policies do not address the scientific community's intolerance of the temporary decline in scientific productivity associated with childbearing and child rearing. At present, a woman's desires to have a professional career conflict with her desires to raise children. The years twenty to thirty-five, crucial for determining a scientist's success in the profession, coincide with a woman's childbearing years. Moving women into science may require more than restructuring the institutions of science alone since, as Hilary Rose explains, this change has implications for the relationship between the reproductive sphere and the productive sphere, and the value we attach to both.[58]

S. Spilerman, "Geographic Constraints on Women's Careers in Academia," *Science* 205 (1979): 1225–31; Judith Ramaley, ed., *Covert Discrimination and Women in the Sciences* (Boulder, Colo.: American Association for the Advancement of Science, 1978); Barbara Reskin, "Scientific Productivity, Sex, and Location in the Institution of Science," *American Journal of Sociology* 83, no. 5 (1978): 1235–43; L. Hargens, J. McCann, and B. Reskin, "Productivity and Reproductivity: Professional Achievement among Research Scientists," *Social Forces* 57 (1978): 154–63; B. Reskin and L. Hargens, "Scientific Advancement of Male and Female Chemists," in *Discrimination in Organizations*, ed. R. Alvarez and K. Lutterman (San Francisco: Jossey-Bass, Inc., 1979), 100–122; Jonathan Cole, *Fair Science: Women in the Scientific Community* (New York: Free Press, 1979); F. Pepitone-Rockwell, ed., *Dual Career Couples* (Beverly Hills, Calif.: Sage Publications, 1980); Rosenfeld (n. 47 above), 89–127. For a comparison by sex of publication records and professional status, see Cole, 79; for critical reviews of this book see Karen Oppenheim, "Sex and Status in Science," *Science* 208 (1980): 277–78; and Margaret Rossiter, "Fair Enough?" *Isis* 72 (1981): 99–103.

57. Barbara Reskin, "Sex Differentiation and Social Organization of Science," in *Sociology of Science*, ed. Jerry Gaston (San Francisco: Jossey-Bass, Inc., 1978): 6–37.

58. Elizabeth Fee, "A Feminist Critique of Scientific Objectivity," *Science for the People* 14, no. 4 (1982): 5–8, 30–32; Hilary Rose, "Hand, Brain, and Heart: A Feminist Epistemology for the Natural Sciences," *Signs* 9, no. 1 (1983): 73–90, esp. 83.

Naturally Few? Biology as Destiny

The literature on scientific definitions of woman's nature can be broken down into three dominant strands. The first view recognizes and emphasizes differences between men and women and traces these differences to inevitable biological differences. The second view minimizes differences between men and women and uses biological similarity to argue for social equality of the sexes. A third standpoint recognizes and appreciates gender based differences but maintains that these differences have been molded by history and environment and are not necessarily inevitable.[59]

By emphasizing intellectual differences between men and women, the first view attempts to explain why, in fact, so few women have made substantial contributions to science over the past two thousand years. Rather than looking at the subtle forms of discrimination built into social and scientific institutions, biological determinists have argued that an inability to do science is built into women. This literature, cast in the guise of science, presents women as naturally incapable of doing science.

The attempt to trace woman's social inferiority to her supposed biological inferiority is an old one, dating back at least to Aristotle. In the ancient world, for example, Hippocrates, Aristotle, and Galen argued that woman's weaker nature justified her inferior social status. Aristotle argued that women were colder and weaker than men and that women did not have sufficient heat to boil the blood and thus purify the soul.[60] The medical and scientific assumptions of these ancients were incorporated into medieval thinking with few revisions and came to dominate

59. See, e.g., Ruth Bleier, *Science and Gender: A Critique of Biology and Its Theories on Women* (New York: Pergamon Press, 1984); The Brighton Women and Science Group, *Alice through the Microscope: The Power of Science over Women's Lives*, ed. Lynda Birke et al. (London: Virago, 1980); Maryanne Horowitz, "Aristotle and Woman," *Journal of the History of Biology* 9 (1976): 183–213; Ruth Hubbard and Marian Lowe, eds., *Genes and Gender II: Pitfalls in Research on Sex and Gender* (New York: Gordian Press, 1979); Ruth Hubbard, Mary Sue Henifin, and Barbara Fried, eds., *Women Look at Biology Looking at Women: A Collection of Feminist Critiques* (Boston: G. K. Hall & Co., 1979); Hilda Smith, "Gynecology and Ideology in Seventeenth-Century England," in *Liberating Women's History: Theoretical and Critical Essays*, ed. Berenice Carroll (Champaign: University of Illinois Press, 1976); Caroline Whitbeck, "Theories of Sex Difference," *Philosophical Forum* 5 (1973/74): 54–80; "Special Issue: Women, Science, and Society," *Signs*, vol. 4, no. 1 (1978); "Special Issue: Les femmes et la science," *Pénélope*, vol. 4 (1981); "Special Issue: Les femmes dans les sciences de l'homme," *Nouvelles questions féministes*, vol. 3 (1982).

60. For Aristotle's views on women, see G. E. R. Lloyd, *Science, Folklore and Ideology: Studies in the Life Sciences in Ancient Greece* (Cambridge: Cambridge University Press, 1983), 58–111; Sarah Pomeroy, *Goddesses, Whores, Wives, and Slaves: Women in Classical Antiquity* (New York: Schocken Books, 1975); M. C. Horowitz, "Aristotle and Woman," *Journal of the History of Biology* 9 (1976): 183–213; Lynda Lange, ed., *The Sexism of Social and Political Theory: Women and Reproduction from Plato to Nietzsche* (Toronto: University of Toronto Press, 1979).

much of Western medical literature until well into the seventeenth century.[61] A fundamental shift in the definition of sex differences emerged in the course of the eighteenth and early nineteenth centuries. The doctrine of humors, which had long provided an explanation for differences in women's physical and moral character, was replaced by a search for sex differences using the methods of modern science. Beginning in the 1750s, a body of literature emerged in England, France, and Germany calling for a finer delineation of sex differences.[62] In the late eighteenth and nineteenth centuries, craniologists tried to account for sex differences in intellectual achievement by measuring the skull. Anatomists asserted that the larger male skull held a heavier and more powerful brain.[63] In the mid-nineteenth century, social Darwinists invoked evolutionary biology to argue that a woman was a man whose evolution—both physical and mental—had been arrested in a primitive stage.[64] In this same period, doctors used their authority as scientists to discourage women's attempts to gain access to higher education. Women's intellectual development, it was argued, would proceed only at great cost to reproductive development. As the brain develops, so the logic went, the

61. On the continuity in medical views from the ancient to the medieval worlds, see Vern L. Bullough, "Medieval Medical and Scientific Views of Women," *Viator* 4 (1973): 487; Julia O'Faolain and Lauro Martines, *Not in God's Image: Women in History from the Greeks to the Victorians* (London: M. T. Smith, 1973), 130–39; Lester S. King, "The Transformation of Galenism," in *Medicine in Seventeenth-Century England: A Symposium Held at UCLA in Honor of C. D. O'Malley*, ed. Allen Debus (Berkeley: University of California Press, 1974), 7–32; Ian Maclean, *The Renaissance Notion of Woman: A Study in the Fortunes of Scholasticism and Medical Science in European Intellectual Life* (Cambridge: Cambridge University Press, 1980).

62. Londa Schiebinger, "Skeletons in the Closet: The First Illustrations of the Female Skeleton in Eighteenth-Century Anatomy," *Representations* 14 (Spring 1986): 42–82.

63. The first critique of craniology came from Hedwig Dohm, *Die wissenschaftliche Emancipation der Frauen* (Berlin: Wedekind & Schwieger, 1874). See also Elizabeth Fee, "Nineteenth-Century Craniology: The Study of the Female Skull," *Bulletin of the History of Medicine* 53, no. 3 (1979): 415–33; Stephen Jay Gould, *The Mismeasure of Man* (New York: W. W. Norton & Co., 1981), esp. chap. 3.

64. The first critique of this literature came from Antoinette Brown Blackwell, *The Sexes throughout Nature* (New York: G. P. Putnam's Sons, 1875). See also Eliza Burt Gamble, *Evolution of Woman: An Inquiry into the Dogma of Her Inferiority to Men* (New York: G. P. Putnam's Sons, 1894); Jill Conway, "Stereotypes of Femininity in a Theory of Sexual Evolution," *Victorian Studies* 14 (1970): 47–62; Elaine Morgan, *The Descent of Woman* (New York: Stein & Day, 1972); Evelyn Reed, *Woman's Evolution: From Matriarchal Clan to Patriarchal Family* (New York: Pathfinder Press, 1975); Stephanie Shields, "Functionalism, Darwinism, and the Psychology of Women," *American Psychologist* 30 (1975): 739–54; Elizabeth Fee, "Science and the Woman Problem: Historical Perspectives," in *Sex Differences: Social and Biological Perspectives*, ed. Michael Teitelbaum (Garden City, N.Y.: Anchor Press, 1976), 175–223; Sarah Hrdy, *The Woman that Never Evolved* (Cambridge, Mass.: Harvard University Press, 1981); Ruth Hubbard, "Have Only Men Evolved?" in Hubbard, Henifin, and Fried, eds. (n. 4 above), 17–46.

ovaries shrivel.[65] In the twentieth century, scientists have given modern dress to these prejudices. Arguments for women's different (and inferior) nature have been based on hormonal research,[66] brain lateralization,[67] and sociobiology.[68]

Though science and the authority of science have often been turned against women, science has also been enlisted in the liberation of women. This argument became especially important with the rise of liberal political theory in the seventeenth century. In 1673, the Frenchman François Poullain de la Barre argued that women's anatomy proved their equality. Women, he wrote, have the same sense organs as men—their eyes see as clearly, their ears hear with the same degree of accuracy, their hands are as dexterous. Their heads, he continued, are the same as men's: "The most exact anatomy has not discovered any difference in that part [the

65. Edward Clarke, *Sex in Education* (Boston: J. R. Osgood, 1873), 39. For critical reflections on the medical case against higher education for women, see Janet Sayers, *Biological Politics: Feminist and Anti-Feminist Perspectives* (London: Tavistock, 1982); Joan Burstyn, "Education and Sex: The Medical Case against Higher Education for Women," *Proceedings of the American Philosophical Society* 117 (1973): 79–80; George Bernstein and Lottelore Bernstein, "Attitudes toward Women's Education in Germany, 1870–1914," *International Journal of Women's Studies* 2, no. 5 (1979): 473–89; Louise Newman, ed., *Men's Ideas/Women's Realities: Popular Science, 1870–1915* (New York: Pergamon Press, 1985).

66. Diana Long Hall, "Biology, Sex Hormones and Sexism in the 1920s," in *Women and Philosophy: Toward a Theory of Liberation*, ed. C. C. Gould and M. W. Wartofsky (New York: G. P. Putnam's Sons, 1976); Bleier (n. 59 above); Estelle Ramey, "Sex Hormones and Executive Ability," *Annals of the New York Academy of Sciences* 208 (1973): 237–45; Ethel Tobach and Betty Rosoff, eds., *Genes and Gender I: First in a Series on Hereditarianism and Women* (New York: Gordian Press, 1978), and *Genes and Gender III: Genetic Determinism and Children* (New York: Gordian Press, 1980).

67. For critical studies of brain lateralization, see Susan Leigh Star, "Sex Differences and the Dichotomization of the Brain: Methods, Limits, and Problems in Research on Consciousness," in Hubbard and Lowe, eds. (n. 59 above); Jannette McGlone, "Sex Differences in Human Brain Asymmetry: A Critical Survey," *Behavioral and Brain Sciences* 8 (1980): 215–63; Sayers (n. 65 above), chap. 6; Bleier (n. 59 above), chap. 4.

68. Lila Leibowitz, *Females, Males, Families: A Biosocial Approach* (North Scituate, Mass.: Duxbury Press, 1978); Marian Lowe and Ruth Hubbard, "Sociobiology and Biosociology: Can Science Prove the Biological Basis of Sex Differences in Behavior?" in Hubbard and Lowe, eds. (n. 59 above); Donna Haraway, "Animal Sociology and a Natural Economy of the Body Politic," *Signs* 4, no. 1 (1978): 21–60, "The Biological Enterprise: Sex, Mind, and Profit from Human Engineering to Sociobiology," *Radical History Review* 20 (1979): 206–37, and "In the Beginning Was the Word: The Genesis of Biological Theory," *Signs* 6, no. 3 (1981): 469–82; Elizabeth Adkins, "Genes, Hormones, Sex and Gender," in *Sociobiology: Beyond Nature/Nurture? Reports, Definitions, and Debates*, ed. George Barlow and James Silverberg (Washington, D.C.: American Association for the Advancement of Science, 1980), 385–415; Sayers (n. 65 above); Helen Longino and Ruth Doell, "Body, Bias, and Behavior: A Comparative Analysis of Reasoning in Two Areas of Biological Science," *Signs* 9, no. 2 (1983): 206–27; Bleier (n. 58 above); Richard Lewontin, Steven Rose, and Leon Kamin, *Not in Our Genes: Biology, Ideology and Human Nature* (New York: Pantheon Books, 1984), 63–81.

head] between men and women; the brain is the same in both, as are memory and imagination."[69] By minimizing differences between the sexes, Poullain de la Barre hoped to overturn Aristotelian arguments that sexual differences justified social inequalities between the sexes. This critique of biological determinism continued throughout the nineteenth century. In America and England, Antoinette Brown Blackwell and Eliza Burt Gamble refuted the social Darwinists' claim that womankind is an arrested, primitive stage of human evolution.[70]

More recently, feminists continue to confront biological determinists. On the one hand, feminists deconstruct misrepresentations of women's nature that have been used to limit women's rights. On the other hand, this effort builds a fuller understanding of the reality of women's bodies and abilities, sorting out, as it were, myth from fact. This kind of critical reevaluation of women's biology comes from neurophysiologist Ruth Bleier and biologists Ruth Hubbard, Marian Lowe, Anne Fausto-Sterling, and Lynda Birke, for example—scientists in their own right who evaluate bias in science in scientific terms.[71] Birke, Bleier, Hubbard, Lowe, and Fausto-Sterling are particularly careful to point out that the mystifications in sociobiology in no way undermine valid biological inquiry; but they are equally careful to point out what is not scientific about the claims of biological determinists.

These feminist biologists (each in her own way) demonstrate that biology is not the static bedrock of organic life that biological determinists would like it to be, and they show instead how cultural factors shape biology. At the same time Birke and Fausto-Sterling caution against any attempt to trace behavior to any single root cause.[72] They, like Lowe, emphasize a dialectic between society and biology: biology does not explain the shape of society so much as society explains the shape of biology. Though biology may condition behavior, biology is constantly being molded and changed by diet, occupation, quality of health care, income levels, stress, and exercise.

Lowe shows, for example, that female weakness—the smaller, frailer female frame (long an explanation for women's frail social powers)—has been molded, at least in part, by social factors. Though women do not

69. François Poullain de la Barre, *De l'égalité des deux sexes discours physique et moral* (Paris: Chez Jean du Puis, 1676), 112. For a similar argument see *An Essay in Defence of the Female Sex* (London: A. Roper & E. Wilkinson, 1696), 12–13 (attributed to Mary Astell or Judith Drake).

70. Blackwell (n. 64 above); and Gamble (n. 64 above).

71. Bleier (n. 59 above); Hubbard and Lowe, eds. (n. 59 above); Anne Fausto-Sterling, *Myths of Gender: Biological Theories about Women and Men* (New York: Basic Books, 1985); Lynda Birke, *Women, Feminism and Biology: The Feminist Challenge* (New York: Methuen, Inc., 1986).

72. Birke, chap. 5; Fausto-Sterling, 7–8.

have the same type of musculature as men, women may have the same potential for strength as men of comparable size. Lowe shows that as exercise becomes more vigorous so does physical strength. Lowe concludes that women's legendary frailness has as much to do with social expectations of ladylike behavior as an inherent weakness in women.[73]

This project of fighting science with science confronts two central contradictions. First, in order to refute sociobiologists, some arguments minimize sex differences, thus retaining the assumption that social equality depends on biological similarity. Second, as Bleier points out, an attempt to fight science with science must accept the idea that there exists a body of "neutral" knowledge that can be refined until we have the "facts" right.[74] Over one hundred years ago, German feminist Hedwig Dohm maintained that variations on the biological argument could be infinite and that one must therefore reject the biological argument altogether.[75] Dohm and those following her lead—philosopher Janet Sayers, biologist Richard Lewontin, historians of science Elizabeth Fee, Steven Rose, and Donna Haraway (Hubbard, Birke, and Bleier also contribute to this work)—reevaluate the function of science in Western society. By lifting the argument about sex differences out of the realm of "pure" science and placing it within its social context, this strand of the biological debate has relieved feminists of the task of minimizing sex differences.

In particular, recent works by Elizabeth Fee, Janet Sayers, Richard Lewontin, and Ruth Bleier attempt to understand what the debates about woman's nature (focused though they are on women) tell us about science and scientific discourse in Western society. Why have biological arguments against women's ability to do science seemed so persuasive to so many? Why did the study of sex differences become a priority of scientific investigation? How have sex differences been used to assign men and women to particular roles in the social hierarchy?

Fee, Lewontin, Rose, and Kamin, for example, have explained the role science has played in reaffirming societal attitudes toward race and sex.[76] They examine how the rise of a belief in meritocracy and individual ability inaugurated and maintained the complementary belief that social inequalities resulted not from systematic discrimination but from intrinsic inabilities within certain groups. If women have not made outstanding contributions to the sciences, so the logic goes, perhaps women are natu-

---

73. Marian Lowe, "Dialectic of Biology and Culture," in Lowe and Hubbard, eds., *Woman's Nature: Rationalizations of Inequality* (Elmsford, N.Y.: Pergamon, 1983), 39–62.

74. Bleier (n. 59 above), 13.

75. Dohm (n. 63 above). On Dohm, see Renate Duelli-Klein, "Hedwig Dohm: Passionate Theorist," in *Feminist Theorists: Three Centuries of Women's Intellectual Traditions*, ed. Dale Spender (London: Women's Press, 1983), 165–83.

76. Fee (n. 63 above); Lewontin, Rose, and Kamin (n. 68 above).

rally incapable of doing science. As Fee and Sayers argue, it became the task of science to define the natural abilities (and inabilities) of identifiable social groups—women and blacks among others.[77] In the course of the eighteenth and nineteenth centuries, scientists mustered evidence that seemed to show that human nature was not uniform, that it differed according to age, race, and sex. The differences they discovered between the male and the female body then were used to justify a social agenda that privileged men, economically, socially, and politically.

It is important to understand that this science was assumed to be neutral, assumed to have no social agenda of its own. Many believed that science could reveal the "nature of human beings" and could play an important role as a "neutral arbiter" in social debates.[78] Yet, the nature and capacities of women were vigorously investigated by a scientific community from which women (and the feminine) were (and are) almost entirely absent. As Bleier points out: "While . . . [a scientific] discourse appears to be the *uncovering* of truth, it [in fact] rests upon and conceals the struggle between those who have the power to discourse and those who do not. Both by their practices of exclusion and their definitions of what *is*, what is to be discussed, what is false or true, discourses *produce* rather than reveal truth."[79]

The Impact of the Few? Gender Distortions in Science

Hitherto institutions, laws, sciences, philosophy have chiefly borne the masculine imprint; all of these things are only half human; in order that they may become wholly so, woman must be associated in them ostensibly and lawfully. [Jenny d'Héricourt, 1864][80]

While many may concede that the institutions of science have been sexist and that sexist institutions may have distorted the content of some sciences, few are willing to concede that gender-based patterns of exclusion may have distorted the norms and methods of scientific practice. In the last decade, a growing number of feminists have begun to define and

77. Fee (n. 64 above); Sayers (n. 65 above).

78. The phrases are Robert N. Proctor's. See his "The Politics of Purity: Origins of the Ideal of Neutral Science" (Ph.D. diss., Harvard University, 1984). As Elizabeth Fee has pointed out, interest in sex differences surges in response to feminist challenges. See also Fee (n. 64 above); and L. J. Jordanova, "Natural Facts: A Historical Perspective on Science and Sexuality," in *Nature, Culture and Gender*, ed. Carol P. MacCormack and Marilyn Strathern (Cambridge: Cambridge University Press, 1980), 42–69.

79. Bleier (n. 59 above), 194.

80. Jenny d'Héricourt, *A Woman's Philosophy of Woman; or, Woman Affranchised* (New York: Carleton Press, 1864), 239. I thank Karen Offen for calling d'Héricourt's work to my attention.

discuss gender-based distortions in the norms and practices of science.[81] These authors emphasize a central tenet of the social critique of science— that scientific knowledge cannot be neutral because it is structured by power relations—and they examine the masculinist values practiced within the context of the norms of scientific inquiry.[82]

Explanations of the development of this bias in the methods of modern science are many and varied; some emphasize psychological dimensions, and others emphasize social and historical dimensions.[83] Evelyn Fox Keller, for example, focuses on the psychological dynamics of family relations as the source of gender distinctions in cognitive skills. Using the psychological theory of object relations, Keller argues that the link in the ideology of modern science between objectivity (a cognitive trait), autonomy (an affective trait), and masculinity (a gender trait) derives from the asymmetries of child care. Social asymmetries in child rearing, Keller holds, create asymmetries in the cognitive skills of girls and boys. As long as caretakers are most often women, children will associate love and relatedness with woman as mother. A girl will thus associate her femaleness with "merging"—what Keller identifies as the close identity of two subjects. A boy, however, will develop a male identity only by distinguishing himself from mother or femaleness. Thus, a boy

---

81. In addition to the literature discussed below, see also Rita Arditti, "Feminism and Science," in *Science and Liberation*, ed. R. Arditti, P. Brennan, and S. Cavrak (Boston: South End Press, 1980); Gabriele Dietze, ed., *Die Überwindung der Sprachlösigkeit* (Darmstadt: Luchterhand, 1979); Hilary Rose and Steven Rose, eds., *Ideology of/in the Natural Sciences* (Boston: G. K. Hall & Co., 1980); Brian Easlea, *Witch-Hunting, Magic and the New Philosophy: An Introduction to Debates of the Scientific Revolution, 1450–1750* (Brighton, Sussex: Harvester Press, 1980), *Masculinity, Scientists and the Nuclear Arms Race* (London: Pluto Press, 1983), and *Science and Sexual Oppression* (London: Weidenfeld & Nicolson, 1981); Luce Irigaray, *Speculum de l'autre femme* (Paris: Editions de Minuit, 1974); Marielouise Janssen-Jurreit, *Sexism: The Male Monopoly on History and Thought*, trans. Verne Moberg (New York: Farrar, Straus & Giroux, 1982); Evelyn Fox Keller, "Women in Science: An Analysis of a Social Problem," *Harvard Magazine* (October 1974), 14–19; Michelle Le Doeuff, "Women and Philosophy," *Radical Philosophy* 17 (1977): 2–10; Teri Perl, "The Ladies' Diary or Woman's Almanack, 1704–1841," *Historia Mathematica* 6 (1979): 36–53; Emmanuel Rudolph, "How It Developed that Botany Was the Science Thought Most Suitable for Victorian Young Ladies," *Children's Literature* 2 (1973): 92–97; Ann Shteir, "With Bliss Botanic: Women and Plant Sexuality" (paper presented at the American Society for Eighteenth-Century Studies, San Francisco, April 1980); Lieselotte Steinbrügge, "Vom Aufstieg und Fall der gelehrten Frau," *Lendemains*, vols. 25/26 (1982); Helen Longino, "Scientific Objectivity and Feminist Theorizing," *Liberal Education* 67, no. 3 (1981): 187–95; Ruth Wallis and Peter Wallis, "Female Philomaths," *Historia Mathematica* 7 (1980): 57–64.

82. For the relation of feminist critique of science to other social critiques of science, see Rose (n. 58 above); and Elizabeth Fee, "Critiques of Modern Science: The Relationship of Feminism to Other Radical Epistemologies," in *Feminist Approaches to Science*, ed. Ruth Bleier (New York: Pergamon Press, 1986), 42–56.

83. For a detailed analysis and critique of feminist philosophies of science, see Sandra Harding, *The Science Question in Feminism* (Ithaca, N.Y.: Cornell University Press, 1986).

learns to differentiate sharply between subject and object—between himself and others—and Keller argues that it is precisely this male mind set that has been written into the norms and methods of modern science.[84]

Others locate the development of the individual psyche within the division of intellectual labor in modern industrial society. Advocates of this approach—Elizabeth Fee, Donna Haraway, Nancy Hartsock, Keller (in her more historical work), Carolyn Merchant, and Hilary Rose, among others—claim that the historical construction of gender roles accompanied the rise of modern market economies. This social division of labor spawned divisions in intellectual labor, and in the course of the eighteenth century, specific intellectual skills became attached to increasingly separate spheres of public and private life. Skills such as reason and objectivity became required for participation in the public spheres of government, commerce, science, and scholarship. At the same time, feeling and subjectivity became skills confined to the private sphere of hearth and home.

As Hartsock points out, intellectual qualities often associated with gender do not have a biological attachment to women or to men. Rather, intellectual qualities are more closely associated with activities than with individuals. Thus particular intellectual skills are developed as a response to the requirements of the public or domestic spheres of society. Western masculinity in science is a bias produced by the overrepresentation of both science and men in the public sphere. Hartsock argues that science and its conceptual tools—in the past constructed almost exclusively by men—represent primarily male cultural experience.[85]

A vast literature has demonstrated how the Western philosophical traditions out of which modern science developed have embedded divisions in intellectual labor within a series of dualisms.[86] The basic categories

84. For other versions of Keller's argument, see her "Gender and Science," *Psychoanalysis and Contemporary Thought* 1 (1978): 409–33, and "Feminism and Science," *Signs* 7, no. 3 (1982): 589–602, esp. 597. For her most recent work, see Keller, *Reflections on Gender and Science* (New Haven, Conn.: Yale University Press, 1985), pt. 2.

85. Nancy Hartsock, "The Feminist Standpoint: Developing the Ground for a Specifically Feminist Historical Materialism," in *Discovering Reality: Feminist Perspectives on Epistemology, Metaphysics, Methodology, and Philosophy of Science*, ed. Sandra Harding and Merrill Hintikka (Boston: Kluwer Academic Press, 1983).

86. See Karin Hausen, "Die Polarisierung der 'Geschlechtscharaktere,' " in *Sozialgeschichte der Familie in der Neuzeit*, ed. Werner Conze (Stuttgart: Klett Cotta Verlag, 1976); Susan Moller Okin, *Women in Western Political Thought* (Princeton, N.J.: Princeton University Press, 1979); S. Bovenschen, *Die imaginierte Weiblichkeit* (Frankfurt am Main: Suhrkamp Verlag, 1979); Zillah Eisenstein, *The Radical Future of Liberal Feminism* (New York: Longman, Inc., 1981), esp. chap. 4; Jean Elshtain, *Public Man, Private Woman: Women in Social and Political Thought* (Princeton, N.J.: Princeton University Press, 1981); Genevieve Lloyd, *The Man of Reason: "Male" and "Female" in Western Philosophy* (London: Methuen & Co., 1984); "Naturwissenschaftlerinnen: Einmischung statt Ausgrenzung," *Feministische Studien*, vol. 4, no. 1 (1985).

of modern thought have thus taken shape as a series of dualities: reason has been opposed to feeling, fact to value, culture to nature, science to belief, the public to the private. One set of qualities—reason, fact, object— came to represent constituents of rational discourse and scientific knowledge. The other set of qualities—feeling, value, subject—have been defined as unpredictable and irrational. When the dualism of masculinity and femininity was mapped onto these categories, masculinity became synonymous with reason and objectivity—qualities associated with participation in public spheres of government, commerce, science, and scholarship. Femininity became synonymous with feeling and subjectivity—qualities associated with the private sphere of hearth and home.

Despite nineteenth-century insistence on the complementarity (and equality) of the sexes, the feminine has consistently been subordinate to the masculine. Carolyn Merchant has argued that the deep-seated dualisms of Western culture have encouraged and maintained a hierarchical domination of subject over object, male over female, and culture over nature. In *The Death of Nature*, Merchant explores how the historical association between science and masculinity has the practical effect of justifying the domination and exploitation of both nature and women.[87]

One task of those who study gender in science has been to reverse traditional values and throw a positive light on the feminine side of the opposition: to promote women's values as an essential aspect of human experience and to seek a new vision of science that would incorporate these values.[88] Recent studies on mothering and caring focus attention on positive aspects of women's experience. Carol Gilligan has found that women speak about moral decisions in a different voice. This voice has been described variously as a voice tempered with caring (Hilary Rose and Nel Noddings) or maternal thinking (Sara Ruddick).[89]

The study of the historical construction of gender differences offers an important opportunity to understand what scientists have devalued and why. However, as Sandra Harding cautions, values ascribed to women originated in the historical subordination of women. In her view, the much touted "feminine" is merely the obverse of the culturally dominant "masculine." Harding also warns that the feminine qualities celebrated by feminists—cooperation, caring, holism—do not accurately

87. Merchant, "Isis' Consciousness Raised" (n. 6 above), and *The Death of Nature: Women, Ecology, and the Scientific Revolution* (San Francisco: Harper & Row, 1980).

88. Fee, "A Feminist Critique of Scientific Objectivity" (n. 58 above), 6–7.

89. Carol Gilligan, *In a Different Voice: Psychological Theory and Women's Development* (Cambridge, Mass.: Harvard University Press, 1982); Rose (n. 58 above); Nel Noddings, *Caring: Feminist Approaches to Ethics and Moral Education* (Berkeley and Los Angeles: University of California Press, 1984); Sara Ruddick, "Maternal Thinking" (paper presented at the Boston Area Colloquium for Feminist Theory, Northeastern University, Spring 1984).

reflect the social experience of all women (women of different races, ages, and classes) and disenfranchised peoples.[90]

Though traditional feminine values alone may not serve well as an epistemological base for new philosophies of science, they provide a powerful critique of science by throwing into bold relief prevailing distortions in the norms and methods of science. The feminist critique of science promotes feminine values as an essential aspect of human experience and envisions a science that would integrate all aspects of human experience into our understanding of the natural world. The task at hand is to refine the human effort to understand the world by restoring to science a "lost dimension"—the feminine—whose loss has distorted human knowledge.[91] In attempting to overcome the parochialisms of science by uncovering the influence of gender ideology on the practices of science, feminists join with other social activists in calling for a more socially responsible science. The task of making science less masculine is also the task of making it more completely human.

Program in Values, Technology, Science, and Society
Stanford University

90. Harding (n. 83 above), 26–27, and chap. 6.
91. Fee, "A Feminist Critique of Scientific Objectivity" (n. 58 above). See also Elizabeth Fee, "Women's Nature and Scientific Objectivity," in Lowe and Hubbard, eds. (n. 73 above), 22–25; and Bleier (n. 59 above), chap. 8.

Sexual Segregation in the Sciences: Some Data and a Model

Margaret W. Rossiter

I would like to extend *Signs*'s previous discussion of occupational segregation[1] by presenting a new model for assessing occupational segregation in the sciences.[2]

Several years ago Henry W. Menard, currently director of the U.S. Geological Survey, in a provocative book, *Science: Growth and Change*,[3] showed that the "growth rate" of a science, however measured, has a strong effect on the careers of scientists in that field. Menard described how advancement comes early to a man *(sic)* in a fast-growing field so that after a few years he can become a full professor, a spokesman for the field, and even president of a very young professional society. In a slowly growing field, however, the same person would have to be older, and have published more, before becoming a full professor, winning a certain prize, or being elected to a major professional office.

One can extend Menard's reasoning beyond the timing of scientific careers. Growth forces may also affect the employment opportunities of women and minority groups and attitudes of other scientists toward them. Menard's model can be used to explain why, for instance, some fields have been more "open" to women and minority members, offering them jobs even during the depression of the 1930s when others did not.

1. *Signs: Journal of Women in Culture and Society,* vol. 1 no. 3, pt. 2 (Spring 1976).

2. Based on Margaret W. Rossiter, "Women Scientists in America: An Historical Survey" (forthcoming).

3. (Cambridge, Mass.: Harvard University Press, 1971).

This essay originally appeared in *Signs,* vol. 4, no. 1, Autumn 1978.

This extended Menard model (referred to hereafter as the Menard-Rossiter model) can lead to several testable hypotheses about the flow of trained womanpower. For example, in a rapidly growing field, with a shortage of highly qualified people, women would be tolerated and even sought out, especially since they would typically be paid lower salaries than men. But in a crowded field where the growth rate had slowed, as in some fields in the 1930s, women, especially married women, would be the first people laid off, or would not be hired. Thus one might expect a correlation between growth rates in various sciences in a given period and the percentage of women entering each field. If other factors were equal, the fastest growing fields should contain the highest percentage of women.

Table 1 presents data on fifteen fields for the period 1920–38. Some confirmation of the hypothesis was obtained for nine of these fields (nutrition, psychology, biochemistry, microbiology, zoology, medical sciences, geology, physics, and botany), by means of Spearman's statistical method of rank correlation. Despite this, however, the correlation across all fifteen fields for the period was not statistically significant ($r = .26$, $\alpha <$ .10) because of the large discrepancies in the six other fields. The cases where the hypothesis apparently failed were helpful in illuminating the other considerations in examining the flow of women into the various sciences. Thus, although the number of women entering engineering and chemistry from 1921 to 1938 was high compared with the past, there were so many men already in those fields that the percentage of women in them barely rose. In the case of mathematics, the method of rank correlation obscured a real change—the percentage of women in the field increased from 5.8 in 1921 to 10.6 in 1938, but the rank rose only from eighth to seventh. In agricultural sciences, despite a moderately rapid growth rate, no women entered the field, and its percentage remained at zero. Possibly the barriers to women were particularly strong in this field, or perhaps a minimum momentum or rate of growth is needed to start the process of integration. Another factor that defies the basic hypothesis can be observed in the fields of anthropology and astronomy, which women entered in relatively large numbers, although they should have found the going very difficult and felt pressure to leave the field to men. In anthropology appearances are misleading because women tended to get fellowships and other temporary appointments in the 1930s, while men got the few available permanent jobs. In astronomy women entered the field and accepted bleak employment prospects, while men left the field or did not enter it at all. Thus women were not only likely to enter and be welcomed into rapidly growing fields but they were, at the same time, more willing than men to endure the hardships of a stagnant or shrinking field. They were relatively less attracted to fields undergoing average growth, where normal competitive and discriminatory practices prevailed. A revised conclusion might be, there-

Table 1

Percentage of Women in Field in 1938 versus Growth Rates, 1920–38

Field	1920 Women	1920 Total	1938 Women	1938 Total	1938 %	1938 Rank	Rate of Increase, 1920–38 Women Only %	Women Only Rank	Both Sexes %	Both Sexes Rank
Agriculture	0	396	0	950	...	15	...	15	139.9	10
Anthropology	8	116	29	254	11.4	6	262.5	8	119.0	12
Astronomy	20	326	36	236	15.3	3	80.0	14	−27.6	15
Biochemistry	7	259	129	1,029	12.5	5	1,742.9	2	297.3	3.5
Botany	84	948	256	2,806	9.1	9	204.8	11	196.0	7
Chemistry	28	1,378	163	4,838	3.4	12.5	482.1	5	251.1	6
Engineering	0	882	8	3,508	.2	14	Infinite	1	297.7	3.5
Geology	23	653	60	1,485	4.0	11	160.9	13	127.4	11
Mathematics	42	726	151	1,426	10.6	7	259.5	10	96.4	14
Medical science	41	1,373	186	3,511	5.3	10	353.7	7	155.7	9
Microbiology	18	216	109	809	13.5	4	505.6	4	274.5	5
Nutrition	20	20	164	389	42.4	1	720.0	3	1,845.0	1
Physics	21	885	63	1,888	3.3	12.5	200.0	12	113.3	13
Psychology	60	294	227	1,277	21.7	2	361.7	6	334.4	2
Zoology	78	1,014	281	2,881	9.8	8	260.3	9	184.1	8
Total	450	9,486	1,912	27,287	(7.0)	...	(324.9)	...	(187.8)	...

SOURCE.—*American Men of Science*, 3d ed. (Garrison, N.Y.: Science Press, 1921), and 6th ed. (New York: Science Press. 1938).

NOTE.—Figures in parentheses are averages. No. of women based on actual count; no. of men on samples of fields of every twenty-fifth male fore, that the women's comparative advantage lies at the two extremes in the growth curve rather than in the middle or at just one end.

Unemployment can also be used to test the validity of the Menard-Rossiter model. Although some authorities have deplored the lack of unemployment data on scientists before the 1970s,[4] such information can be obtained from the older editions of the *American Men of Science*, which retained the names of scientists who lost their jobs. Of the 1,912 women scientists in the 1938 edition of *American Men of Science*, 181 or 9.5 percent did not hold jobs in 1938, compared with 1.3 percent of the over 27,000 men. Some of these women may have stayed out of the work force voluntarily, but others had held precarious positions in the early 1930s before finally losing even that vestige of employment by 1938. Many listed themselves as "guest investigator" or "fellow by courtesy" at a university, and others claimed to be in "private research" or doing "ecological investigations" with no institution specified. They wished others to consider them actively involved in science, although they did not hold jobs. Curiously, these entries in the *American Men of Science* show that unemployment in 1938 was not evenly spread across the sciences but varied among fields, with high unemployment in anthropology, geology, and astronomy, low unemployment in nutrition, and none in engineering.[5] This variation seems strange, because there is no reason why a woman botanist should be more eager to stay at home than a woman geologist, unless there were important differences in her marital status or employment prospects.

One difference among the sciences was in the percentage of women in the field who were married. Although the *American Men of Science* is not a totally reliable source of data on the marital status of its women (the thrice-married Margaret Mead listed herself as "Miss" in 1938), it does provide a rough guide. A breakdown of the unemployment figures for 1938 by marital status shows a large difference between rates for married women (24.0 percent) and single women (4.3 percent), confirming the contemporary attitude that justified discrimination against married women who took good jobs away from family men. Under such hostile conditions it is perhaps less surprising that 24 percent of the married women scientists were unemployed than that 76 percent did hold jobs!

Assuming that married women were a "safety valve" in the scientific labor market of the 1930s (first fired, last hired), the employment rate of these married women scientists should be a barometer of opportunities in a field. A second hypothesis to test the Menard-Rossiter model would be a comparison of the growth rate of women in a field from 1920 to 1938 with the rate of *employment* of married women in the same field in

4. Carolyn Shaw Bell, "Women in Science: Definitions and Data for Economic Analysis," in *Successful Women in the Sciences: An Analysis of Determinants*, ed. Ruth B. Kundsin (New York: William Morrow & Co., 1972), pp. 134–42.

5. Taken from *American Men of Science* (New York: Science Press, 1938).

1938 (calculated from the data on *unemployment* described earlier). Thus, in a fast-growing field the percentage of married women employed should be high, almost 100 percent, but in a slowly growing or stagnating field it should be low. These predictions were confirmed, as shown in table 2, where the rank correlation was statistically significant ($r = .71$, $\alpha < .01$). However, as before, the anomalies were interesting. In anthropology and botany married women were doing better than one would have predicted. In fact, all the married women anthropologists were employed, and it was the single ones who lacked jobs, a surprising finding for a slowly growing science. This peculiarity probably stemmed from the traditional role of wealthy married women in anthropology. They made do with titles such as "collaborators" or "associates" since they wished to be considered legitimate investigators. However, they were not in direct competition for salaried jobs, which the younger women sought. The good showing of the married women botanists is surprising and may indicate that conditions at the U.S. Department of Agriculture, one of their principal employers, were better than is often assumed. In two fields, chemistry and nutrition, married women did less well than predicted by the hypothesis. However, women chemists in industry were expected to quit their positions on marriage. The relatively poor showing of the married women nutritionists is inexplicable, especially since the field of home economics has generally been depression and war proof because of its close ties to two governmental interests, social welfare and war planning. It could be, however, that married

Table 2

Growth Rates versus Employment Rates of Married Women, by Field, 1938

Field	Growth Rates, 1920–38 (Women in Field)		Employment Rates, 1938 (Married Women)	
	Rank	Rate	Rank	Rate (%)
Engineering	1	...	1.5	100.0
Biochemistry	2	1,742.9	3	90.3
Nutrition	3	720.0	7	79.4
Microbiology	4	505.6	4	86.7
Chemistry	5	482.1	10	73.7
Psychology	6	361.7	8	78.0
Medical science	7	353.7	5	86.3
Anthropology	8	262.5	1.5	100.0
Zoology	9	260.3	11	69.1
Mathematics	10	259.5	9	75.0
Botany	11	204.8	6	79.6
Physics	12	200.0	14	58.3
Geology	13	160.9	13	68.4
Astronomy	14	80.0	12	68.7

Sources.—See table 1.
Note.—$r = .71$ (significant at .01 level).

home economists were a self-selected group, more domestically inclined than other women scientists.

Thus the Menard-Rossiter model which claims that the rate of growth of a field has some importance for the percentage of women entering it and for the percentage of married women employed in it has received some confirmation. Other factors, such as the vast predominance of men, rigid entrance barriers, availability of government employment, or other peculiarities, prevent the model—which assumes that sexual discrimination is a function of crowding—from working more accurately in all fields. Its effectiveness in some fields, however, implies that there is some feedback mechanism that channels women scientists into those fields most open or receptive to them. But the process by which women scientists choose fields, and the conservative function of written or spoken "vocational guidance" which advises women at every level to head for safe, familiar, "feminine" fields and to avoid pioneering in risky territory, remains to be explored.

Office for the History of Science and Technology
University of California, Berkeley

Images of Female Medical Students at the Turn of the Century

Sandra L. Chaff

Many nineteenth-century supporters of the cause of women in medicine felt that women, if given a chance, would bring a new dimension to health care. These photographs capture the image of women engaged in that endeavor. The scenes pictured here were as familiar around Philadelphia at the turn of the century as they were controversial. The photographs are reproduced from the Archives and Special Collections on Women in Medicine of the Medical College of Pennsylvania. This school, formerly known as the Woman's Medical College of Pennsylvania, was founded in 1850 as the Female Medical College of Pennsylvania, the first medical school regularly organized for the exclusive purpose of training women to be physicians. It is the only one of the seventeen medical schools for women founded in the latter half of the nineteenth century to have survived into modern times. In 1969 the school began admitting male medical students, and a year later it changed its name to the Medical College of Pennsylvania.

The Special Collections on Women in Medicine, a component of the College's Archives and Special Collections, is one of the most comprehensive collections of material on women physicians. It consists of institutional collections, personal papers, books, reprints, and a file of over 2,000 photographs relating to women not only as physicians but also as visiting nurses, administrators, teachers, and students of medicine.

Florence A. Moore Library of Medicine
Medical College of Pennsylvania

This article originally appeared in *Signs*, Vol. 4, No. 1, Autumn 1978.

41

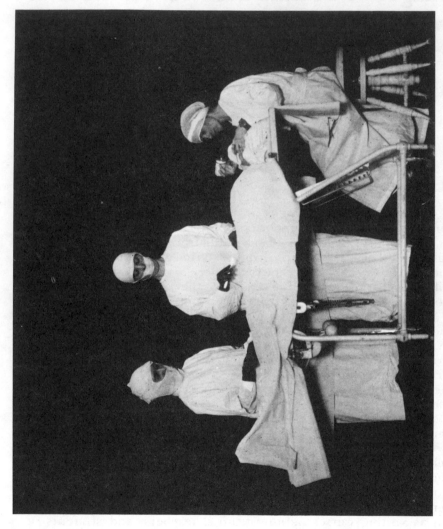

Fig. 1.—Administering ether in an operating room, Woman's Medical College of Pennsylvania, ca. 1920s.

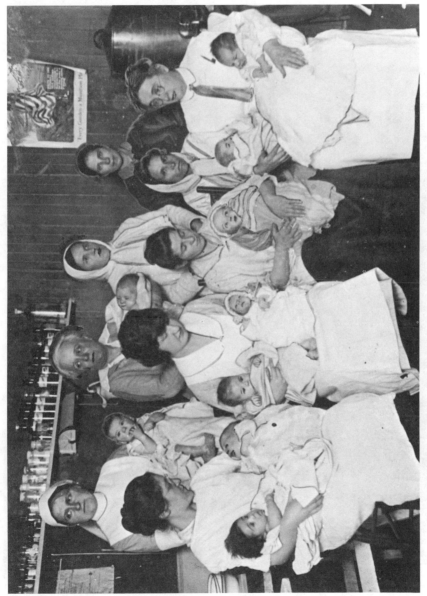

Fig. 2.—Barton Dispensary of the Woman's Medical College of Pennsylvania, ca. 1918

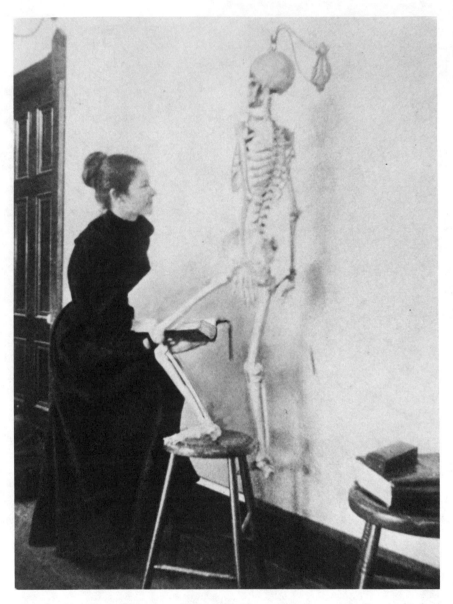

Fig. 3.—Mary Irvin Thompson, M.D., 1898. Photo taken from student's scrapbook.

FIG. 4.—Matilda A. Evans, M.D., 1897. Reportedly the first black physician in South Carolina.

Women and the History of American Technology

Judith A. McGaw

Only six years ago, when preparing a paper for the bicentennial meeting of the Society for the History of Technology (SHOT), Ruth Schwartz Cowan puzzled over a new question: "Was the female experience of technological change significantly different from the male experience?" Answering in the affirmative, Cowan accepted the challenge of charting what was largely terra incognita. She identified four areas of special concern for a feminist history of technology: technology's impingement

My thanks to Estelle Freedman, Brooke Hindle, Elizabeth Hitz, Margo Horn, Thomas Hughes, Henrika Kuklick, Alejandra Laslo, Russell Maulitz, Rosemary Stevens, and Arnold Thackray for reading and commenting on an earlier draft of this essay and to the authors of several recent or forthcoming publications for making manuscript copies available to me. Joan N. Burstyn, Ruth Schwartz Cowan, Laurence F. Gross, Joseph C. Miller, Leila J. Rupp, Joan W. Scott, and Helena Wright helped me clarify some of the ideas set forth in the first section.

EDITORS' NOTE: Technological innovation has in our time the centrality—and the moral, intellectual, and social significance—that theological speculation had in the medieval period. Judith McGaw addresses the ways in which women serve as potential and actual contributors to constantly changing technologies.

This essay originally appeared in *Signs,* vol. 7, no. 4, Summer 1982.

on women's activities as bearers and rearers of children; technology's relationship to the segregated and exploitative world of women's employment; technology's role in woman's place, the home; and technology's peculiar relationship to women in a society simultaneously celebrating "Yankee ingenuity" and "systematically training more than half of our population to be 'un-American' by socializing women to be unskilled in mathematics and mechanics."[1]

Six years later Cowan's categories still encompass most of the history of American women and technological change. Moreover, as in 1976, we can offer only tentative answers to most of her provocative questions. It also remains generally true that "the history of the uniquely female technologies"—those that affect women because they "menstruate, parturate and lactate"—"is yet to be written."[2]

Yet measured against Cowan's bicentennial benchmark, the scholarship available today reflects prodigious effort in a new and promising field. The bicentennial meeting at which Cowan spoke witnessed the organization of Women in Technological History (WITH), which has encouraged feminist scholars, promoted greater attention to women and technology at SHOT meetings and in the society's journal, *Technology and Culture,* and disseminated primary and secondary bibliographic information through the *WITH Newsletter.* At the same time, scholarship in two sister disciplines, the history of science and the history of medicine, has explored the roles of women as scientific technologists and studied the development of specialized medical technology for women. The years since 1975 have also been the first in which "the study of women's work both inside and outside the home has attracted a truly broad range of inquiry."[3] Thus, many women's historians, although not primarily concerned with technology, have nonetheless increased our awareness of technology's relationship to women's industrial, commercial, and domestic labor.

The resulting rich, diverse, scattered, uneven, and amorphous body of literature makes this review essay exploratory rather than definitive. As in most developing fields, there is little explicit agreement on what the central questions are, but I will suggest some emerging themes im-

1. Ruth Schwartz Cowan, "From Virginia Dare to Virginia Slims: Women and Technology in American Life," *Technology and Culture* 20 (January 1979): 51–63.

2. Ibid., pp. 51–52.

3. Mary Beth Norton, "Review Essay: American History," *Signs: Journal of Women in Culture and Society* 5, no. 2 (Winter 1979): 324–37, esp. 328. Eleanor Maas, WITH bibliographer for the organization's first five years, is currently preparing a review of her findings for *Technology and Culture.* I am grateful to her for sharing her card catalog and an early version of her essay with me. Her successor, Helena Wright, Librarian, Merrimack Valley Textile Museum, will continue to collect bibliographic information and pass on her findings through the *WITH Newsletter.* Those acquainted with useful, but relatively obscure, primary or secondary works or with dissertations or works in progress can assist by calling such materials to her attention.

plicit in the literature. The enormous number of potentially relevant works has prompted the tactical decision to concentrate on two of Cowan's categories, the technology of homemaking and the technology of the nondomestic work place, because the scholarship is fullest and most developed on these topics. Moreover, scholars in related historical specialties have produced much of the best research on Cowan's remaining categories, technology influencing women because of their distinctive physiology and women's limited participation as inventors, developers, and innovators of new technology. Earlier review essays in this journal already make much of this work accessible to women's studies scholars.[4] I will, therefore, confine myself to a few brief remarks on the existing literature in these areas and on continuing needs and opportunities for study. Finally, since Cowan's essay was explicitly "suggestive, but not definitive"[5] and a few scholars have explored areas outside her four fields, I will conclude with a brief discussion of several of these: technology and women's work in predominantly agricultural communities—those of the colonial era, the American West, and the South; technology as a tool for enhancing sex differences and reinforcing sex-role stereotypes through clothing, cosmetics, and hairdressing; the technological preconditions for and consequences of women's increasing importance as consumers; and the differential impact on women of technologies generally examined only from a masculine perspective—manufacturing technologies in industries from which women have been excluded, municipal technologies, and transportation and communications technologies.

Before proceeding, I must pause to discuss two special characteristics of the history of technology which transcend subject-matter divisions and may be unfamiliar to specialists in other women's studies disciplines. First, a number of fine scholars in the history of technology work in museums and have embodied their research in museum exhibits and living historical villages rather than in books or articles. Indeed, museums are often strongest where published work is weakest. For example, although scholarly publications give little sense of women's skills and tools in agricultural communities, living historical villages such as Plymouth Plantation, Old Bethpage Village Restoration, Old Stur-

4. Relevant review essays that have appeared in *Signs: Journal of Women in Culture and Society* include Nancy Schrom Dye, "History of Childbirth in America," 6, no. 1 (Autumn 1980): 97–108; Michele L. Aldrich, "Women in Science," 4, no. 1 (Autumn 1978): 126–35; Barbara Hayler, "Abortion," 5, no. 2 (Winter 1979): 307–23; M. Louise Fitzpatrick, "Nursing," 2, no. 4 (Summer 1977): 818–34; June A. Kennard, "The History of Physical Education," 2, no. 4 (Summer 1977): 835–42; Dolores Hayden and Gwendolyn Wright, "Architecture and Urban Planning," 1, no. 4 (Summer 1976): 923–34; Martha H. Verbrugge, "Women and Medicine in Nineteenth-Century America," 1, no. 4 (Summer 1976): 957–72; and Rayna Green, "Native American Women," 6, no. 2 (Winter 1980): 248–67. I will make only limited reference to primary sources and works published before 1970.

5. Cowan, p. 51.

bridge Village, Old Economy, Hancock Shaker Village, Old World Wisconsin, and Conner Prairie Pioneer Settlement offer a clear and detailed look at women's technological activities and the annual cycles of their work. Similarly, museums implicitly address the question, How did clothing technology help define woman and limit her activities? Museum collections usually have an upper-class bias, the First Ladies' gown collection of the National Museum of American History being only the most obvious example; even so, changes over time in the emphasis given to secondary sex characteristics invite the viewer acquainted with American women's history to draw conclusions about clothing's articulation of sex-role stereotypes. In addition, some scholars at living historical villages have searched the artifactual, pictorial, and literary record for information on common women's clothing. They "publish" their research in the dress of village guides and the manufacture of clothing by village personnel. Similarly, the collection of wigs and the manufacture of hairpieces at Colonial Williamsburg afford unique glimpses of hairdressing technique, a subject on which we lack published scholarship. The National Museum of American History supplements our meager knowledge of female inventors by exhibiting women's inventions displayed at the Centennial Exposition, the Peabody Institute offers a rare examination of women who went to sea, and the Wisconsin Historical Society's fine exhibit on the evolution of clothes-washing technology presents the fullest exposition of that subject.[6]

Museums also provide an essential complement to the more ample published accounts of work-related technology. Merrimack Valley Textile Museum's excellent display of both craft and industrial textile mechanisms makes clear, in ways that the extensive literature on textile mechanization does not, the changes occurring in production technology, the physical demands that excluded women from mule spinning, and the skill spinsters needed to produce uniform and unbroken threads. Old Slater Mill makes a similar contribution and, juxtaposed to the associated Wilkinson Machine Shop, provokes reflection about differences between men's and women's work in the early industrial era. The Harriet Beecher Stowe House kitchen renders graphically and faithfully the ideals and innovations of Stowe and her sister Catharine Beecher, two of America's earliest domestic reformers. The National Museum of American History's recent exhibit, "Perfect in Her Place," offers a more general survey of women and technology in paid employment and in the home. The exhibit is intended to travel, and an accom-

6. Published works on most of these subjects are discussed below, but see Deborah J. Warner, "Women Inventors at the Centennial," in *Dynamos and Virgins Revisited: Women and Technological Change in History*, ed. Martha Moore Trescott (Metuchen, N.J.: Scarecrow Press, 1979), pp. 102–19; and Julia C. Bonham, "Feminist and Victorian: The Paradox of the American Seafaring Woman of the Nineteenth Century," *American Neptune* 37 (July 1977): 203–18.

panying slide collection also makes it accessible to classrooms distant from the Washington, D.C., area.[7]

My second general observation is *caveat emptor:* ideally the social history of technology offers a nice balance of technical and social analysis, but the field remains underdeveloped and most works fall far short of that ideal. Unlike many earlier scholars who stressed the internal history of technology—the evolution of particular tools, machines, and techniques—more recent historians of technology define technology broadly as the system of tools, skills, and knowledge needed to make or do things. They stress that technology reflects social decisions and cannot be understood outside its social, political, economic, and cultural context. But they also recognize that social decisions cannot be understood without looking closely at technology itself. Without understand-

7. Similarly, Susana Torre, ed., *Women in American Architecture: A Historic and Contemporary Perspective* (New York: Whitney Library of Design, 1977), and Bernard Rudofsky, ed., *Now I Lay Me Down to Eat* (Garden City, N.Y.: Anchor Press, 1980), preserve for scholars' continued use earlier exhibits on residential and domestic technology mounted by the Archive of Women in Architecture and the Cooper-Hewitt Museum. Closer to home, most teachers of women's history will probably find that one of the numerous and widespread smaller museum collections includes the household furnishings women purchased and kept clean; the decorative arts they produced; the clothes they made, wore, washed, ironed, and mended; or the industrial and office machines they tended. Especially useful collections may be found at the Georgia Agrirama, Tifton; Ford Museum and Greenfield Village, Dearborn, Michigan; Mercer Museum, Doylestown, Pennsylvania; Museum of the City of New York; Watkins Mill, Lawson, Missouri; Oakland Museum, Oakland, California; Valentine Museum, Richmond, Virginia; Greenville County Museum of Art, Greenville, South Carolina; Gaston County Art and History Museum, Dallas, North Carolina; Eleutherian Mills-Hagley Foundation and Winterthur Museum, Wilmington, Delaware; Oklahoma Territorial Museum, Guthrie, Oklahoma; Kansas State Historical Museum, Topeka; Milwaukee Public Museum, Milwaukee, Wisconsin; Eckley Miners' Village, Eckley, Pennsylvania; and Margaret Woodbury-Strong Museum, Rochester, New York. Like published works, museums vary greatly in standards of historical excellence, depending upon their personnel, financial resources, and relative commitment to research, education, and entertainment. Publications helpful in understanding and evaluating museums include *History News* (American Association for State and Local History), *Museum News* (American Association of Museums), and museum reviews published in *Technology and Culture.* Those wishing to use historical villages for teaching should see Warren Leon, "Fear of Tripping, or Designing a College Field Trip That Works," *AHA Newsletter* 19 (January 1981): 7–8. Finally, material culture preserved outside museums is also a rich resource in the social history of technology. A useful introduction to several important federal agencies producing guides to the built environment is Ted Sande, "The National Park Service and the History of Technology: The New England Textile Mill Survey," *Technology and Culture* 14 (July 1973): 404–14. One guide especially rich in material relevant to women's industrial work is Historic American Engineering Record [Gary Kulik and Julia C. Bonham], *Rhode Island: An Inventory of Historic Engineering and Industrial Sites* (Washington, D.C.: Government Printing Office, 1978). These remarks are limited both by the extent of my travels and by my status as a museum visitor rather than a museum professional. I have, however, benefited greatly from the comments and suggestions of several historians of technology working in museums, including Elizabeth Hitz, Laurence F. Gross, Gary Kulik, Patrick C. Malone, Theodore Z. Penn, Caroline Sloat, Deborah J. Warner, and Helena Wright.

ing the tools, skills, and knowledge at a society's disposal, we cannot judge what technical options a society possessed and, thus, how a particular technological decision reflected the preferences of particular classes, individuals, or institutions. The cost of ignoring technology is the implicit technological determinism of most social historians who adduce "technological factors" as one "cause" of social change or most economic historians who treat technology as an "exogenous factor."[8]

Because the social history of technology is still in its infancy, much of the literature I cite either treats technology without examining its social context or, more commonly, explores technologically based social change without examining the technology itself. One clear example comes from the history of contraception—a central topic for a feminist history of technology. Currently, Norman Himes's classic survey of contraceptive technology details technique but does not analyze its social context. By contrast, Linda Gordon's insightful social and political survey *Woman's Body, Woman's Right* discusses contraceptive technology only in a general way, thus missing the chance to demonstrate convincingly that social institutions shaped women's technological options. James Reed provides a fuller discussion of the professionals, entrepreneurs, and philanthropists who promoted and financed birth control, but by failing to examine why elite decision makers favored investment in developing and distributing particular contraceptive techniques over others that were available or potentially available, he likewise shortchanges the issue of technological choice.[9]

Similar objections could be raised to virtually all of the works I cite. They offer useful approaches to the subject of women and technological

8. Thomas Parke Hughes, "Emerging Themes in the History of Technology," *Technology and Culture* 20 (October 1979): 697–711, reviews outstanding recent work in the history of technology and suggests the promise and current limitations of the social history of technology. The best general bibliographies are Eugene S. Ferguson, *Bibliography of the History of Technology* (Cambridge, Mass.: MIT Press, 1968); and Brooke Hindle, *Technology in Early America: Needs and Opportunities for Study* (Chapel Hill: University of North Carolina Press, 1966). For more recent works, see *Technology and Culture*'s annual bibliography in the history of technology. Although it treats male workers and technological change, the best example of a study that examines technological options closely to draw conclusions about social choice is David F. Noble, "Social Choice in Machine Design: The Case of Automatically-controlled Machine Tools, and a Challenge for Labor," in *Case Studies in the Labor Process,* ed. Andrew Zimbalist (New York: Monthly Review Press, 1980), pp. 18–50.

9. Norman Himes, *Medical History of Contraception* (New York: Gamut Books, 1963); Linda Gordon, *Woman's Body, Woman's Right: A Social History of Birth Control in America* (New York: Penguin Books, 1976); James Reed, *From Private Vice to Public Virtue: The Birth Control Movement and American Society since 1830* (New York: Basic Books, 1978). Other important works include Margaret Sanger's autobiography (*Margaret Sanger: An Autobiography* [New York: Dover Publications, 1971]) and David Kennedy's *Birth Control in America: The Career of Margaret Sanger* (New Haven, Conn.: Yale University Press, 1970), which also depict influential actors, but neglect or obscure what technical options they recognized, why they preferred some to others, and what medical and social consequences particular choices entailed.

change in history, but potential readers must approach them critically, aware both that technologies affect human experience and that people choose technologies and decide how to use them.

Technology and Women's Work Outside the Home

Scholarship on technology and women in the labor force has concentrated on two periods: the years from about 1790 to 1850, when manufacturing moved from the household and workshop to the factory, sometimes entailing women's movement outside the home to work; and the years from about 1870 to 1920, when larger percentages of women sought and obtained employment outside the home and the percentage of women in white-collar work rapidly increased. At present, the role of technology remains largely unexplored for a third critical period: the years since 1940, when married women's labor force participation has grown dramatically.[10] By concentrating on areas of obvious change, historians of technology, like most other historians, have neglected colonial women; rural women; black, hispanic, and native American women; and women living outside the Northeast.

Most early scholarship depicted technology as an agent of change, drawing women into paid employment and transforming tasks so that men's jobs were reassigned to women. By contrast, recent studies present the story of technology and women's work as a tale of continuity, despite superficial change. As Cowan has noted, three "economic facts of life for women" predated industrialization and persisted through the revolutionary technological developments of the nineteenth and twentieth centuries: women were almost always paid less than men even for the same work, women rarely performed the same tasks as men, and women considered themselves and were considered by others to be transitory members of the labor force.[11] Leslie Woodcock Tentler's study of early twentieth-century wage-earning women demonstrates how these constraints could reinforce one another, creating a distinctively passive, dis-

10. Women's historians will immediately notice that each of the three periods mentioned above also produced articulate and innovative feminist movements. The same three periods saw great increases in the percentage of the population enrolled in elementary, secondary, and college education, respectively. Few have explored directly how experience with the technology of the work place influenced feminist thought or how women's labor force participation accommodated changing educational requirements for coping with technology or, alternatively, facilitated the development of human capital as a prerequisite for technological innovation. These larger issues clearly call for sustained and sophisticated new scholarship. See Elizabeth Faulkner Baker, *Technology and Woman's Work* (New York: Columbia University Press, 1964); and Valerie Kincaid Oppenheimer, *The Female Labor Force in the United States* (Berkeley, Calif.: Institute of International Studies, 1970).

11. Cowan, p. 53.

ciplined, noncompetitive feminine response to a work-place technology designed to require rapid performance of repetitive tasks. She also shows how work-place technology and discipline underscored for young women the thought that they deserved less pay than did men and that they should marry and return to the home.[12]

Sex-segregated labor markets, the assumption that female workers are transient, and the persistence of lower pay for women encouraged the conceptualization of "women's work" according to preexisting sex-role stereotypes and permitted the continued employment of women in less mechanized or industrialized occupations. Thus, substantial historical continuities also exist in women's job descriptions. When we examine the tasks women performed before and after industrialization we find that, except in textiles, women often did the same tasks, whereas men's work became mechanized. Even when machines or the emergence of new industries created apparently novel jobs, employers consistently assigned women to jobs that were relatively monotonous and did not call for rapt attention, were interruptible and easily resumed, and were not visibly hazardous. The same characteristics describe women's work in both contemporary and historic preindustrial societies, suggesting that society's definition of women's work has remained essentially unchanged. One especially interesting recent study showing that women entered technologically new jobs because of sex-role stereotypes, not despite them, is Joseph Corn's analysis of female pilots. He musters convincing evidence that the aviation industry hired women to allay public fears by portraying flying as safe and easy. The women served their employers by always appearing well-groomed and feminine and by understating their mechanical skills and the dangers they encountered.[13]

Although it antedates the best new scholarship, any discussion of the industry-specific relationship of women to technology must begin with Elizabeth Faulkner Baker's *Technology and Woman's Work,* because it remains the only comprehensive survey. It also provides a useful summary of the earlier literature, primarily the work of contemporary social commentators, social scientists, and government statisticians. Covering

12. Leslie Woodcock Tentler, *Wage-earning Women: Industrial Work and Family Life in the United States, 1900–1930* (New York: Oxford University Press, 1979).

13. Baker; ibid., pp. 26–57; Judith A. McGaw, "Technological Change and Women's Work: Mechanization in the Berkshire Paper Industry, 1820–1885," in Trescott, ed., pp. 78–99; Susan L. Hirsch, *The Roots of the American Working Class: The Industrialization of Crafts in Newark, 1800–1860* (Philadelphia: University of Pennsylvania Press, 1978); Judith K. Brown, "A Note on the Division of Labor by Sex," *American Anthropologist* 72 (October 1970): 1073–78; Judith A. McGaw, "Historians and Women's Work: Insights and Oversights," in *Women and the Workplace: Conference Proceedings,* ed. Valerie Gill Couch (Norman: University of Oklahoma Press, 1979); Rosalyn Baxandall, Elizabeth Ewen, and Linda Gordon, "The Working Class Has Two Sexes," *Monthly Review* 28 (July/August 1976): 1–9; and Joseph Corn, "Making Flying 'Thinkable': Women Pilots and the Selling of Aviation, 1927–1940," *American Quarterly* 31 (Fall 1979): 556–71.

the decades from 1790 to 1960, Baker shows women "gradually re-capturing lost positions in a man's world—pushed and pulled by the driving forces of science and technology, wars, and education." Thus, she implicitly treats simple numerical gains in employment as evidence of progress and accords technology an active role. Although she does not ignore the influence of lower wages for women, union discrimination, and protective labor legislation, Baker's greatest weakness is her ten-dency to cite biological differences and sex stereotypes to explain the ways in which new technology facilitated women's movement into new occupations. Occasionally she acknowledges that sex differences were in the eye of the beholder, as when she notes that power-driven sewing machines reduced women's employment "because the pace set . . . was considered too great for women to maintain." More frequently, how-ever, she merely cites women's "light touch," "greater manual dexterity and softer hands," "keener appreciation of appearance," and limited strength as reasons for their employment. She also states repeatedly that women moved into unskilled jobs created by technological change but does not examine critically employers' designation of most women's tasks, even hand sewing, as unskilled. Nonetheless, Baker still offers the best and often the only historical treatment of technological change and women's work in many enterprises, including those of the electrical and electronics, chemical, and airplane manufacturing industries; data pro-cessing; knit goods and synthetic textiles; and tobacco and food pro-cessing.[14]

The manufacturing sector has never employed a majority of work-ing women, and the percentage employed in textiles has constituted a continuously diminishing minority. Like Baker's study, however, much of the recent literature on technology and women's work gives primary attention to women in manufacturing, especially textile manufacturing. But the new textile history scholarship has moved beyond Baker's em-phasis on technologically determined job shifts and rates of female em-ployment to bring increasingly complex analysis to the relationship be-tween technology and society. Thomas Dublin's thorough reexamination of the Lowell experience argues persuasively that women workers' tra-ditional values as "daughters of freemen," their desire for social and economic independence, and the solidarity fostered by their communal residence made them active, articulate critics of their employers' attempt to make them tend more and faster-moving machines while earnings remained constant or declined. The women generally accepted the dis-cipline of the factory and the boardinghouse, but they rejected exploita-tion. Dublin also explores the ways in which women's early industrial experience influenced their later marriage and residence choices. Focusing on Manchester, New Hampshire, another Lowell-system textile

14. Baker, pp. 27, 128, 140, 200, 425, and passim.

center, Tamara Hareven traces the continued interrelationship of home and workplace in the early twentieth century.[15]

By abandoning earlier scholars' preoccupation with Lowell-system mills and devoting attention to the more representative small mills outside northern New England, to woolen mills, and to carpet mills, a large and growing number of additional studies contribute to an emerging portrait of women's active and varied roles in early industrialization. David Jeremy's study of both cotton and woolen textile technology indicates that women's employment in American mills was one factor shaping the transfer of early technology from Britain to America and, increasingly, from America to Britain. In contrast to Jeremy, however, Helena Wright discovers that "unskilled" female textile operatives brought considerable skill to the factories. Anthony F. C. Wallace broadens the role of women in industrialization to include the contribution to work discipline made by mill owners' wives and daughters who organized Sunday schools for workers, especially female workers. Similarly, Barbara Tucker integrates women at home into the process of textile mechanization by showing that the southern New England mills' early success depended on child labor made tractable, in part, by maternal teachings and household discipline.[16]

15. Thomas Dublin, *Women at Work: The Transformation of Work and Community in Lowell, Massachusetts, 1826–1860* (New York: Columbia University Press, 1979); Tamara Hareven, "Family Time and Industrial Time: Family and Work in a Planned Corporation Town, 1900–1924," *Journal of Urban History* 1 (May 1975): 365–89.

16. David Jeremy, *Transatlantic Industrial Revolution: The Diffusion of Textile Technologies between Britain and America, 1790–1830s* (Cambridge, Mass.: MIT Press, 1981); Helena Wright, "Acres of Girlhood: Women in the Textile Industry, 1790–1860," in *New Work: Women's Industrial Employment in 19th Century America*, ed. Helena Wright (forthcoming, 1982); Anthony F. C. Wallace, *Rockdale: The Growth of an American Village in the Early Industrial Revolution* (New York: Alfred A. Knopf, 1978); Barbara Tucker, "The Family and Industrial Discipline in Ante-Bellum New England," *Labor History* 21 (Winter 1979–80): 55–74; John W. Lozier, "The Forgotten Industry: Small and Medium Sized Cotton Mills South of Boston," in *Working Papers from the Regional Economic History Research Center*, ed. Glenn Porter and William Mulligan, Jr. (Wilmington, Del.: Eleutherian Mills-Hagley Foundation, 1979), pp. 101–24; Elizabeth Hitz, "A Technical and Business Revolution: American Woolens to 1832" (Ph.D. diss., New York University, 1978); Susan Levine, "Ladies and Looms: The Social Impact of Machine Power in the American Carpet Industry," in Trescott, ed., pp. 67–76. The scholarship on the American textile industry is immense and growing rapidly. The best recent review and critique is Gary Kulik, "The Historiography of the American Textile Industry: Overview and Analysis" (paper presented at the Second Textile History Conference, Merrimack Valley Textile Museum, 1980). For a sense of the diversity of recent approaches, see John F. Kasson, *Civilizing the Machine: Technology and Republican Values in America, 1776–1900* (New York: Penguin Books, 1976), pp. 55–106; Tamara Hareven and Randolph Langenbach, *Amoskeag: Life and Work in an American Factory City* (New York: Pantheon Books, 1978); Carl Gersuny, " 'A Devil in Petticoats' and Just Cause: Patterns of Punishment in Two New England Textile Factories," *Business History Review* 50 (Summer 1976): 131–52; Daniel J. Walkowitz, *Worker City, Company Town: Iron and Cotton Worker Protest in Troy and Cohoes, New York, 1855–1884* (Urbana: University of Illinois Press, 1978).

Despite the fascination of both contemporaries and historians with female textile machine tenders, such workers represented a minority of women in manufacturing, and their experience of technological change was atypical. Most women did not tend machines, and most remained in traditional sex-typed occupations. Substantially lower female wages allowed manufacturers to employ women to perform the numerous hand operations that persisted in American manufacturing despite mechanization, tasks which women had performed in mills or shops before the advent of machines. Men usually operated the new machines that fabricated products, while women gave raw materials a preliminary screening and sorting, put finishing touches on the final product, scanned finished goods for flaws, sorted and graded them, washed them, folded them, labeled them, or packed them for shipping. True, mechanization affected such women's work indirectly. By speeding and multiplying the work of men in primary processing activities, machines created relatively more demand for female workers. For example, by 1900, American makers of glass and of nonferrous metal products each employed about thirty-five times as many women as they had in 1850. The female glass factory employees worked at finishing, decorating, and packing the product, and the female metalworkers at polishing, filing, soldering, weighing, and packing. Similarly, women did not tend papermaking machines, but machine production required increased numbers of women to prepare rags, sort and count paper, and operate less productive machines that applied polished surfaces or decorative imprints to the finished product and transformed flat sheets into envelopes. Electronic technology enabled post–World War II manufacturers to automate some women's tasks, such as scanning finished goods for defects and assembling electronic equipment, but historians have yet to assess whether automation had a disproportionate impact on female manual labor.[17]

Among women's manual occupations, sewing jobs have consistently led the list, but only a few studies have added much to Baker's discussion of women in the needle trades. As Rachel Maines demonstrates, the sewing industries offer peculiar challenges to scholars, because needlewomen produced a variety of goods, including garments, hats, shoes, gloves, corsets, umbrellas, bags, collars, artificial flowers, upholstery, home furnishings, horse goods, sails, awnings, suspenders, and numer-

17. Baker, pp. 51, 194–211, and passim; Hirsch, pp. 27–39; McGaw, "Technological Change"; and Tentler. In the printing trades, feminine occupations such as type rubber (finisher) and proofreader conform to the general pattern of women performing unmechanized work. The exclusion of women from mechanized work (especially operating steam presses and linotype machines) and their increasing relegation to peripheral tasks and marginal shops in the course of the nineteenth century resulted primarily from male trade unionists' sustained efforts. See Betsey W. Bahr, "The Smell of Printer's Ink: Women's Occupations in Printing and Publishing," in Wright, ed.

ous others; they labored in a variety of settings, including the home, the custom shop, the large factory, and the sweatshop; and they were influenced by a host of new technologies, including "the development of standard sizes, the invention of printed patterns, [and] improvements in cutting tools," as well as the introduction of the sewing machine.[18] Claudia B. Kidwell analyzes another new nineteenth-century technology, dressmakers' drafting systems, and attributes the highly specialized character of the devices to the fact that less skilled professionals or amateurs generally produced women's clothing, just as they had in the colonial era. Similarly, Maines, Mary H. Blewett in her study of the shoe industry, and Susan L. Hirsch in her treatment of Newark craft workers each find women performing analogous tasks before and after the advent of new sewing technologies, their exclusion from the most highly skilled work being one pervasive element of continuity. In part by speeding women's traditional work, the introduction of the sewing machine reduced the relative employment of women in the needle trades—a contrast with the impact of mechanization in the glass, paper, and nonferrous metal industries. As in the case of textiles, mechanization and industrialization eventually drove clothing and shoe manufacture from the home and progressively reduced female operatives' skill requirements. But apparel manufacture, like the glass, paper, and metal trades, remained far less mechanized than textiles, so that labor costs continued to be manufacturers' principal expense. Consequently, to cut costs apparel makers took advantage of women workers' limited employment alternatives, restricted geographical mobility, and competition from the large numbers of women who knew how to sew. Throughout the nineteenth and early twentieth centuries, extremely low wages, poor working conditions, and seasonal unemployment persisted despite, and often because of, more efficient technology.[19]

Historians also need to devote far more attention to technology's relationship to women's clerical and sales work, the largest current employment sector for women. As with sewing occupations, the number and variety of activities, settings, and relevant technological innovations

18. Rachel Maines, " 'Laboring as Men for the Pay of Children': Women in the Needle Trades, 1820–1880," in Wright, ed.

19. Claudia B. Kidwell, *Cutting a Fashionable Fit: Dressmakers' Drafting Systems in the United States* (Washington, D.C.: Smithsonian Institution Press, 1979); Mary H. Blewett, "From Their Kitchen to the Shops: The Role of Women in the Growth and Mechanization of the New England Shoe Industry, 1800–1865," in Wright, ed.; Hirsch (above, n. 13); Tentler (above, n. 12); Mark Aldrich and Randy Albelda, "Determinants of Working Women's Wages during the Progressive Era," *Explorations in Economic History* 17 (October 1980): 323–41. Good studies also exist tracing the development of the sewing machine and examining its early mass production and mass marketing. See Grace Rodgers Cooper, *The Invention of the Sewing Machine* (Washington, D.C.: Smithsonian Institution Press, 1968); Ruth Brandon, *A Capitalist Romance: Singer and the Sewing Machine* (Philadelphia: J. B. Lippincott, 1977).

make analysis of white-collar work especially challenging. Nonetheless, several scholars have made a significant start, moving the field well beyond Baker's emphasis on "new opportunities" and nimble fingers suited to typing. Margery Davies's "Woman's Place Is at the Typewriter" adduces three simultaneous developments—increased demand for low-level, dead-end, but educated clerical employees; relatively limited alternative opportunities for the growing number of female high-school graduates; and the emergence of typewriting, a new and non-sex-typed occupation—to account for the simultaneous introduction of large numbers of women and typewriters in American offices. She also shows that journalists and businessmen quickly adapted contemporary patriarchal values to redefine low-level clerical work as women's work and to legitimate the sexual hierarchy of the office. Maurine Greenwald's chapter on the early Bell System finds the situation similar for telephone operating. Scholars still need to assess how a wide array of office technologies—ranging from paper clips and carbon paper, to inventory controls and accounting standards, to word processors and facsimile transceivers—have affected women and their work.[20]

Moreover, despite the continuous development of increasingly sophisticated office machines, only $3,000 in capital supports the contemporary office worker, as compared to $35,000 per factory operative and $50,000 per farmer. As in manufacturing, then, the relationship of women's clerical employment to the limited industrialization of the office invites exploration. In the late nineteenth- and early twentieth-century office, Davies finds that personalized work relations and the employment of women permitted and encouraged relatively limited industrialization. By contrast, Greenwald shows that the far larger number of workers per manager in the Bell System allowed it to rationalize and speed work by applying scientific management techniques borrowed from industry. As in office work, the importance of personal interaction in sales work set limits to the rationalization of department store clerking, but, by treating spatial arrangement and merchandise displays as technologies, Susan Porter Benson demonstrates that managers increasingly used technology to control and speed the activities of both saleswomen and their predominantly female customers. But women have been active creators of clerical technology, not just passive objects of male manipulation.

20. Margery Davies, "Woman's Place Is at the Typewriter: The Feminization of the Clerical Labor Force," *Radical America* 8 (July/August 1974): 1–28; Maurine Weiner Greenwald, *Women, War and Work: The Impact of World War I on Women Workers in the United States* (Westport, Conn.: Greenwood Press, 1980), pp. 185–232. Cindy S. Aron, " 'To Barter Their Souls for Gold': Female Clerks in Federal Government Offices, 1862–1890," *Journal of American History* 67 (March 1981): 835–53, indicates that Davies's portrait of rapid redefinition of clerical work as appropriate for women must be qualified in view of the lengthy debate over the issue when small numbers of women first entered government offices.

Scholarship on filing systems especially might illuminate how women abetted or resisted the rationalization of office work.[21]

A final focus for recent revisionist scholarship on women and technology is the influence of modern warfare on women in the work force. As with most new scholarship on technology and women's employment, new studies accent continuity, whereas older studies emphasized novel opportunities. Greenwald analyzes a variety of industries and shows that World War I merely accelerated long-term employment trends that began as early as 1870 and that most new industrial, transportation, and clerical workers had already been employed, but in lower-paying, feminized occupations such as domestic service. In addition, the technical expertise required in most occupations opened to women was quite similar to that needed in the work women had previously performed. Likewise the double burden of sex and race discrimination confined most black women to tasks analogous to their prewar work. Although a few overcame earlier barriers to their employment in industrial and office work, they were largely confined to the dirtiest, most dangerous occupations. Most remained essentially domestic servants, although they received higher pay when they cleaned railroad stations and distributed linen on Pullman cars than they would have in a private home. All women found that their traditional association with lowered pay and reduced skill provoked union hostility, skilled workers' fears, and campaigns to return them to traditional women's work after the war. Their response and ultimate fate depended on the government's role, the local situation, and the earlier history of their employer, a complex array of factors which Greenwald examines through a series of diverse case studies. No similar detailed examination exists of how other wars affected women in the labor force, but Leila Rupp's close analysis of World War II propaganda makes clear that it explicitly minimized the potential novelty of or increased competence required by women's wartime work. It underscored the femininity of women in new occupations and showed them as merely extending their traditional maternal or supportive roles.[22]

21. "Device Makers Dream of Electronic Office, but Obstacles Remain," *Wall Street Journal* (March 13, 1981), pp. 1, 12; Davies; Baxandall, Ewen, and Gordon (above, n. 13), p. 4; Greenwald; Susan Porter Benson, "Palace of Consumption and Machine for Selling: The American Department Store, 1880–1940," *Radical History Review* 21 (Fall 1979): 199–221; Aron.

22. Greenwald; Leila Rupp, *Mobilizing Women for War: German and American Propaganda, 1939–1945* (Princeton, N.J.: Princeton University Press, 1978). Given the contemporaneous increases in married women's employment, the existence of Title VII guarantees, and the presence of an articulate feminist movement, women's work during the Korean and Vietnam wars especially invites attention. We also know little about the impact of the Civil War on female employment, although Maines finds that much new military sewing went on in traditional domestic settings, and increased mechanization and

We still know little about the management of female workers, women's protective labor legislation, and the hazards of women's work, although a few studies suggest the possibilities for further research. For example, we know that piecework incentives have long been the principal technique for managing female workers and that scientific managers attempted to refine traditional piecework systems in the late nineteenth and early twentieth centuries. Greenwald finds that women played an important part in the development of personnel administration, especially of techniques for managing female workers, but that the sexist assumptions and conservative labor relations theories of these professional women often blunted their sensitivity to working women's needs and preferences. J. Stanley Lemons's study of "social feminists" in the 1920s treats several of the same women as advocates of women's protective labor legislation, although he emphasizes their benevolent motives rather than the discriminatory consequences of such laws. He paints a less sympathetic portrait of their articulate opponents in the National Woman's Party (NWP). Echoing the NWP's early critique, however, several recent studies question whether women's protective labor legislation was intended to protect women, men's jobs, or unborn or even unconceived children. Although historically women have been excluded from visibly dangerous jobs, Tentler suggests that the study of obvious hazards can be misleading. Lower fixed capital investments, characteristic of industries employing women, reduced manufacturers' incentives to guard against fire and, combined with the ability of such industries to convert old structures for industrial use, subjected female workers especially to hidden fire hazards. Disproportionate numbers of women also labored near hazards such as radiation, benzene, lead, contagious diseases, and airborne fiber, substances less visibly linked to workers' deaths than falling rock in coal mines or molten metal in steel mills, but often fatal nonetheless. Thus, we must critically examine repeated assertions that Americans preferred to keep women in safe occupations. In some instances employers and the general public may not have perceived invisible dangers, but Vilma R. Hunt's study of worker, manager, and public awareness of occupational health risks indicates that this was not always the case.[23]

---

industrialization in the needle trades simply accelerated earlier trends. Female clerks in government offices represented a more visible departure from tradition, but, as Davies notes (pp. 6–7), female clerical workers were not unprecedented, their numbers continued to be small, and their wages were held well below those of men.

23. Tentler; Greenwald, pp. 46–86; J. Stanley Lemons, *The Woman Citizen: Social Feminism in the 1920s* (Urbana: University of Illinois Press, 1973); Rosalind Petchesky, "Workers, Reproductive Hazards and the Politics of Protection: An Introduction," *Feminist Studies* 5 (Summer 1979): 233–46; and, in the same issue of *Feminist Studies,* Ann Corinne Hill, "Protection of Women Workers and the Courts: A Legal Case History," pp. 247–73; Vilma R. Hunt, "A Brief History of Women Workers and Hazards of the Workplace," pp. 274–85; Carolyn Bell, "Implementing Safety and Health Regulations for Women in the

Finally, Daniel Rodgers breaks important ground by extending to feminist thinkers his analysis of changes in the work ethic as America industrialized. Kenneth L. Ames's account of Victorian hall furnishings and women's calling-card customs implicitly links new etiquette codes to industrialization. And fragmentary evidence from a number of studies indicates that household manufacturing continued throughout the nineteenth century as an important employer of women. Clearly scholarship on manufacturing technology and women cannot be confined within the walls of the factory.[24]

Technology and Domestic Work

Scholars treating technology and the American home have emphasized three related questions: To what extent did new technologies "industrialize" the home and alter domestic labor? How have new construction techniques and spatial arrangements reshaped "woman's place" and her work there? What were the origins, content, and results of the movement to make housekeeping "scientific"? Like scholarship on workplace technology, research on domestic technology has concentrated on a period of apparently rapid change: the years between 1870 and 1930, when home economics became professionalized; architects, builders, and reformers redesigned the house; and middle-class households purchased an increasing array of domestic conveniences. As a result, white, middle-class, urban, northeastern women figure prominently in the literature.

Although it focuses on those most influenced by new household technology and science, the new scholarship finds fundamental continuities in woman's domestic experience, continuities akin to those she encountered in the work place. Domestic technology made housework

---

Workplace," pp. 286–301; and Michael J. Wright, "Reproductive Hazards and 'Protective' Discrimination," pp. 302–9.

24. Daniel Rodgers, *The Work Ethic in Industrial America, 1850–1920* (Chicago: University of Chicago Press, 1974), pp. 182–209; Kenneth L. Ames, "Meaning in Artifacts: Hall Furnishings in Victorian America," *Journal of Interdisciplinary History* 9 (Summer 1978): 19–46. At present the only general study of household manufacture is Rolla Milton Tryon, *Household Manufactures in the United States, 1640–1860* (Chicago: University of Chicago Press, 1917), which is dated and heavily dependent on the limited data collected in the federal agricultural censuses. Fortunately, Thomas Dublin's work in progress on household manufacturing promises to revise and expand our information on this subject. See also Elizabeth Ramsey, *The History of Tobacco Production in the Connecticut Valley* (Northampton, Mass.: Smith College, 1930), pp. 126–27; Margaret Richards Pabst, *Agricultural Trends in the Connecticut Valley Region of Massachusetts, 1800–1900* (Northampton, Mass.: Smith College, 1940), pp. 40–41, 58–59, 80–82; Howard S. Russell, *A Long, Deep Furrow: Three Centuries of Farming in New England* (Hanover, N.H.: University Press of New England, 1976), pp. 299, 355, 380, 406.

less arduous but was not used to make it less time-consuming. Housing reform altered interior and exterior spatial arrangement and decor but kept homemakers relatively isolated and inefficient. Well-trained specialists in home economics sought to professionalize, industrialize, and standardize America's domestic work but relied on unpaid generalists serving single families to implement their suggestions. Thus, substantial changes in household technology left the sex, hours, efficiency, and status of the household worker essentially unaltered.

Most research on technological change in American households indicates that, Victorian rhetoric to the contrary notwithstanding, the home was inseparable from the larger society. Two surveys provide especially extensive evidence that this was the case. The classic study of industrialization's impact on the domestic environment is Siegfried Giedion's *Mechanization Takes Command*. His account, published in 1948, gives fullest coverage to American technology during the preceding hundred years and examines changes in household furnishings and housing design, the mechanization of most household tasks, and the evolution of the bathroom. His generous selection of photographs and patent office drawings is another asset. Like the text, however, the illustrations feature prototypes and new products, leaving the dissemination and influence of novel technology unclear.[25]

By contrast, Susan Strasser's *Never Done* relies on a variety of sources, particularly mail-order catalogs, to determine which inventions were produced commercially and when they became widely available.[26] Written for a popular audience and richly illustrated, her work provides a history of housework since the colonial era that is well suited for introductory women's studies courses. Another strength is Strasser's imaginative reconstruction of tasks to recapture easily overlooked details of nineteenth-century housework. She reminds us, for example, that each bucket used to haul the minimum four hundred pounds of water needed for clothes washing added eight or nine pounds of weight; that most housewives used heavy cast-iron cookware requiring considerable routine maintenance; and that prior to the late nineteenth century even women who purchased food at the market continued to perform processing tasks such as plucking and eviscerating chickens, roasting coffee, and seeding raisins.

Strasser also supplies the best generalizations about technology's role in the nineteenth-century household. She finds that only two innovations, the cast-iron cookstove and the Dover eggbeater, measurably

25. Siegfried Giedion, *Mechanization Takes Command: A Contribution to Anonymous History* (New York: Oxford University Press, 1948). See also Alan Gowans, *Images of American Living: Four Centuries of Architecture and Furniture as Cultural Expression* (New York: Harper & Row, 1976).

26. Susan Strasser, *Never Done: A History of American Housework* (New York: Pantheon Books, in press).

lightened women's domestic labor. Two additional new technologies, oil lamps and heat stoves, accustomed Americans to warmer, brighter homes, but the many fires in Victorian homes increased the burdens of cleaning and made the housewife's work environment dry and stuffy. Other nineteenth-century inventions, such as washing machines and gas stoves, remained too expensive for the average family to afford and required utilities and plumbing that were even more costly.

Most other scholarship devoted to nineteenth-century domestic technology features prescription rather than practice. Dolores Hayden, Kathryn Kish Sklar, Clifford E. Clark, Jr., and William D. and Deborah C. Andrews concur that model Victorian homes, made possible by new building practices, designed to promote efficient housekeeping, and incorporating new cooking, plumbing, and heating technologies, were promulgated as part of a system that also encouraged women to devote increased attention to child rearing, especially religious education, and to consumption. Thus, they were not intended to shorten women's workday but to raise American living standards. Moreover, assessing such prescription's influence on the Victorian housekeeper requires knowledge of which model houses and efficiency systems consumers adopted. In their technology, Catharine Beecher and Harriet Beecher Stowe placed the kitchen at the center of the dwelling and kept women in touch with other household activities while at work; significantly different were housing plans by romantic revival architects which relegated the kitchen to the basement or back of the house, isolating women's activities. Beecher and Stowe also tied the housewife's increased efficiency to the elimination of domestic servants, whereas other designers assumed that efficient housewives would manage hired help. Despite the wide dissemination of Beecher's and Stowe's manuals, Gwendolyn Wright's study of model homes in late nineteenth-century Chicago indicates that most houses kept the kitchen peripheral and were designed with servants in mind.[27]

27. Dolores Hayden, *The Grand Domestic Revolution: A History of Feminist Designs for American Homes, Neighborhoods, and Cities* (Cambridge, Mass.: MIT Press, 1981); Kathryn Kish Sklar, *Catharine Beecher: A Study in American Domesticity* (New York: W. W. Norton, 1973); Clifford E. Clark, Jr., "Domestic Architecture as an Index to Social History: The Romantic Revival and the Cult of Domesticity in America, 1840–1870," *Journal of Interdisciplinary History* 7 (Summer 1976): 33–56; William D. Andrews and Deborah C. Andrews, "Technology and the Housewife in Nineteenth-Century America," *Women's Studies* 2, no. 3 (1974): 309–28; Gwendolyn Wright, *Moralism and the Model Home: Domestic Architecture and Cultural Conflict in Chicago, 1873–1913* (Chicago: University of Chicago Press, 1980), pp. 36–39. Recent works indicate that similar developments occurred in colonial America and early modern England. Carole Shammas "uses artifacts to uncover work patterns . . . in the absence of time allocation studies" and concludes that, between the sixteenth and eighteenth centuries, houses grew larger and became more differentiated, furnishings increased in number and variety, levels of household consumption rose, families spent more time at home together, and women's domestic work became more specialized and less arduous ("The Domestic Environment in Early Modern England and

Claudia L. Bushman's detailed portrait of Harriet Hanson Robinson's housekeeping offers a good test case for the influence of prescription on practice and a useful complement to Strasser's generalized account of nineteenth-century housekeeping. Robinson, who had read
Beecher and Stowe, cooked in the back of her house and managed her
mother, daughters, and hired servants while doing her own work. In
general she employed new technology to raise her family's standard of
living rather than to shorten her workday. She used her season railroad
ticket to journey frequently from Malden to Boston to shop. Aided by
her sewing machine, she spent many hours making and remaking
clothing for herself and her children so as to conform to changing styles
and keep up the family's outward appearance. Her typically Victorian
multiroom dwelling, with its more and more ample furnishings, required a month of spring cleaning and a shorter stint in the fall. Other
tasks that Robinson found far more time-consuming than had her
mother included entertaining her friends and those of her husband and
children, flower and vegetable gardening, carpet cleaning and repair,
and planning and supervising frequent improvements on her house. On
the other hand, Robinson certainly found much of her work interesting
and rewarding, especially newly emphasized tasks such as shopping,
child rearing, sewing, and gardening. To find out whether she was representative, scholars will need to abandon their assumption that Beecher
and Stowe were traitors to their sex who consigned women to tedious
housework. We need to examine the possibility that many middle-class,
nineteenth-century housewives had more varied, interesting, rewarding
work and better working conditions than virtually all other contemporary female workers and most contemporary male employees.[28]

Scholarship on technology and twentieth-century housework finds
that domestic labor continued to be time-consuming and that it lost
much of its creativity and individuality. Two articles by Ruth Schwartz
Cowan discuss how the shift from muscle power to electrical power,

America," *Journal of Social History* 14 [Fall 1980]: 3–24, esp. 5). Alice E. Messing finds
evidence of increasing rationalization of domestic tasks in eighteenth-century American
cookery literature (" 'A Bundle of Sweet Herbs': The Transcription of Women's Cooking
into Culinary Literature in Eighteenth-Century America," in *Proceedings of the Society for the
History of Technology, 1981* [Milwaukee: Milwaukee Public Museum, in press]).

28. Claudia L. Bushman, *"A Good Poor Man's Wife": Being a Chronicle of Harriet Hanson
Robinson and Her Family in Nineteenth-Century New England* (Hanover, N.H.: University Press
of New England, 1981), chap. 7. Indeed, most of the literature published since its writing
contradicts Gerda Lerner's classic assertion that the early nineteenth century offered
working-class women new opportunities while restricting middle-class women to un-
rewarding idleness and display. Bushman's study is especially illuminating on this point
because she discusses Robinson's early Lowell experiences as well as her domestic life. See
also Harriet H. Robinson, *Loom and Spindle, or Life among the Early Mill Girls* (Kailua,
Hawaii: Press Pacifica, 1976); and Gerda Lerner, "The Lady and the Mill Girl: Changes in
the Status of Women in the Age of Jackson," in *Our American Sisters: Women in American Life
and Thought,* ed. Jean E. Friedman and William G. Shade (Boston: Allyn & Bacon, 1973),
120–32.

from coal and wood cooking to gas and electric stoves, from gas and oil to electric lighting, from stoves to central heating, and from outdoor pumps to running water affected middle-class housewives. Most new technologies reduced household dirt and lightened individual tasks but did not create the sociologists' mythic idle, neurotic housewife, in part because advertisers and psychologists successfully manipulated guilt and embarrassment to enforce higher standards of cleanliness and greater attention to child care and shopping. Simultaneously, domestic servants and unpaid helpers grew less common, and middle-class wives found themselves to be household workers, rather than household managers. Contrasting technology's role in twentieth-century homes with its role in nineteenth-century factories, Cowan argues that the industrialization of the home reduced specialization and differentiation in the work force, decreased managerial functions, and "heightened the emotional context of the work."[29] Baxandall, Ewen, and Gordon concur that manufacturers and marketers proletarianized housework, but they emphasize parallels between home and factory under "monopoly capital," including the demotion of artisan craftsworkers to machine tenders, the separation of the conception of work from its execution, and the substitution of simplified direction for the understanding of work processes. Strasser's chapters on the twentieth century substantiate their model and underscore as well the increasing isolation of the domestic worker.[30]

Joann Vanek's work complements that of Cowan. She analyzes a number of temporally and geographically diverse twentieth-century surveys of time spent on housework, finding that only labor force participation greatly reduced household activity, but that even employed women labored long hours at home. With statistical controls for work outside the home, Vanek finds that, by the 1960s, after poor and rural women had obtained the technologies that middle-class women had acquired in the 1920s, no significant class differences existed in length of housework day or distribution of time among tasks. By contrast, rural women in the 1920s and 1930s lacked labor-saving technologies but spent fewer hours on housework than their urban, middle-class contemporaries. They devoted more time to household maintenance and food and clothing production, but considerably less time to child care and shopping.[31]

29. Ruth Schwartz Cowan, "The 'Industrial Revolution' in the Home: Household Technology and Social Change in the 20th Century," *Technology and Culture* 17 (January 1976): 1–23, esp. 23; and "Two Washes in the Morning, and a Bridge Party at Night: The American Housewife between the Wars," *Women's Studies* 3 (1976): 147–72.

30. Baxandall et al. (above, n. 13), pp. 5–8; Strasser. See also Joan Rothschild, "Technology, 'Women's Work,' and the Social Control of Women," in *Women, Power and Political Systems,* ed. Margherita Rendel (London: Croom Helm, 1981).

31. Joann Vanek, "Household Technology and Social Status: Rising Living Standards and Status and Residence Differences in Housework," *Technology and Culture* 19 (July 1978): 361–75, and "Time Spent on Housework," *Scientific American* 231 (November 1974): 116–20.

As Vanek notes, working-class women had a different experience of domesticity. S. Jay Kleinberg offers a broadly conceived analysis of working-class households in late nineteenth-century Pittsburgh that links industrialization and urbanization to longer hours of housework. Steel mill owners, who controlled the resources of the city, invested primarily in order to rationalize and expand steel production and to provide municipal services to middle- and upper-class neighborhoods. As a result, working-class wives lacked both the capital to invest in new appliances and the utilities and work space that would have made appliances usable. At the same time, their husbands, sons, and male boarders performed extremely dirty work so that laundry became a more arduous chore, and they worked several different shifts so that meals had to be cooked around the clock. Heavy traffic on unpaved roads in working-class districts churned up dust and mud that, added to the soot belching from nearby factories, made household cleaning an endless task. Inadequate water mains and the competition of mills for water forced women to haul water long distances at very early hours. The sewage system did not extend to workers' neighborhoods, and the municipal sewers and mills polluted workers' drinking water, increasing demands on working-class wives' nursing skills. When widowed by high rates of illness and industrial accidents, women found that local industry offered little employment, except indirectly by creating a large market for poorly paid laundresses.[32]

Daniel E. Sutherland and David M. Katzman examine how technology influenced and was influenced by another group of working-class women who represented the largest percentage of female employees in nineteenth- and early twentieth-century America: domestic servants. They find the relationship complex. For example, after 1910 washing machines and commercial laundries virtually eliminated the need for laundresses outside the South but added to the duties of maids-of-all-work, who now assumed laundry duties as well. As was the case in factories employing poorly paid women, the availability of cheap domestic servants generally slowed technological innovation in the household. Because middle-class families usually chose either to buy new appliances and houses or to hire servants, domestics worked with inadequate tools and in poorly designed kitchens. Gwendolyn Wright concurs, noting that as long as kitchens were the province of domestics, architects and builders paid little attention to kitchen design. Sutherland and Katzman also find that maids tolerated inferior sleeping and bathing facilities, despite their exhausting, dirty work. Frequent complaints by servants and reformers about employers' lack of planning and failure to establish set schedules of tasks provide added evidence that the availability of

32. S. Jay Kleinberg, "Technology and Women's Work: The Lives of Working Class Women in Pittsburgh, 1870–1900," in Trescott, ed. (above, n. 6), pp. 185–204, and *The Shadow of the Mills: Working Class Family Life, Pittsburgh, 1870–1900* (Champaign: University of Illinois Press, in press).

servants retarded the industrialization of the home. Both authors also discuss the social status servants conferred, indicating that middle-class households turned to household technology only because servants became scarce, not because new technology superseded servants. Interregional comparisons strengthen Katzman's analysis. In the South, race discrimination and low wages for black men made a large pool of black wives available for domestic service. Well supplied with cheap domestic labor, Southern white households waited longest to adopt new household technology. Contradictory and less persuasive is his argument that work in middle-class households helped modernize working-class women by instilling time discipline, introducing them to some new technology, and, especially in the North, enhancing their efficiency through training. By contrast, Sutherland chronicles the limited success of reformers' repeated attempts to train workers and encourage more efficient management by their employers.[33]

Hayden and Wright's 1976 review of the literature on women, architecture, and urban planning allows me to limit my discussion of housing form and women's work to a few salient themes and important works. Wright's study of how architects, builders, and reformers simplified and standardized the home between 1873 and 1913 merits special attention for its treatment of an enormous array of relevant technologies, including domestic appliances, utilities, building and woodworking techniques, printing and papermaking methods, arts and crafts traditions, home economists' advice, and public health measures. She relates housing reform to contemporary demographic trends, increases in female employment, feminist thought, settlement house activity, and professionalization of architecture. She concludes that housing reformers failed to ease domestic work, or solve the social problems they addressed, because they preferred ideal and symbolic architectural solutions to a realistic analysis of and attack upon the problems. As Wright points out, "there is something peculiarly American about" proposing model houses in order to extend "the moral influence of the home" and preserve family stability, a thesis she extends to earlier and later eras in *Building the Dream,* a broad social history of housing in America. Clark finds similar goals and symbolic rather than realistic solutions proffered by romantic revival architects between 1840 and 1870, and Robert C. Twombley discovers in Frank Lloyd Wright's seemingly radical prairie

33. Daniel E. Sutherland, *Americans and Their Servants: Domestic Service in the United States from 1800 to 1920* (Baton Rouge: Louisiana State University Press, 1981); David M. Katzman, *Seven Days a Week: Women and Domestic Service in Industrializing America* (New York: Oxford University Press, 1978); Wright, pp. 36–39. Alice E. Messing's research in culinary literature suggests that the availability of slave women and freedwomen for domestic service encouraged inefficient cooking methods and the retention of antiquated, regionalized recipes (" 'From Cooking to Dancing, from Singing to Serving': Continuity in Southern Plantation Cooking, 1824–1930," unpublished paper [Philadelphia: University of Pennsylvania, Department of American Civilization, 1981]).

house yet another conservative, symbolic attempt to "save the family." Again we find essential continuity beneath changing facades.[34]

Given their continuous commitment to relatively uniform buildings to house nuclear families in suburban settings, it is hardly surprising that architects, builders, and their customers have come up with only a narrow choice of spatial arrangements formalized within the home and imposed upon the landscape. While woman's place as service worker for her husband and children remained unquestioned, the industrialization of the home inevitably remained partial and superficial. New household appliances entered the home, but, unlike industrial machines, they did not achieve economies by operating continuously or on a large scale. Both housewives and servants had to be generalists, and thus were limited in their skill and efficiency. As Dolores Hayden demonstrates in her monographs, *Seven American Utopias* and *The Grand Domestic Revolution*, feminists and socialists have repeatedly recognized the inefficiency of conventional residences. They have drawn plans for kitchenless houses or apartments, built and lived in utopian communities, operated cooperative housekeeping enterprises, and used tracts and novels to argue their case for reform. Yet even among these visionaries, relatively few questioned the traditional sexual division of labor. Moreover, the activities most commonly chosen for collectivization were those for which working commercial models already existed: laundry, baking, and cooking. Successful alternative societies, such as the Shakers and Oneida communitarians, generally created institutions for collective, evolutionary spatial design and shared a commitment to eliminating conventional families and to overcoming women's isolation from other women. They also achieved high rates of innovation in domestic technology, a subject that merits further study.[35]

The emergence and professionalization of home economics and the influence of the domestic efficiency movement also call for additional research. Thus far most scholarship has merely chronicled the activities of leading practitioners and has focused on formal education and official

34. Hayden and Wright (above, n. 4); Wright, *Moralism and the Model Home* (above, n. 27), p. 293; and *Building the Dream: A Social History of Housing in America* (New York: Pantheon Books, 1981); Clark (above, n. 27); Robert C. Twombley, "Saving the Family: Middle Class Attraction to Wright's Prairie House, 1901–1909," *American Quarterly* 27 (March 1975): 57–72.

35. Margaret Benston, "The Political Economy of Women's Liberation," *Monthly Review* 21 (September 1969): 13–27; Cynthia Rock, Susana Torre, and Gwendolyn Wright, "The Appropriation of the House: Changes in House Design and Concepts of Domesticity," in *New Space for Women,* ed. Gerda R. Wekerle, Rebecca Peterson, and David Morley (Boulder, Colo.: Westview Press, 1980), pp. 83–101; Dolores Hayden, *Seven American Utopias: The Architecture of Communitarian Socialism, 1790–1975* (Cambridge, Mass.: MIT Press, 1976), and *The Grand Domestic Revolution* (above, n. 27). Also warranting more attention is the differential influence of suburbanization on women, especially in view of recent evidence that suburbanized families have differed at different historical periods; see, e.g., John Modell, "Suburbanization and Change in the American Family," *Journal of Interdisciplinary History* 9 (Spring 1979): 621–46.

pronouncement. Existing studies show that informal educational activities, such as Chatauqua, World's Fairs, and agricultural extension work, as well as ties between home economists and the food-processing industry and continued complaints about the scarcity of good domestic tools and the surfeit in the American diet, all were common enough to merit systematic consideration. We also need a good collective biography analyzing the emerging profession. And we need to look more closely at the movement's clients and how their training affected their lives, especially in light of Wright's observation that home economists did not consider the housewife's real needs and experiences when they relied on industrial analogies. Linda Marie Fritschner's interregional comparison of home economics education argues that the movement responded to different social needs in the East, where shortages of trained servants existed; in the Midwest, where farm households wanted to preserve their way of life; and in the South, where the legacy of black servitude stigmatized domestic work and divided and weakened the movement. Katzman's observation that midwestern cities experienced the greatest servant shortages raises questions about Fritschner's interpretation of the midwestern situation, but her emphasis on the movement's social contexts remains provocative and deserves further exploration. Given the professionals' tendency to expand work to fill the time saved, her general contention that the movement sought to preserve a passing order and control change certainly rings true.[36]

Technology of Women, Technology by Women, and Technology for Women

Other than works on birth control (considered earlier) and studies of medical technologies impinging on reproduction (surveyed in review essays cited above), scholarship on the technology of women as bearers and rearers of children remains much as Cowan reported it six years

36. Most of the works cited earlier in this section give some attention to home economics or domestic reformers. See also Emma Seifrit Weigley, *Sarah Tyson Rorer: The Nation's Instructress in Dietetics and Cookery* (Philadelphia: American Philosophical Society, 1977), and "It Might Have Been Euthenics: The Lake Placid Conferences and the Home Economics Movement," *American Quarterly* 26 (March 1974): 79–96; Ruth Schwartz Cowan, "Ellen Swallow Richards: Technology and Women," in *Technology in America: A History of Individuals and Ideas,* ed. Carroll W. Pursell, Jr. (Washington, D.C.: Voice of America, 1979), pp. 157–66; Harvey Levenstein, "The New England Kitchen and the Origins of Modern American Eating Habits," *American Quarterly* 32 (Fall 1980): 369–86; Robert Clarke, *Ellen Swallow: The Woman Who Founded Ecology* (Chicago: Follett Publishing Co., 1973); Barbara Ehrenreich and Deirdre English, *For Her Own Good: 150 Years of Experts' Advice to Women* (Garden City, N.Y.: Anchor Press, 1978); Wright, *Moralism and the Model Home,* pp. 272–73; Linda Marie Fritschner, "Women's Work and Women's Education: The Case of Home Economics, 1870–1920," *Sociology of Work and Occupations* 4 (May 1977): 209–34; Lynn Nyhart, "Efficiency in Home Economics," unpublished paper (Philadelphia: University of Pennsylvania, Department of History and Sociology of Science, 1981).

ago. The baby bottle, baby carriage, and toilet training before and after water closets and disposable diapers still await their historians, as do the breast pump, the pacifier, and the technologies of teething. Several interesting accounts of child-rearing technique exist, but they study professional recommendation, not parents' behavior. Among these, Michael Zuckerman's analysis of Benjamin Spock's *Baby and Child Care* and Nancy Pottishman Weiss's comparison of Spock's book to the Children's Bureau's *Infant Care* deserve special mention for their careful attention to changing social circumstances. Strasser's chapter entitled "When the Bough Breaks" provides a more general overview of child-rearing advice and child-care institutions. Carroll W. Pursell, Jr., offers evidence that toymakers promoted both modern technology and conventional sex roles through the toys they marketed in the 1920s and 1930s. His study also suggests the possibilities for similar research on the toys of earlier and later generations.[37]

The literature on the technologies of menstruation and breast-feeding is even scantier. Janice Delaney, Mary Jane Lupton, and Emily Toth bring together information on changes in menstrual technology and the persistence of psychological, rather than technological, appeals in product advertising. But they do not question whether the sanitary napkin and tampon industry's emphasis on more absorbent products was in women's best interest, and they do not understand that quality control seeks a commercially acceptable product, not optimal quality. Vern L. Bullough also supplies some useful material, especially on patent activity, for a variety of products for women, although his analysis repeatedly treats breast-feeding, menstruation, and conception as "problems" to be overcome. Thus, like the authors of the early literature on the technology of women's work, he sees technology as liberating women. Much remains to be done![38]

Female technologists and technology created by women have also received far less attention than the technology of women's work. Most studies reviewed by Michele Aldrich for the *Signs* special issue on science (Autumn 1978), as well as essays by Sally Gregory Kohlstedt and Margaret W. Rossiter in the same issue, concentrate on women engineers and scientific technicians, underscoring the limits professionalization set

37. See nn. 4 and 9 above. See also Michael Zuckerman, "Dr. Spock: The Confidence Man," in *The Family in History*, ed. Charles E. Rosenberg (Philadelphia: University of Pennsylvania Press, 1975), pp. 179–207; Nancy Pottisham Weiss, "Mother, the Invention of Necessity: Dr. Benjamin Spock's *Baby and Child Care*," *American Quarterly* 29 (Winter 1977): 519–46; Strasser (above, n. 26), chap. 12; Jay Mechling, "Advice to Historians on Advice to Mothers," *Journal of Social History* 9 (Fall 1975): 44–63; Carroll W. Pursell, Jr., "Toys, Technology and Sex Roles in America, 1920–1940," in Trescott, ed. (above, n. 6), pp. 252–67.

38. Janice Delaney, Mary Jane Lupton, and Emily Toth, *The Curse: A Cultural History of Menstruation* (New York: New American Library, 1976), pp. 101–19; Vern L. Bullough, "Female Physiology, Technology and Women's Liberation," in Trescott, ed. (above, n. 6), pp. 236–51.

on women's participation. Biographies, especially of home economists and architects, make up most of the remaining works in the field. Two other approaches that scholars have pursued are more problematic. One is the attempt to uncover ways in which female relatives contributed to the technological achievements of notable inventors and entrepreneurs—an extremely difficult and often speculative enterprise, as Martha Moore Trescott has shown. Equally speculative are the studies suggesting that women's limited participation in the technical professions has deprived Americans of more humane technical options. Women have certainly figured prominently in crusades against the adverse social and environmental consequences of sophisticated twentieth-century technology, but documenting a uniquely female perspective or disproportionate female presence and influence in protest movements will be extremely difficult. More promising might be the study of women's innovative use of household tools, modification of standardized housing plans, and experimental gardening inside and outside the home. Whereas female diarists had little reason to record routine domestic activities, they probably took pride in and recorded their innovations. Claudia Bushman's evidence from Robinson's diaries and Nannie T. Alderson's published memoirs indicate that this was the case. May Thielgaard Watts's imaginative reconstruction of intergenerational changes in a single house and garden suggests the possibility of gleaning added insights from the records of garden club meetings and from the landscape itself.[39]

Four additional subject areas deserve brief discussion: agriculture, clothing, consumption, and technologies influencing both sexes, but differently. A growing number of studies touch on women's relationship to agricultural technology and women's technology in rural America. As in agricultural history generally, most attention has gone to the West and Midwest. Like women's work throughout the nation, farm women's

---

39. Aldrich (above, n. 4); Sally Gregory Kohlstedt, "In from the Periphery: American Women in Science, 1830–1880," *Signs: Journal of Women in Culture and Society* 4, no. 1 (Autumn 1978): 81–96; and, in the same number of *Signs,* Margaret W. Rossiter, "Sexual Segregation in the Sciences: Some Data and a Model," pp. 146–51. On women home economists, see n. 34 above. Torre, ed. (above, n. 7), pp. 51–132, includes biographies of female architects by Judith Paine, Susan Fondiler Berkon, Sara Boutelle, Mary Otis Stevens, Doris Cole, Jane McGroaty, and Susana Torre. See Martha Moore Trescott, "Julia B. Hall and Aluminum," in Trescott, ed. (above, n. 6), pp. 149–79; essays in Torre, ed., pp. 136–202; Joan A. Rothschild, "A Feminist Perspective on Technology and the Future," *Women's Studies International Quarterly* 4, no. 1 (1981): 65–74; Carolyn Merchant, *The Death of Nature: Women, Ecology and the Scientific Revolution* (San Francisco: Harper & Row, 1980); Philip Sterling, *Sea and Earth: The Life of Rachel Carson* (New York: Thomas Y. Crowell, 1970); Rock et al. (above, n. 35); Bushman (above, n. 28), chap. 7; Nannie T. Alderson and Helena Huntington Smith, *A Bride Goes West* (Lincoln: University of Nebraska Press, 1969); May Thielgaard Watts, *Reading the Landscape of America* (New York: Collier Books, 1975), pp. 320–46. An interesting fictional speculation about a feminist technology is Sherwood Anderson, *Perhaps Women* (Mamaroneck, N.Y.: Paul P. Appel, 1970).

labor compensated for limited capital. The parallels with Pittsburgh's working-class women are especially striking. Virtually all scholars concur that men made investment decisions and women coped with the consequences. Because agricultural capital, such as machines and irrigation systems, took priority over better houses and domestic appliances, farm women devoted long hours to arduous housework, including cooking and washing for hired hands, and they spent their time in cramped houses that became oppressively hot in the summer and uncomfortably cold in the winter. They sewed by poor light and struggled against impossible odds to keep sod houses clean and to exterminate omnipresent vermin. Moreover, the larger farms promoted by mechanization and specialization increased rural women's traditional isolation. As John Mack Faragher shows in his study of women and men on the Overland Trail, that isolation deepened for women who migrated west. Such women also had to part with at least some of the inadequate housekeeping tools they had used in the Midwest and, after each day on the trail, they had to cook and sometimes launder with the few tools and fuels available. Despite increased research, scholars remain divided on two important questions: How common was women's field labor, especially among homesteaders? How influential was the eastern cult of domesticity, with its attendant pressure for higher housekeeping standards than most western women could maintain? Answers to both of these questions are essential if we are to assess how confined western women were and what skills they needed.[40]

We know less about technology's relationship to three other large groups of rural women: colonial women, southern women, and female migrant laborers. This especially limits our knowledge of the technology used by black, hispanic, and native American women. For the colonial period, Mary Beth Norton's observation that the "standard books . . . are now badly outdated" is especially applicable to works on technology for women. The classics, by Alice Morse Earle and Mary Earle Gould, were published in 1899 and 1949, respectively. They assume that every

40. The potentially relevant literature is massive, although treatment of technology is generally descriptive, rather than analytic. The most thorough study, John Mack Faragher, *Women and Men on the Overland Trail* (New Haven, Conn.: Yale University Press, 1979), provides access to many important works through its bibliography. Mary W. M. Hargreaves, "Homesteading and Homemaking on the Plains: A Review," *Agricultural History* 47 (April 1973): 156–63, reviews three useful published diaries and memoirs and supplies an excellent list of additional published primary sources. See also Mary W. M. Hargreaves, "Women in the Agricultural Settlement of the Northern Plains," *Agricultural History* 50 (January 1976): 179–89; Christine Stansell, "Women on the Great Plains," *Women's Studies* 4 (1976): 87–98; Frances W. Kaye, "The Ladies Department of the *Ohio Cultivator:* A Feminist Forum," *Agricultural History* 50 (July 1976): 414–23. Treatments of women who played unusual roles in agriculture include Lonnie E. Underhill and Daniel F. Littlefield, "Women Homeseekers in Oklahoma Territory, 1889–1901," *Pacific Historian* 17 (Fall 1973): 36–47; Gladys L. Baker, "Women in the U.S. Department of Agriculture," *Agricultural History* 50 (January 1976): 190–201.

household owned a vast array of domestic equipment and that every housewife possessed an enormous number of manufacturing skills, whereas recent scholarship has found little evidence of these implausible self-sufficient households. It seems likely that colonial housewives, especially those in newly settled areas, labored under many of the same technological disadvantages as nineteenth-century western women, but we need to examine the historical record. Likewise, we need to know how colonial women developed recipes using unfamiliar New World plants and how they adapted European cooking technology to the far greater abundance of meat, poultry, and fish in early America.[41]

By all accounts southern women have taken a more active part in farm labor than any other group of rural American women, but the accounts remain fragmentary and scattered. Gavin Wright finds that slave owners' ability to force black women to do field work was an integral part of the political economy of the cotton South, a system that discouraged investment in improved agricultural technology. Both contemporary observers and recent scholars document the massive post-bellum withdrawal of black women from agricultural labor, yet throughout the late nineteenth century the proportion of southern black women who performed agricultural work far exceeded the proportion of white women doing such work in any region. Sociologist Margaret Jarman Hagood also found field work common among white southern tenant women in the 1930s. As in the antebellum era, then, we find women's labor associated with limited investment in new agricultural technology. Similarly, the possibility of employing poorly paid female migrant laborers discouraged innovation by fruit and vegetable growers. On the other hand, the role of migrant women stemmed in part from technological change for, prior to the advent of automobiles, migrant labor had been almost exclusively the province of men who rode the rails.[42]

41. Norton (above, n. 3), p. 334; Alice Morse Earle, *Home Life in Colonial Days* (New York: Macmillan Co., 1899), and *Colonial Dames and Goodwives,* reprint (New York: Frederick Ungar, 1962); Mary Earle Gould, *The Early American House: Household Life in America, 1620–1850* (Rutland, Vt.: Charles E. Tuttle Co., 1949); Carole Shammas, "The Pre-Industrial Consumer in England and America," forthcoming (Department of History, University of Wisconsin—Milwaukee); Laurel Thatcher Ulrich, " 'A Meet Help': Social Dimensions of Housework in Northern Colonial New England" (paper presented at the Berkshire Conference of Women Historians, August 1978); Judith A. McGaw, "New York Colonial Kitchens and Dining Rooms: Questions and Answers" (Department of History and Sociology of Science, University of Pennsylvania, 1972). Studies of America's remarkably successful agriculture must also address its consequences for maternal health, infant nutrition, and fertility. For an excellent review of relevant literature, see Jane Menken, James Tressell, and Susan Watkins, "The Nutrition Fertility Link: An Evaluation of the Evidence," *Journal of Interdisciplinary History* 11 (Winter 1981): 425–41.

42. Gavin Wright, *The Political Economy of the Cotton South: Households, Markets, and Wealth in the Nineteenth Century* (New York: W. W. Norton, 1978), pp. 82–101, 162; Lynda Morgan, " 'To Taste the Blessedness of Intelligent Womanhood': Black Women and the

Whereas rural technology has affected a declining proportion of women, clothing is a technology that has defined woman's place and restricted her movements throughout American history. But scholarship has generally been preoccupied with professional design, high fashion, and stage costume, so that we know relatively little about how the average woman dressed, how her clothing impeded her movements, or how much creativity domestic clothing manufacture exhibited. The best study of the common American's dress is Claudia B. Kidwell and Margaret Christman's, which emphasizes the standardization and democratization of style since the eighteenth century. Its numerous illustrations, besides showing changes in style, portray women manufacturing clothing in homes, shops, tenements, and factories. Dublin and Tentler provide evidence that women's work experience also encouraged more uniform dress. Because outraged contemporaries left a vast literary record for scholars to peruse, two feminine clothing styles have especially fascinated social historians: Victorian dress, which exaggerated secondary sex characteristics, and the flapper costume, which minimized them. The flapper literature also addresses hairdressing and cosmetic technique, subjects long overdue for study. Kenneth A. Yellis's "thoughts on the flapper" are especially provocative. He links the new fashion to a host of technological changes, including the development of standard sizes, which, because they fit a larger percentage of young women, encouraged designers to cater to youthful figures; the emergence of popular cinema, which made movie stars, rather than the royal family, arbiters of fashion; central heating and auto transport, which made heavy clothing less necessary; improved transportation and communication, which facilitated the dissemination of a radically new style; and new fabrics, which were lighter, cheaper, more durable, and easier to care for. He also examines the role of the enlarged market created by female white-collar workers. Scholars treating women's work make occasional reference to new work clothes, clothing as a justification for women's exclusion from some jobs, and clothing as occupational hazard, but no systematic treatment of any of these subjects exists. Likewise, students of domestic reform indicate that improved technol-

---

Transition from Slave to Free Labor, 1865–1900" (Department of History, University of Virginia, 1980); Margaret Jarman Hagood, *Mothers of the South: Portraiture of White Tenant Farm Women*, reprint (New York: W. W. Norton, 1977); Joseph Interrante, "You Can't Go to Town in a Bathtub: Automobile Movement and the Reorganization of Rural American Space, 1900–1930," *Radical History Review* 21 (Fall 1979): 157, 162–63; Everett Dick, *The Dixie Frontier* (New York: Capricorn Books, 1964), pp. 293–309. Research in numerous southern white women's diaries, conducted by undergraduate women's history students at the University of Virginia, has uncovered rich resources for studying the role of southern female farm managers during the Civil War. For a brief discussion indicating that the role of female agricultural workers in World War I would also reward scrutiny, see Lemons (above, n. 23), p. 17.

ogy eliminated the need for women to dress differently when at work than when in society but do not develop the subject.[43]

Just as traditional scholarship depicted female workers as passive victims or beneficiaries of technological change, historical studies have generally treated consumption, a task increasingly assigned to women, as a passive response to industrialization, and they have held modern technology responsible for the family's withdrawal from society. But recent work by Carole Shammas finds early modern consumers laying the basis for industrial development and creating households comfortable enough to keep families home. She also traces the feminization of consumption to the preindustrial era, when general stores developed and made shopping more egalitarian and less time-consuming. Her approach offers great potential for scholarship on the role of consumers in industrializing and industrialized America. In particular, her demonstration that consumption is real work should help us analyze phenomena other scholars have documented: the increased time women have devoted to shopping, home economists' attempts to train consumers, and women's important roles in advertising and in fighting for consumer protection.[44]

Finally, several studies show that technologies affecting both men and women often had different consequences for the two sexes. In Pittsburgh, municipal and industrial technologies designed to promote

43. Claudia B. Kidwell and Margaret Christman, *Suiting Everyone: The Democratization of Clothing in America* (Washington, D.C.: Smithsonian Institution Press, 1974); Claudia B. Kidwell, *Women's Bathing and Swimming Costume in the United States* (Washington, D.C.: Smithsonian Institution Press, 1968); Dublin, *Women at Work*, pp. 81–82 (above, n. 15); Tentler (above, n. 12), pp. 73–74; Kenneth A. Yellis, "Prosperity's Child: Some Thoughts on the Flapper," *American Quarterly* 21 (Spring 1969): 44–64; Gerald Critoph, "The Flapper and Her Critics," in *"Remember the Ladies": New Perspectives on Women in American History,* ed. Carol V. R. George (Syracuse, N.Y.: Syracuse University Press, 1975), pp. 145–60; John S. Haller, Jr., and Robin M. Haller, *The Physician and Sexuality in Victorian America* (New York: W. W. Norton, 1974), pp. 141–87; Helene E. Roberts, "The Exquisite Slave: The Role of Clothes in the Making of the Victorian Woman," *Signs: Journal of Women in Culture and Society* 2, no. 4 (Spring 1977): 554–69; David Kunzle, "Dress Reform as Antifeminism: A Response to Helene E. Roberts's 'The Exquisite Slave: The Role of Clothes in the Making of the Victorian Woman,' " ibid., pp. 570–79; Fred Shroeder, "Feminine Hygiene, Fashion, and the Emancipation of American Women," *American Studies* 17 (Fall 1976): 101–10; Greenwald (above, n. 20), pp. 28–31, 117, 125–27; Bahr (above, n. 17); Weigley, *Rorer* (above, n. 36); Faragher (above, n. 40), pp. 105–7.

44. Shammas, "Pre-Industrial Consumer" (above, n. 41); Mary P. Ryan, *Womanhood in America: From Colonial Times to the Present* (New York: New Viewpoints, 1979), pp. 151–82; Joseph J. Corn, "General Electric's *The Home Electrical* and Early Corporate Promotional Films" (paper presented at the Society for the History of Technology, October 1980; available from Department of American Civilization, Stanford University); Ronald Lawson, Stephen Barton, and Jenna Weissman Joselit, "From Kitchen to Storefront: Women in the Tenant Movement," in Werkele, Peterson, and Morley, eds. (above, n. 35), pp. 255–71; Lemons (above, n. 23), pp. 117–52. On shopping and home economists, see studies cited in the preceding section, and Interrante (above, n. 42).

more efficient masculine production heightened the difficulty of women's work. On the farm, gasoline-powered vehicles increased men's productivity and broadened their wives' horizons, but lengthened women's workdays by adding to the time they spent shopping. In both theaters and homes, new communications media beamed different messages at men and women and structured special programming around women's traditional work schedules. Yet, especially for transportation and communications technologies, historians' oversights far outnumber their insights. For example, no work yet addresses such questions as whether telephones relieved homebound women's isolation or discouraged visits by making them less necessary, how improved transportation affected women's employment options, or why automobile driving was viewed as masculine.[45]

Nonetheless, even if historically women have not come a long way, the history of women and American technology certainly has. Its sophistication and scope today are impressive when contrasted with the state of the art only six years ago. Although many issues still await serious inquiry, the fine studies at hand will make it easier to frame and respond to questions earlier scholars left unasked or unanswered.

Department of History and Sociology of Science
University of Pennsylvania

45. Kleinberg (above, n. 32); Interrante (above, n. 42); June Sochen, *"Mildred Pierce* and Women in Film," *American Quarterly* 30 (Spring 1978): 3–20; Mary P. Ryan, "The Projection of a New Womanhood: The Movie Moderns in the 1920s," in Friedman and Shade, eds. (above, n. 28), pp. 366–84; Carol Lopate, "Daytime Television: You'll Never Want to Leave Home," *Feminist Studies* 3 (Spring/Summer 1976): 69–82; Russell Merritt, "Nickelodeon Theaters 1905–1914: Building an Audience for the Movies," in *The American Film Industry*, ed. Tino Balio (Madison: University of Wisconsin Press, 1976), pp. 59–79.

Outrunning Atalanta: Feminine Destiny in Alchemical Transmutation

Sally G. Allen and Joanna Hubbs

> Quite in contrast to Freud's assumption that "penis envy" is a natural phenomenon in the constitution of the woman's psyche, there are good reasons for assuming that even before male supremacy was established there was a "pregnancy envy" in the male, which even today can be found in numerous cases. In order to defeat the mother, the male must prove that he is not inferior, that he has a gift to produce. Since he cannot produce with a womb, he must produce in another fashion; he produces with his mouth, his word, his thought.[1]

In the seventeenth-century alchemical treatise *Atalanta fugiens* [Atalanta in flight], Michael Maier, one of the most prominent adepts of the hermetic art in his time, uses the image of the Greek heroine Atalanta to suggest the symbolic relationship of alchemical transmutation to the conquest of nature, represented by him as woman. The myth concerns the virgin who complies with her father's order to choose a husband, but who refuses to consider any suitor incapable of defeating her in a race; the losers are to be killed by her own hand. Only the clever Hippomenes—his name means "horse might," and he derives his wiliness from Venus, goddess of love and of animals[2]—succeeds in overtaking the fiercely independent girl. The victory is due not to his skill but

---

1. Erich Fromm, *The Forgotten Language* (New York: Grove Press, 1956), p. 233.

2. In the Homeric "Hymn to Aphrodite" she is presented as a mistress of animals: "After her came gray wolves, fawning on her, and grim-eyed lions, and bears, and fleet leopards, ravenous for deer; and she was glad in heart to see them, and put desire in their breasts, so that they all mated, two together, about the shadowy vales" (Hesiod, *The Homeric Hymns and Homerica*, trans. H. G. Evelyn-White [Cambridge, Mass.: Harvard University Press, 1920], p. 41).

This essay originally appeared in *Signs*, vol. 6, no. 2, Winter 1980.

to the goddess's gift of three golden apples which he throws in Atalanta's path, distracting her attention from the race. In Maier's treatise, Atalanta, like nature, must be brought into union with human needs. "Hippomenes and Atalanta, united in the temple of the Mother of the Gods (Cybele, here synonymous with Venus as goddess of fertility and of animal life) are transformed into lions."[3] But this union, which is designed to harness the energies of nature for the benefit of humanity, involves mastery over the feminine (the primordial empowerer) as revealed in a series of illustrations of the stages through which the alchemical opus must proceed. These emblems (reproduced at the end of the article), or visual symbols, which constitute a major portion of *Atalanta fugiens,* imply that the written word alone is insufficient to communicate the secret knowledge experienced by the alchemist.[4]

The first image—and stage—shows us the muscular and bearded figure of the phallic god Hermes Trismegistus, the mythical founder of alchemy.[5] His powerful body is drawn in a manner which suggests that he is not simply a messenger of the gods from heaven to earth but, rather, that in him earth and heaven are united. The caption, "Portavit eum ventus in vetre sui" [The wind carried him in his womb],[6] refers to the lightly etched outline of an infant in fetal position placed at the base of his belly, indicating the inseminating role of the male and, even more explicitly, the desire to appropriate the function of maternity. This remarkable image is followed by that of the earth depicted as a globe, representing the torso of a woman suckling a child; it is entitled: "Nutrix eius terra est" [His nurse is the earth].[7] In these two introductory engravings, then, we see the birth of a child from the male; the female's function is reduced to that of nurse, a notion consistent with the Aristotelian view of woman as the passive "nurse" or carrier of the active male seed, the *form* of the foetus.[8]

3. Michael Maier, *Atalanta fugiens* (Oppenheim, 1618), p. 9.

4. Mario Praz, in his *Studies in Seventeenth Century Imagery* (London: Warburg Institute, 1939), p. 12, ascribes the following purpose to emblems: "In need as he was of certainties of the sense, the seventeenth century man did not stop at the purely fantastic cherishing of the image: he wanted to externalize it, to transpose it into a hieroglyph, an emblem."

5. N. O. Brown links Hermes' trickery with love and seduction and points to his phallic aspect in the union with Athrodite (*sic*)—the Hermaphroditus (see *Hermes the Thief* [New York: Random House, 1967], pp. 19–20). Guthrie links his phallic fertility aspects to his function as *psychopompos* leading souls to Hades. The relationship between fertility and the underworld—the cycle of life and death—is common to Mother Goddesses (see W. K. C. Guthrie, *The Greeks and Their Gods* [Boston: Beacon Press, 1968], p. 89).

6. Maier, p. 13; Maier in the commentary associates the wind with Hermes, the messenger of the gods to earth.

7. Ibid., p. 17.

8. "The contribution which the female makes to generation is the matter used therein; and this is to be found in the substance constituting the menstrual fluid. . . . There are some who think that the female contributes semen during coition because women sometimes derive pleasure from it comparable to the male and also produce a fluid secre-

The third drawing, however, is intended to convey the beginning of the alchemical process, that is, to explicate how the child born of man, not of woman, can be conceived. A woman stands between a well and a hearth pouring water into vessels, while the caption instructs the reader: "Vade ad mulieram lavantum pannos. Tu fac similiter" [Go to the woman washing pans. You do so in the same manner].[9] As philosophers, anthropologists, archaeologists, and psychoanalysts have frequently suggested, fire, water, and vessel are all feminine symbols.[10] The Jungian psychoanalyst Neumann makes the link explicit: "At the center of the mysteries over which the female group presides stood the guarding and tending of the fire. As in the house round about, female domination is symbolized in its center, at the fireplace, the seat of warmth and food preparation, the hearth which is also the original altar."[11] Elsewhere Neumann writes: "The natural elements that are essentially connected with vessel symbolism include both earth and water. This containing water is the primordial womb of life from which, in innumerable myths, life is born."[12] The notions of containment and transformation—of earth, fire, water—reinforce the idea in the initial engraving that "male pregnancy" can be accomplished only through imitation of feminine life. The alchemist essentially becomes a "container" of life: "Tu fac similiter!"

The imagery in *Atalanta fugiens* is not an anomaly in the literature

---

tion. This fluid, however, is not seminal . . . the female in fact, is female on account of an inability of a sort, viz., it lacks the power to concoct semen out of the final state of the nourishment (this is either blood, or its counterpart in bloodless animals because of the coldness of its nature). . . . Now what happens is what one would expect to happen. The male provides the 'form' and the 'principle of the movement,' the female provides the body, in other words the material. . . . Taking then the wildest formulation of each of these two opposites, viz., regarding the male qua active and causing movement, and the female qua passive and being set in movement, we see that the one thing which is formed is formed from the carpenter and the wood or a ball from the wax and the form" (Aristotle, *Generation of Animals*, trans. A. L. Peck [Cambridge, Mass.: Harvard University Press, 1943], bk. 1, pp. 101, 103, 109, 113).

9. Maier, p. 21.

10. See G. Van der Leeuw, *Religion in Essence and Manifestation* (trans. J. E. Turner [New York: Peter Smith, 1963], 1: 59–64), in which he implies the relationship of water and fertility with Demeter, the Mother Goddess (p. 59), and associates fire and female fertility (p. 61). O. Nahodil ("The Mother Cult in Siberia," in *Popular Beliefs and Folklore Tradition in Siberia*, ed. V. Dioszegi [The Hague: Mouton, 1968], pp. 459–77) points to the widespread belief among the Siberians of a Mother of fire and to the concomitant custom of female guardianship of the hearth. See also G. Bachelard, *La Psychoanalyse du feu* (Paris: Gallimard, 1949). Mircea Eliade comments on the belief in New Guinea that women have fire hidden in their genitals—and use it in cooking! (*Myths, Rites, Symbols: A Mircea Eliade Reader*, ed. W. C. Beane and W. G. Doty [New York: Harper & Row, 1976], 1:196.) The association of women with water is explored by Erich Neumann, *The Great Mother* (Princeton, N.J.: Princeton University Press, 1963), p. 47.

11. Neumann, p. 284.

12. Ibid., p. 47.

and language of alchemical transmutation. Alchemical symbolism, rich in both mythological and biological allusion, presents the image of the opus as the wresting of an embryo from the womb of the earth, embodied in woman, a birth from a man-made alembic.[13] This recurrent symbolism in alchemical works suggests an obsession with reversing, or perhaps even arresting, the feminine hegemony over the process of biological creation. More broadly, alchemical practice itself signals the continuous effort to gain control over a recalcitrant nature for the benefit of humanity. This desired mastery is also depicted in such images as that of Zeus giving birth to Athena from his head (see Emblem XXIII) or of Adam being delivered of Eve from his chest. The alchemist, who exemplifies the primordial striving for control over the natural world, seeks nothing less than the magic of maternity conferred on the "lesser" half of the species. Thus the great alchemist Paracelsus gives an affirmative answer to the question: "Whether it was possible for art and nature that a man should be born outside a woman's body and a natural mother's." And the culmination of the alchemical process is frequently depicted through the image of the birth of a male child who, according to Paracelsus again, "by art received life, through art . . . received a body, flesh, bones, and blood and through art . . . was born."[14]

The origins and evolution of alchemy contain further evidence of an identification with the maternal feminine as a source of creative power, what we would call today "womb envy." In *The Forge and the Crucible,* Mircea Eliade argues that the source of alchemical operations should be sought in shamanic practices, which also aim to control the forces of nature: "By conquering nature through physio-chemical sciences, man can become nature's rival," writes Eliade. More specifically, "The alchemist becomes the master of time when with his various apparatus he symbolically reiterates the primordial chaos and the cosmogony or when he underwent initiatory death and resurrection. Every initiation was a victory over death, i.e., temporality. The initiate proclaimed himself immortal, that is, he had forged for himself a post mortem existence which he claimed to be indestructible."[15]

Like the alchemist, the shaman in his *rite de passage* submits to ritual death and rebirth, a process of reintegration with a primordial totality followed by a repetition of the creation of a cosmogony through the

13. Mircea Eliade, *The Forge and the Crucible,* trans. S. Corrin (New York: Harper & Row, 1971) hereafter cited as *The Forge;* see the chapter, "Terra Mater, Petra Genitrix," pp. 43–52. "Indeed, metallurgy, like agriculture—which also presupposes the fecundity of the Earth-Mother—ultimately gave to man a feeling of confidence and pride. Man feels himself able to collaborate in the work of Nature, able to assist the process of growth within the bowels of the earth. He jogs and accelerates the rhythm of these slow chthonian maturations. In a way he does the work of Nature" (p. 47).

14. Paracelsus as quoted in H. Silberer, *Hidden Symbolism of Alchemy and the Occult Arts,* trans. S. E. Jelliffe (New York: Dover Publications, 1971), pp. 139–141.

15. Eliade, *The Forge,* p. 175.

breaking up of primitive unity. Out of the dismemberment which he undergoes, the shaman receives a new body from spirits who in early lore were in the service, or even in the shape, of the Mother of the Universe.[16] The knowledge imparted to the shaman of death and rebirth, of curative powers and transformational abilities, comes to him in the earliest versions from a mistress of animals whose metamorphosis into a master, a number of scholars argue, marks a later development in shamanic practices.[17] Moreover, although the shaman is considered the "master of fire," fire is originally a feminine element; thus among Siberian peoples who practice shamanism still, the Fire Mother is central to the life of the clan.[18] And for the Greeks Hestia was the goddess of the hearth until she was later displaced by Dionysius. The identification with the feminine in order to assume her power is most strikingly illustrated in the widespread custom among male shamans of wearing feminine attire.[19] Indeed, peoples who practice shamanism believe that the first shamans were women and that women could be shamans by nature while men had to learn the rites.[20]

16. See Mircea Eliade, *Shamanism: Archaic Techniques of Ecstasy*, trans. W. R. Trask (Princeton, N.J.: Princeton University Press, 1972), pp. 37 ff., 79–81.

17. The shaman's "tool" for communication with spirits, his *dru*, is cut from a tree of life linked with feminine spirits; the skin which he stretches over the wooden base is given to him by the mother of all wild things (see S. F. Anisimov, "Shamanskii chum u evenkov i problema proizhozhdenie shamanskogo oriada," *Trudy instituta ethnografii akademii nauk SSSR* 18 [1952]: 212–14). On the notion of the priority of a mistress of animals in shamanic rites, see Eliade, *Shamanism: Archaic Techniques of Ecstasy*, p. 81. There he speculates that her image may represent "matriarchal" conceptions which postdate those of a male "Uranian Being." Anisimov argues the opposite case; he suggests that the notion of the "Master of Animals" bears within it later patriarchal conceptions and for that reason the image of the "Master" is vaguer and less compelling than that of the earlier "Mistress" (see A. F. Anisimov, *Kosmologicheski predstavlenia narodov-severa* [Moscow-Leningrad: Akademia Nauk SSSR, 1959], p. 22). Nahodil insists on the priority of the Mistress of the Universe in Siberian lore (pp. 159–77).

18. Nahodil, p. 468.

19. For the phenomenon of tranvestism in shamanic rites, see L. Ia. Shternberg, "Shamanism and Religious Election," in *Introduction to Soviet Ethnography*, ed. S. Dunn and E. Dunn (Berkeley and Los Angeles: University of California Press, 1974), 1:77. For documentation of dress and sex change among the Siberians, see M. A. Czaplicka, *Aboriginal Siberia: A Study in Social Anthropology* (Oxford: Oxford University Press, 1914), pp. 244–45. For the belief in the origin of shamanism among women of the Siberian Enets, see Anisimov, "Shamanskii chum u evenkov i problema proizhozhdenie shamanskogo oriada," p. 213.

20. N. K. Chadwick, "Shamanism among the Tartars of Central Asia," *Journal of the Royal Anthropological Institute* 66 (January–June, 1936): 81. R. Briffault in his monumental *The Mothers* (London: G. P. Allen & Unwin, 1927), 2:531–32, gives a number of examples which suggest the notion of the priority of female shamans in the beliefs of Siberian as well as other peoples who practice shamanism. D. Zelenine (*Le Culte des idoles en Siberie*, trans. G. Welter [Paris: Payot, 1952], p. 244) seems convinced that shamanism was linked with the matriarchal clans and that the process of initiation whereby the male shaman received his spirit was originally thought of in the same terms as pregnancy. It is perhaps the experience of the women themselves, continues Zelenine, which suggested this analogy.

The primordial importance of the feminine as transmitter of the shaman's and later the alchemist's power is also underscored by shamanic initiation rites which necessitate a "dangerous" reentry into the paradigmatic body of the mother—cave, earth, water, fire—in order to learn her secrets and, thus, no longer be subject to her control. This parallels the alchemical vision of the absorption of the initiate into the body of the monster or into the womblike darkness called *nigredo* in order to begin the process of rebirth.[21] Eliade connects the shaman and the alchemist through the figure of the smith, who burrows into Mother Earth to extract her children, the metals,[22] in order to fashion them to his needs. In this act, he repeats, like Prometheus, the dangerous entry into a forbidden world: "In the symbols and the rites accompanying metallurgic operations, there comes into being the idea of an active collaboration of man and nature, perhaps even the belief that man by his own work is capable of superseding the processes of nature. The act, par excellence, of the cosmology, starting from a living primal material, was sometimes thought of as a cosmic embryology: the body of Tiamat was in the hands of Marduk a foetus. . . . And in all creation and all construction reproducing the cosmogonic model, man in constructing or creating, imitated the work of the demiurge."[23]

In the Babylonian myth of creation the original demiurge was the Mother—Tiamat. In fact, the alchemists' work on the *prima materia*—*materia* being etymologically traceable to *mater* or mother—suggests not so much a collaboration as an operation, in Aristotelian terms, in order to force that matter to deliver its perfect child, gold. More than a midwife, the alchemist functions as the form maker, the active principle, through his work upon the *materia*. We find here the Aristotelian equation of male activity and female passivity-materiality. The alchemist, the shaman, and the smith, draw their creativity from the feminine realm, identified by the shaman with the mistress of animals, by the smith with Mother Earth, and by the alchemist with the *prima materia*.

In *Atalanta fugiens*, the collaboration or union between man and nature is achieved by denying the independent status of the feminine and by containing and arrogating her creative powers. This tendency, already apparent in the first two illustrations of Maier's treatise, which describe the assumption of maternity by the male, deepens in subsequent images where woman is progressively divested of any creative function. The first images showing woman as nursing earth rather than mother denigrate the specificity of the feminine role, as Maier's commentary in the discourse of the second emblem reveals: "It makes no difference whether the nest is set in place by the hen or by the farmer, for the generation of the egg is the same."[24] This leads logically to the

21. Eliade, *The Forge*, p. 41; and *Shamanism: Archaic Techniques of Ecstasy*, pp. 46 ff.
22. Eliade, *The Forge*, pp. 53–64.
23. Ibid., p. 75.
24. Titus Burckhardt, *Alchemy* (Baltimore: Penguin Books, 1971), pp. 149–51.

third emblem, which indicates that the alchemist imitates woman's work, and even more to the fourth image, where we see the king (as the alchemist himself) in the act of joining man and woman, brother and sister in an act of love.[25] Although the emblem appears to show the equal conjunction of male and female as a prerequisite of the beginning of the alchemical opus, this conjunction is clearly under the aegis of the male alchemist himself.

In the fifth emblem, the nurturing aspect of the mother is drawn from the female by the king who holds a toad at her breast. "Place a toad at the breasts of a woman," instructs the motto, "that she may nurse it / And the woman dies, and the toad grows big from the milk." The epigram is even more explicit about the toad's use of the milk: "Put an ice-cold toad on the breast of a woman / so that like a child it may drink from the cup of milk. / Let it grow large by emptying the teat / and let the woman become ill and die. / Then make a noble medicine from it, / Which drives the poison from the human heart and relieves pestilence."[26] The noble medicine from the woman's breasts is the natural power of creation and healing which the toad, as intermediary, takes for the male. Symbolically, the toad has long been associated with the sphere of the feminine; already in neolithic times it was thought to possess mysterious powers over life. This appears as a motif in innumerable artifacts from cultures of the sixth millennium and later in Europe and the Mediterranean world. Gimbutas explains:

> Toad's meat was eaten as an aphrodisiac and dry toads were hung up to protect the house against evil. Such beliefs suggest a benevolent goddess, but the toad as a nocturnal and mysterious creature can cause madness, can take away milk and suck the blood from humans while they are asleep. In Baltic and Slavic mythologies, she is the main incarnation of the chthonic magician goddess. In the Indo-European mythologies she is basically an incarnation of a goddess of death while in the South . . . the most prevalent beliefs concerning the toad are those connected with birth, pregnancy or the womb (uterus).[27]

Moreover, perhaps as far back as the paleolithic age, the toad, as a "wandering uterus," was believed to cause pregnancy.[28] Like a woman's belly and like the moon to which it has been assimilated, the toad as a creature can go from flatness to fullness. Given these associations the significance of the fifth emblem becomes clear: The man in the illustration draws into the toad the maternal milk or the nurturing function of

25. Ibid., p. 25.
26. Ibid., p. 29.
27. M. Gimbutas, *The Gods and Goddesses of Old Europe* (Berkeley and Los Angeles: University of California Press, 1974), pp. 177–78.
28. Ibid.

the mother, which is in his possession. From woman's dark nature is taken the poison that can be transformed by the toad into the spiritual child of the male.

The following emblems reflect on the progress of masculine empowerment in the process of fertilization from which life emerges. Like a sky god, man is shown consistently to inseminate from above. Women no longer appear in the next three emblems: in Emblem VI the male is planting gold coin-seeds into the ground like the farmer;[29] and in Emblem VII he is identified with an eagle which descends from the sky to impregnate his mate and then builds his nest upon the summit of a phallic-shaped mountain.[30] Emblem IX, however, is harder to decipher: the king is seated inside a womblike pavilion eating apples from a tree, and the epigram advises that he "learn to be young before he becomes old."[31] Does this mean that the king-alchemist must return again to the womb of woman in order to master the operation of birth giving himself? This seems to be the case in Emblem X, where the king is shown returning to the fire, a symbol of the transformational power of the womb.[32] In the following emblem, XI, he is led to water, represented by the illustration of the goddess Latona or Leto, mother of Apollo and Artemis, sun and moon, astral twins who refer us back to those conjoined by the alchemist in Emblem VI. Since Leto, the goddess associated with hard labor, is permitted by a jealous Hera to give birth only on the water, the astral twins appear to have their origin from uterine water. The reference to this mother of the day and night implies that the alchemist must begin anew his cycle of generation by returning to the first source of all creation, the mother goddess herself, out of whose body the inseminating heavenly spheres themselves emerge. For in Emblem XIII the king is presented like a naked child by the river and is instructed in the motto to wash seven times (a reference to the Creation) to be rid of disease, thus suggesting the rebirth of the king who passes again through the uterine water of the original creatrix.[33] Rebirth is further underscored in the following emblem which shows the uroboric image of the dragon devouring its tail, a theme of self-regeneration through the symbolism of the womb.[34] An illustration of a craftsman-potter in Emblem XV who forms a womblike container out of the earth emphasizes again the beginning of another cycle and refers to Emblem III: "Tu fac similiter."[35]

29. Maier, p. 33.
30. Ibid., p. 37.
31. Ibid., p. 45.
32. Ibid., p. 49; for the association of fire with the womb, see the discussion of Emblem III above, and esp. n. 10 above.
33. Maier, p. 61.
34. Ibid., p. 65.
35. Ibid., p. 69.

In later emblems, this cycle of arrogation of feminine procreative powers is accelerated, the imagery becoming increasingly complex as the alchemist approaches his goal. Emblem XXIII represents the birth of Athena from the head of Zeus, a birth aided by the alchemist who brandishes an axe with which he has split the head of the Greek Father God. Athena is greeted by a golden shower from the sky, the heavenly source of life and seed, and the motto reads: "Gold rains down from the sky while Pallas was born in Rhodes, and Sol [Apollo] cohabits with Venus."[36] The sexual act empowered from above (Apollo and the golden shower) generates the birth of woman out of man, that is, reverses the natural relationship of male child emerging out of the body of the female. And in the following emblem, XXIV, we see the king devoured by a wolf only to be reborn out of a fire in which the wolf perishes.[37] Although there is a consistent pattern of birth and rebirth in these emblems, as in all alchemical literature, it must be noted that the male never really perishes, but rather he is resurrected, to underscore his growing mastery over the secrets of generation. Woman, on the other hand, is consistently being killed and not represented as reborn, thus implying a need to eliminate the feminine presence.

Emblem XLIX has as its motto, "The philosophical child acknowledges three fathers, as Orion," and shows Phoebus, Apollo, Vulcan, Haephestus, and Hermes—Olympian sky gods empowered by their mother goddesses, Hera, Maia, and Leto[38]—usurping maternity for themselves. They are presented gathering their sperm into an ox hide. The female maternal role is dismissed in the summary of the discourse that accompanies the emblem: "Lullius . . . attributed this same number of fathers to the Philosophical Child, namely Sol, who is the same as Apollo or the heavenly Sun, as the first begetter of this birth; who by virtue of indescribable force affects matter in a secret way, which is a sign to the Philosophers, just as if that matter were in the womb of a woman. In this way he bears a Son, resembling himself, to whom he later bequeaths his weapons, the signs of his manliness and his paternal rights. . . ."[39] Like the Egyptian patriarchal god Ra, creation is produced by the male through a form of masturbation which negates the female.[40]

36. Ibid., p. 101.
37. Ibid., p. 105.
38. Each of these gods, according to Robert Graves, was the son of former dominant Great Goddesses of fertility whose functions and role they usurped. See his "Introduction" to *The Greek Myths* (Harmondsworth, Middlesex: Penguin, 1970), 1:11–24. The argument that mother goddesses were displaced by their sons and consorts is also made in the recent work of archaeologists such as Gimbutas (n. 27 above), p. 238, who have been doing research on the East European sites which predate those of Greece. See also B. A. Rybakov, "Kosmogenia i mifologiia zemledel 'tsov eneolita,'" *Sovetskaia arkheologiia* 1 (1965): 24–47; and ibid., 2 (1965): 13–33.
39. Maier, pp. 206–7.
40. C. H. Long, *Alpha: The Myths of Creation* (New York: George Braziller, Inc., 1968), p. 187.

In the next and final emblem, where a woman and a dragon-snake are intertwined in a grave, the dispensibility of the female is emphasized with a certain vehemence. "The Dragon kills the woman, and she kills it, and together they steep themselves in the blood," declares the accompanying motto, and Maier writes in commentary: "Have a deep grave dug for the poisonous Dragon / With which the woman should be tightly intertwined: / While it rejoices in the marriage-bed, she dies / Have the Dragon buried with her. / Thereupon its body is abandoned to death and is imbued with blood. / Now this is the true way of your work."[41] In a summary, which states that the woman represents air and water and the dragon earth and fire, Maier insists that the dragon must be united with the woman in the grave and be killed or else it will attack the eggs of the eagle,[42] the sky-inseminating symbol of the male, which rest upon the phallic-shaped mountain.

*   *   *

The aggressive arrogation of the powers of the feminine described in Maier's treatise seems to deny the egalitarian conjunction of opposites upon which the process of alchemical transmutation is theoretically founded. In the *mysterium conjunctionis,* the male is purported to play a role equivalent to that of the female.

However, despite the insistence on the balance of the two sexes, the writings of both critics and adepts reveal, perhaps inadvertently, that the union is in actuality not an equal one. In the Christian framework within which Western alchemy developed, the feminine is relegated to the interior realm of nature, in which the spirit is trapped. Titus Burckhardt, the adept, commented:

> If the distance—and the relationship—between man and God is represented by a vertical line, then the distance between man and woman, or between the two corresponding powers of the soul, is represented by a horizontal line which results in a figure like an inverted T. At the point where the two opposed forces are balanced, that is to say at the center of the horizontal line, the later is the [*sic*] touched by the vertical axis, descending from God, or rising up to God. This corresponds to the supra-formal spirit which unites the soul to God.

> Although following this image the two forces or poles of human nature (the sulphur and quicksilver of the inward alchemical work) lie on the same level, there is nevertheless a difference of rank, similar to that of the right and left hands so that the masculine pole can be said to lie above the feminine. And indeed Sulphur, as the masculine pole, plays the role toward Quicksilver, the feminine

41. Maier, p. 209.
42. Ibid., pp. 210–11.

pole, which is similar to that of the spirit in its action on the whole soul.[43]

Burkhardt's "vertical line" ascends to God and carries the "horizontal line," which has no direct contact with the divine realm and, by its very horizontality, implies opposition were it not for its intersection with the "masculine pole." Active knowledge, he argues, belongs to the masculine; passive being to the feminine. This is, of course, traditional Christian dogma and leads to his conclusion that man saves woman: "The two powers [masculine intellect and feminine being] experience their full development on the plane of the soul, but realize their fulfillment only in the spirit, for only here does feminine receptivity attain its broadest breadth and its purest purity and is wholly united to the victorious masculine act."[44]

According to Jung, alchemy is the corrective underside of Christian consciousness, the unconscious element which "endeavors to fill in the gaps left open by the Christian tension of opposites."[45] It brings back the feminine, the image or memory of the fertility Mother Goddesses of prehistoric and neolithic religions discarded by the Christian Trinity: "Thus the higher, the spiritual, the masculine inclines to the lower, the earthly, the feminine; and accordingly the Mother who was anterior to the world of the father accommodates herself to the masculine principle and with the aid of the human spirit (alchemy or 'The Philosophy') produces a son."[46] Jung continues: "Thus the *filius philosophorum* is not just the reflected image in unsuitable material of the son of God; on the contrary, this son of Tiamat reflects the features of the primordial maternal figure. Although he is decidedly hermaphroditic he has a masculine name, a sign that the chthonic underworld, having been rejected by the spirit and identified with evil, has a tendency to compromise. There is no mistaking the fact that he is a concession to the weight of the earth and the whole fabulous nature of primordial animality."[47] As Jung observes, the equality of masculine and feminine in the alchemical opus is created in order to produce a masculine being, to right a masculine imbalance in masculine terms. The hermaphroditic nature of the *filius philosophorum* is a concession to the maternal by the dominant Father who thus encompasses the feminine (primordial animality) within the purer spiritual realm of the mind, the masculine "womb." The goal of the alchemical opus is to make conscious that which is hidden and to unite consciousness to the unconscious in such a manner as to contain the

43. Burckhardt, pp. 149–51.
44. Ibid., p. 154.
45. C. G. Jung, *Psychology and Alchemy* (Princeton, N.J.: Princeton University Press, 1968), p. 23.
46. Ibid.
47. Ibid., p. 25.

unconscious within the spiritual realm, that is, to masculinize it, by symbolically giving birth to the *filius,* not the *filia.*

The particularly vehement absorption and denial of the feminine by
the masculine which marks Maier's version of the alchemical process
may be related to the particular tenor of his era, a period of scientific
revolution in which man's attempt to control the natural world seemed
to be meeting with unprecedented success. Like the scientists,
seventeenth-century alchemists claimed that they could, through a mastery over the union of opposites (symbolized by the conjugal act) harness
natural energy by themselves. Thus, Atalanta *fugiens,* the free feminine
spirit, the spirit of Mother Nature trapped by the cunning youth who
found the source of creative erotic energy stands as an allegory of
technology triumphing over nature, through the triumph of man over
woman. Maier believed that alchemical power could imitate nature:
"The retort is indeed artificial, but it makes no difference whether the
nest is set in place by the hen or by the farmer, for the generation of the
egg is the same."[48] Since artificial creation is then as good as natural
creation, why keep the natural, which also contains poison, alive when
the man-created may be generated by man's design and at his whim?

Today, the triumph of the test-tube baby seems to replicate the
alchemical attempt to dominate the natural world by a "self-in-
seminating, masturbatory" creation (as in Emblem XLIX of *Atalanta
Fugiens*)[49] and to demonstrate the strong obsession of patriarchy to control the creative aspects of nature. The male vision, in effect, triumphs;
Atalanta is indeed in need of flight. And yet while the test-tube baby
makes the human container almost obsolete—woman and dragon /
serpent as *umbilicus* are "killed"—so does it also suggest that the male
inseminator may be superfluous—woman and serpent as *phallus* intertwined. Remains the alchemist amid his jars.

Seattle, Washington (Allen)

*School of Humanities and Arts
Hampshire College* (Hubbs)

48. Burckhardt, pp. 149–51.
49. Illustrations are reprinted courtesy of the Beinecke Rare Book and Manuscript
Library, Yale University.

Portavit eum ventus in ventre suo.

EPIGRAMMA I.

EMbryo ventosâ BOREÆ qui clauditur alvo,
  Vivus in hanc lucem si semel ortus erit;
Unus is Heroum cunctos superare labores
  Arte, manu. forti corpore, mente. potest.
Ne tibi sit Cæso, nec abortus inutilis ille,
  Non Agrippa, bono sydere sed genitus.

B 3   HER-

Nutrix ejus terra est.

EPIGRAMMA II.

ROmulus hirta lupæ preßiße, sed ubera capræ
  Jupiter, & factis, fertur, adeße fides:
Quid mirum, teneræ SAPIENTUM viscera PROLIS
  Si ferimus TERRAM lacte nutriße suo?
Parvula si tantas Heroas bestia pavit,
  QUANTUS, cui NUTRIX TERREUS ORBIS, erit?
                                C   Apud

Vade ad mulierem lavantem pannos, tu fac similiter.

EPIGRAMMA III.

Abdita quisquis amas scrutari dogmata, ne sis
  Deses, in exemplum, quod juvet, omne trahas:
Anné vides, mulier maculis abstergere pannos
  Ut soleat calidis, quas superaddit, aquis?
Hanc imitare, tuà nec sic frustraberis arte,
  Námque nigri fœcem corporis unda lavat.

C 3   S,

Appone mulieri super mammas bufonem, ut ablactet eum
& moriatur mulier, sitque bufo grossus de lacte.

EPIGRAMMA V.

FOemineo gelidus ponatur pectore Bufo,
    Instar ut infantis lactea pocla bibat.
Crescat & in magnum vacuata per ubera tuber,
    Et mulier vitam liquerit ægra suam.
Inde tibi facies medicamen nobile, virus
    Quod suget humano corde, levétque luem.

                 D 3   TOTA

Aurum pluit, dum nascitur Pallas Rhodi, & Sol concumbit Veneri.

EPIGRAMMA XXIII.

R Es est mira, fidem fecit sed Græcia nobis
    Ejus, apud Rhodios quæ celebrata fuit.
Nubibus Aureolus, referunt, quòd decidit imber,
    Sol ubi erat Cypriæ junctus amore Dea:
Tum quoque, cum Pallas cerebro Jovis excidit, aurum
    Vase suo pluviæ sic cadat instar aquæ.

                                                N 3   AURUM

EPIGRAMMA XXIV.

MUltivorum captare lupum tibi cura sit, illi
    Projiciens Regis corpus, ut ingluviem
Hoc domet, hunc dissone rogo, Vulcanus ubi ignem
    Excitet, in cineres belua quo redeat.
Illud agas iterum atque iterùm sic morte resurget
    Rexq; Leonino corde superbus erit.

               O   QUANTA

EPIGRAMMA XLIX.

FAbula narratur. *Phœbus, Vulcanus & Hermes*
  in pellem bubulam semina quod suerint;
Trésque Patres fuerint magni simul ORIONIS:
  Quin Sobolem Sophiæ sic tripatrem esse ferunt:
SOL *etenim primus, Vulcanus at esse secundus*
  Dicitur, huic præstans tertius arte pater.

Draco mulierem,& hæc illum interimit,simulque
sanguine perfunduntur.

EPIGRAMMA L.

ALtavenenoso fodiatur tumba Draconi,
   Cui mulier nexu sit bene vincta suo:
Ille maritales dum carpit gaudia lecti,
   Hæc moritur,cum qua sit Draco tectus humo.
Illius hinc corpus morti datur,atque cruore
   Tingitur: Hæc operis semita vera tui est.

Dd   DRACO-

Science, Politics, and Race

Inez Smith Reid

I

I confront a serious topic: science, politics, and race. I am not a natural scientist nor a psychologist, but a political scientist and lawyer. I am not a "race woman," but I am a member of a racial minority. My article will reflect that scholastic background and experience as well as the study I have done as a lay person. While this essay is directed primarily toward the effects of bad science and cheap genetics on race, it is also meant to reveal dangers for the female movement for equality.

Although scientists may be perceived as gods, they are not. Many scientists may regard themselves as anointed students of a mysterious realm of knowledge, which only the elitist (also labeled "qualified") can comprehend or command. The fact remains, however, that in a mass culture, such a body of knowledge, whether accurate or inaccurate, becomes public property. Politicians and policymakers capsulate and distort it. It then filters through to the general public with disastrous consequences for the subjects of the scientists' findings.

Let me be more specific. Some geneticists and psychologists pride themselves on articulating theories that propose to answer questions about the hereditary influences of both disease and intelligence. The overworked politician or policymaker, unschooled in esoteric terminology and pressured by an equally unschooled constituency, seeks to distill the essence of scientific findings and reduces a 123 page study, or 587 pages of quantitative, complicated research, into one short sentence. A sophisticated code is broken, but what results is a careless simplicity that the public takes as solid, irrefutable scientific truth. Public response then becomes predictable. Indeed, who can blame the hysterical mothers and

I am grateful for the help of my brother, a biochemist.

This essay originally appeared in *Signs,* vol. 1, no. 2, Winter 1975.

fathers of Boston who rant about the impossibility of their white children sharing an educational experience with black youngsters, when eminent New England and California scientists have labeled them as possibly intellectually inferior by inheritance; perhaps prone to criminality by the presence of an XYY chromosome; or physically diseased because of sickle cell anemia or hypertension? If the scientists are advised that a theory may be misused, they retreat from annoying distractions behind a shield. In turn, policymakers and politicians may make unfair demands upon scientists and misdirect the course of research.

What of the victims of this research? What of the member of a racial minority who is often the object of scientific findings and who learns that he or she is an undesirable genetic mess: dumb, unhealthy, and criminal? Racial minorities, however, are not alone in their victimization via "scientific truth." Their partners on the receiving end may well be the female sex. Women have not escaped inquiries that, on the surface, seem to be objective in nature and that are also eloquently rationalized as honest attempts to ameliorate their lives, to point them in directions that their "fair sex" may tolerate mentally and physically. The effects of science and genetics on race, then, have a parallel in their effect on women. Moreover, because we cannot forget how many blacks are women, to write about harm done to a race is to write about harm done to members of a sex.

These generalizations are the setting for the message I wish to convey about the irresponsibilities of both science and politics.

II

In 1971, the president of the United States called for a national program on a genetic blood disease that affects relatively small segments of the black population: sickle cell disease. In 1972 and 1973, the secretary of Health, Education, and Welfare pronounced the need to tackle a killer disease that disproportionately affects the black population and that many scientists argue is hereditary: hypertension, or high blood pressure. Sickle cell anemia results in death for some 340 black persons a year, compared with hypertension, which takes the lives of some 13,500 blacks a year.[1]

Politicians and policymakers moved the nation onto a wrong course. Scientists and doctors followed. Members of each group sensed a need for education—of a nebulous public, of blacks—but not education of the politician or bureaucrat and the dedicated, qualified scientist and doctor equipped with superior knowledge. So, in his August 16, 1972 speech before the National Medical Association in Kansas City, Missouri, Elliott

1. Barbara J. Culliton, "Sickle Cell Anemia: The Route from Obscurity to Prominence," *Science* 178 (October 1972): 141.

Richardson could state with a clear conscience: "An essential aspect of the new program [meaning sickle cell] will be public education. A recent survey found that only three of ten blacks had ever heard of sickle cell anemia."[2] Clearly, the onus of ignorance was on blacks. What of the do-gooders who asked for screening programs in elementary and secondary schools and on job sites without the faintest recognition of the educational and economic repercussions? What of the doctors who barely were aware of the phrase "sickle cell anemia," let alone the distinction between trait and disease?

Error and misplaced emphasis were compounded in the same 1972 speech. Secretary Richardson moved on to stress repeatedly that hypertension is a killer disease, urging the nation into action to control if not to stamp it out. He said, "This is the moment which must not slip away from us, to go on the attack against the killer hypertension." Again he assumed that scientists and doctors knew (or could learn quickly) what to do—could get the hypertensives to swallow "one drug . . . or drug combination," get "vital information about the disease to both professionals and the public." Not a word about the impediments to pill taking stemming from the customs and fears of people: not a word about the economics of detection and drug taking.[3] Nor were there words of wisdom to educators, employers, insurance companies, or federal and state bureaucrats.

The arousal of public interest in those diseases without proper at-

2. Elliott L. Richardson, "Urgent Health Problems," *Journal of the National Medical Association* 65, no. 3 (May 1973): 189.

3. For a discussion of hypertension—problems of screening, detection, and treatment—see transcript of proceedings of the Miniconsultation on the Mental and Physical Health Problems of Black Women, Black Women's Community Development Foundation, Washington, D.C., March 29–30, 1974. See also the following articles by the Veterans Administration Cooperative Study Group on Antihypertensive Agents: "Effects of Treatment on Morbidity in Hypertension," *Journal of the American Medical Association* 202, no. 11 (December 1967): 116–22; "Effects of Treatment on Morbidity in Hypertension. II. Results in Patients with Diastolic Pressure Averaging 90 through 114 mm Hg," *Journal of the American Medical Association* 213, no. 7 (August 1970): 1143–52; and "Effects of Treatment on Morbidity in Hypertension. III. Influence of Age, Diastolic Pressure, and Prior Cardiovascular Disease: Further Analysis of Side Effects," *Circulation* 45 (May 1972): 991–1004. Other articles include Edward D. Freis, "Age, Race, Sex, and Other Indices of Risk in Hypertension," *American Journal of Medicine* 55 (September 1973): 275–80, and "The Treatment of Hypertension," *American Journal of Medicine* 52 (May 1972): 664–71; Editorial, "Hypertension Is Different in Blacks," *Journal of the American Medical Association* 216, no. 10 (June 1971): 1634–35; Gilbert McMahon et al., "A Study of Hypertension in the Inner City," *American Heart Journal* 85, no. 1 (January 1973): 65–71; Margaret M. Kilcoyne, "Hypertension and Heart Disease in the Urban Community," *Bulletin of the New York Academy of Medicine* 49, no. 6 (June 1973): 501–09; Frank A. Finnerty, Jr., Edward C. Mattie, and Francis A. Finnerty III, "Hypertension in the Inner City. I. Analysis of Clinical Dropouts," *Circulation* 47 (January 1973): 73–75; Joseph A. Wilber, "The Problem of Undetected and Untreated Hypertension in the Community," *Bulletin of the New York Academy of Medicine* 49, no. 6 (June 1973): 510–20.

tention to the education of specific policymakers and politicians left an unclear idea about who should be the beneficiaries of a national assault on sickle cell anemia and hypertension. A case can be made that the beneficiaries were doctors, psychologists, bureaucrats, scientists, and politicians who enjoyed lucrative research grants, remunerative opportunities to penetrate the public arena to lecture, and exposure to the media so that their immediate constituents and the public at large might know of their dedication to the task of wiping out the odious effects of genetic or genetic-based disease. Those afflicted with the disease turned out to be victims, not even secondary beneficiaries.

Early on came the attack on impressionable children. Mandatory screening programs were introduced in some elementary and secondary schools, despite the fact that young children were not the most critical of those potentially afflicted with sickle cell disease or carriers of the disease.[4] Some teachers were quick to conclude that youngsters with sickle cell trait had a central nervous system disease and hence were afflicted with learning and behavior disorders.[5] Some youngsters with high hopes for athletic careers, or just eager to engage in sports in a joyful, carefree manner, were banned from participation because they carried the sickle cell trait.[6] Imagine a child who wants to live each day in playful and vigorous activity, suddenly inactive because of a misguided policymaker. Who knows what effects it may have on the growth and development of that child? As Dr. Helen M. Ranney puts it: "A child who is found to have sickle cell trait is hardly benefited if he is then treated as 'different,' or if this information is utilized to deny him employment or life insurance in the future."[7]

Then came the attack on adults. First, there was the demand to screen them before they get married.[8] Individuals, euphoric about their impending marriage, could be hurtled into depression and anxiety when the revelation came that one or both had the sickle cell trait. Few were told, as reported in the *New England Journal of Medicine*, that "the prevention approach is based on the unproved assumptions that sickle cell anemia is so undesirable that potential trait-bearing parents would be better off not having any children rather than taking a 25 percent chance of having a child with sickle cell anemia, that those who have the trait and wish to have children should not marry another person with

4. "Since the knowledge that a child has sickle cell trait, or other genotypes that are asymptomatic, does not by current standards alter the child's lifestyle or medical care, this age group might be afforded second-order priority" (Donald L. Rucknagel, "The Genetics of Sickle Cell Anemia and Related Syndromes," *Archives of Internal Medicine* 133 [April 1974]: 602).

5. Culliton, "Sickle Cell Anemia: The Route from Obscurity to Prominence," p. 142.

6. Ibid., pp. 141–42.

7. Helen M. Ranney, "Sickle Cell Disease," *Blood* 39, no. 3 (March 1972): 436.

8. Charles F. Whitten, "Sickle-Cell Programming—an Imperiled Promise," *New England Journal of Medicine* 288, no. 6 (February 1973): 318–19.

the trait irrespective of other desirable factors in the prospective marriage, and that those with sickle cell anemia cannot live satisfying lives."[9] Was the right of a man and a woman to association or to freedom of choice in selection of a mate diluted or hidden under a bundle of scientific rationalizations?[10] Not only did the decision to marry become fraught with anxiety. Adult employees often were faced with loss of jobs or with denials of employment after being detected as carriers of the trait. "At least one airline stewardess was grounded after her company found she carried the sickle cell trait," reported Dr. Rudolph E. Jackson, head of the national sickle cell program; and "persons have allegedly been denied jobs because they are carriers."[11]

Moreover, compulsory screening of pregnant women may have implications for the personal freedoms supposedly guaranteed to all. While advocates of pregnancy screening tend to think of the practice not so much as "screening" but as "laboratory tests appropriate to medical care,"[12] the woman who is advised not to have children because she has sickle cell anemia or the sickle cell trait cannot nurture "a laboratory test." What, then, becomes of her right to bear children? Is it psychologically eclipsed?[13] In matters related to fertility, fertility control, and pregnancy, it is of course possible to see an intersection of racial and female concerns.

Most pernicious, perhaps, were the actions of insurance companies with respect to sickle cell carriers. Several specialists attached to the New York State Department of Health in Albany, New York reported in late 1973 that ". . . approximately half of the 45 insurance companies we recently surveyed had increased their premiums for carriers of the sickle cell trait, presumably because the trait was believed to be associated with reduced life expectancy."[14] Similar findings were reported by Joseph Christian of the Indiana School of Medicine who asserted: "We know of

9. Ibid.

10. Premarital screening for sickle cell anemia not only raises the specter of infringement of personal freedoms guaranteed by the First Amendment, or the various "zones of privacy" carved out of the Bill of Rights by Justice William Douglas, but it also can lead to psychological problems for women and men. As Dr. Helen M. Ranney has stated, "There are obvious problems for the individual concerned if screening is carried out at the time of premarital examinations. Genetic counseling could be done, but the psychological problems engendered by premarital screening may make it undesirable at least at the present time" (Ranney, p. 437).

11. Culliton, "Sickle Cell Anemia: The Route from Obscurity to Prominence," p. 141.

12. Ranney, p. 437.

13. What are the probabilities that parents with sickle cell anemia or sickle cell trait will have children without disease or trait? According to Dr. Donald L. Rucknagel, "Should both parents have the trait, the probabilities for each child to have normal hemoglobin only, sickle cell trait, or sickle cell anemia are 25%, 50%, and 25% respectively . . ." (Rucknagel, pp. 595–96).

14. D. T. Janerich et al., "Age Trends in the Prevalence of the Sickle Cell Trait," *Health Services Report* 88, no. 9 (November 1973): 805.

several firms that have a special rating for trait carriers on the books. Some of them charge as much as 150 percent of the usual premium, some added 30 percent."[15] The experience of sickle cell carriers and hypertensives in obtaining life insurance policies at regular or fair premiums, or in getting them at all, is not unlike that of urban property or business owners who operate in the thick of poor economic conditions and whose insurability is almost negligible.

The injustice of the sickle cell passion, with its negative fallout on carriers, is that in the early stages of the popularity of the sickle cell assault, and even today, few resisted the confusion of the trait with the disease. Some with the trait were erroneously classified as having sickle cell disease.[16] Erroneous diagnosis undoubtedly increases the incidence of social and economic discrimination against the victim.[17] This discrimination may even be compounded by those who are ignorant about the nature of the disease. How many people in the population, for example, operate under the mistaken assumption that sickle cell disease and sickle cell trait are contagious or communicable diseases?

Although Dr. Rudolph Jackson reported in 1974 that some state laws requiring screening for sickle cell anemia "are being repealed or amended" and "fewer allegations of job discrimination or other such practices are heard: and companies and airlines are changing their hiring policies relative to sickle cell trait individuals,"[18] the pernicious discrimination of the past four years cannot be erased, nor can one be certain that the subtlety and sophistication of social and economic discrimination against sickle cell carriers does not persist today. Low-cost brainstorming sessions among scientists and educators; doctors, educators, and lawyers; bureaucrats and employers in the private sector; and others might have mapped out an educational campaign designed to minimize or avoid the social injustice of scientific research.

Now hypertension continues to be publicized. The problems of screening and detection are arousing specters similar to those seen in the sickle cell screening and detection process.[19] Hypertension is associated, racially, with the black American populace but really affects the total American population and has a specific impact on females—black and white.[20]

15. Culliton, "Sickle Cell Anemia: The Route from Obscurity to Prominence," p. 142.

16. Whitten, p. 319. For a discussion of the sickle cell disease, see Lloyd H. Smith, Jr., Alan Engelberg, and Michael Messer, "Sickle Cell Disease," *California Medicine, the Western Journal of Medicine* 118 (April 1973): 48–55.

17. See J. E. Bowman, "Mass Screening Programs for Sickle Hemoglobin: A Sickle Cell Crisis," *Journal of the American Medical Association* 222 (December 1972): 1650, and Janerich, p. 806.

18. Rudolph E. Jackson, "A Perspective of the National Sickle Cell Disease Program," *Archives of Internal Medicine* 133 (April 1975): 533.

19. See n. 3, above.

20. Figures vary, but it is estimated that hypertension strikes between 15 and 20

Hypertension in females of reproductive age presents special considerations. Often great fear is engendered in a pregnant woman with hypertension that somehow neither she nor the baby can survive. In reality, though, "in 85% of patients with hypertension and pregnancy, the outcome with good prenatal care is good."[21] Yet, historically, whenever any kind of hypertension was viewed as complicating pregnancy, the general practice was "therapeutic abortion and subsequent sterilization."[22] Now doctors are discovering that chronic hypertension in pregnant women "is poorly defined and may include some patients possessing surgically correctable renal vascular lesions."[23] A case in point is that of a twenty-six-year-old Native American (Indian American) woman, pregnant and hypertensive, who resisted a therapeutic abortion despite the fact that her blood pressure ranged from 170/105 to 230/140. Doctors then decided on surgery and discovered a lesion in the left main renal artery, which "was excised and an end to end anastomosis of the artery . . . performed."[24] What the doctors discovered, then, was a "fibromuscular dysplasia of the renal artery." What is surprising and worth noting is that "standard obstetrical textbooks fail to mention the existence of this potentially correctable lesion as a cause of hypertension, despite the fact that its highest incidence is in the woman of childbearing age."[25] The twenty-six-year-old Native American woman not only sur-

---

percent of Americans, with the black affliction rate being twice that of whites. See David Werdegar, "Hypertension: Insight into the Health Needs of the Community," *Journal of Urban Health* 2, no. 2 (April 1973): 32. Particularly high is the incidence of hypertension in the young black woman. See Editorial, "Hypertension Is Different in Blacks" (n. 3, above), pp. 1634–35, and Medical News, "Hypertension Affects Many Young Black Women," *Journal of the American Medical Association* 215 (February 1971): 876. Black and white alike may not know that they are victims of hypertension. It is postulated, for example, that of the 23 million people suffering from hypertension (160 systolic, 95 diastolic blood pressure; some medical experts recognize high blood pressure beginning at 150/90) in 1972 only about half "knew their condition and of that half only half were receiving treatment." See John B. Stokes III, Gerald H. Payne, and Theodore Cooper, "Hypertension Control—the Challenge of Patient Education," *New England Journal of Medicine* 289 (1973): 1369–70. One study, carried out in Georgia in 1962 and reported on in 1973, revealed that some 35.6 percent of white females knew they had high blood pressure but were not taking medication, while 31.7 percent of the white females had no idea that hypertension had invaded their bodies. By comparison, 60.7 percent of the black males in the same study did not know they were hypertensives and therefore were not under treatment. See Wilber (n. 3 above), p. 513.

21. S. L. Matseoane, "Hypertension in the Harlem Community," in *Mental and Physical Health Problems of Black Women* (Washington, D.C.: Black Women's Community Development Foundation, 1975), pp. 72–76.

22. Yuichi Ito, Newell Falkinburg, Dilworth T. Rogers, and Donald C. Martin, "Renal Vascular Hypertension in Pregnancy: Problems in Diagnosis and Management," *Journal of Urology* 108 (July 1972): 9–11.

23. Ibid., p. 9.

24. Ibid.

25. Ibid., p. 10.

vived surgery but maintained a normal blood pressure until shortly before cesarean delivery, when her pressure climbed to 152/110. After
delivery of a normal baby, the mother's blood pressure dropped to a
range of 120/80 to 132/92.[26] The doctors who treated this patient stress
that "many young women with curable disease are possibly
overlooked."[27] That is, not every hypertensive female who becomes
pregnant is necessarily doomed to therapeutic abortion and eventual
sterilization. The vast majority can produce healthy babies.[28] Thus, like
black males, all women need to be sensitized to hypertension, to know
what are the alternatives in coping with that affliction, to know when and
if childbearing capacities must lie unused or even be taken away.

Many women's organizations could drum up support for, or even
launch, educational programs concerning the special impact of hypertension on the female sex. Studies have revealed that "the physician's
main problem is that he has no time to spend with the patient to educate
him, and he has no one in the office who can do this for him."[29] Rather
than leaving a vacuum, women's groups searching for meaningful projects could take on such a public educational campaign. It could include
basic data on hypertension, information about the treatment of essential
hypertension, and documentation on hypertension and pregnancy.[30]

Yet, although one constantly finds emphasis on the "killer disease,"
little stress is placed on the implications of that disease for policymakers.
Let me single out one policymaking category for emphasis, the Social
Security Administration. If hypertension is a disabling disease, victims
should be able to qualify for disability benefits without a lot of unpleas-
antness. Yet, in two recent cases, hypertensives had to resort to court
action before being declared eligible for disability benefits. The subjects
of the two cases in question are both women, a matter that deserves
further exploration. Are female hypertensives forced to resort to legal
action more often than male hypertensives in order to obtain disability
benefits?

26. Ibid., p. 9.
27. Ibid., p. 11.
28. This is not to deny that some hypertensives, estimated at 10–15 percent by Dr.
S. L. Matseoane, have the kind of hypertension which forebodes badly for the health of
mother and child. As the Ito study also revealed, "Hypertension complicating pregnancy
may have devastating effects upon mother and fetus" (Ito et al., p. 9). For further discussion of hypertension and pregnancy, see N. M. Simon and Frank A. Krumlovsky, "The
Pathophysiology of Hypertension in Pregnancy," *Journal of Reproductive Medicine* 8 (March
1972): 102–5; M. G. Hill, "Pulmonary Hypertension in Pregnancy," *Proceedings of the Royal
Society of Medicine* 65, no. 1 (January 1972): 100 (a woman successfully delivered a child
despite pulmonary hypertension although she elected to be sterilized five months after
delivery); C. P. McCartney, "The Acute Hypertensive Disorders of Pregnancy, Classified
by Renal Histology," *Gynaecologia* 167 (1969): 214; J. A. Tweedie and W. F. Mengert,
"Hypertensive Disease and Pregnancy," *Obstetrics and Gynecology* 25 (1965): 188.
29. Wilber (n. 3 above), p. 518.
30. See n. 3 above.

In the 1960s, prior to the public concentration on hypertension, a fifty-four-year-old woman filed a claim of disability under the Social Security Act. By that time, May L. Hayes had lived her entire life in Roanoke, Virginia, working first in a plant (1933–58) until the company went out of business, and then as a counter attendant or waitress in a drugstore until her physician advised her to stop work in 1962. After several operations, May Hayes tried to return to work in 1964 but by November of that year was forced to file for disability.

She had been under the care of her doctor for hypertension since 1950, when she began to suffer from headaches, blurred vision, substernol pain, blood pressure of 220/110, enlarged heart, and coronary insufficiency. She had been hospitalized also for a series of operations including one for a total hysterectomy. A Social Security Administration doctor examined May Hayes and described her condition as "systolic hypertension benign, without evidence of cardiac disease, no retinophy [*sic*] and no evidence of renal disease." Another doctor for the Social Security Administration did not examine May Hayes, but in looking over her records concluded: "While claimant had some degree of hypertension, there have been normal periods at times." He stated further that he could find "no contraindication why she could not do light forms of work that didn't involve any heavy lifting or straining or involve physical strain." Then came the industrial psychologist, who insisted that there were jobs in Roanoke, Virginia that May Hayes could perform. How did he know? He had looked at a listing of jobs made available to him by the local Chamber of Commerce. When asked whether he was aware of the attitudes of local employers about hiring those with health conditions like May Hayes's, "he was extremely evasive." The hearing examiner at the Social Security Administration turned down May Hayes's claim; the Appeals Council affirmed the rejection, as did the federal district court in Virginia. The federal court of appeals, however, overturned the rejection and concluded that ample evidence of disability from hypertension had been presented and May Hayes should receive disability payments.[31]

Despite the precedent of the Hayes case and alleged aroused public concern for hypertension victims, in the 1970s another case had to be brought to the courts by a hypertensive seeking disability payments. Ruby R. Martin, forty-nine years of age, a weaver in a mill for twenty-three years, formally educated through the eighth grade, had suffered from hypertension and diabetes at least since 1962. As a result, she experienced headaches, shortness of breath, severe dizzy spells with nausea and vomiting, and an occasional loss of orientation. Her own personal physician indicated that she was eligible for disability benefits. An internist, a consultant for the South Carolina Rehabilitation Depart-

---

31. Hayes v. Gardner, 376 F.2d 517 (4th Cir. 1967).

ment, conducted an independent examination and concluded that Ruby Martin's "hypertension is so severe and has been of sufficient duration as to preclude returning her to a semblance of good health." She had even been treated for hypertension at the National Institutes of Health—an indication that her case was exceptional.

The Social Security Administration asked one of its doctors to look over the records of Ruby Martin and to draw some conclusions. The assigned doctor reported that Ruby Martin had had "essential hypertensive vascular disease for at least 10 years, without end organ damage, although she did have some narrowing of the arteries and Grade II retinophy [*sic*]." Still, he felt that Ruby Martin could do "light or sedentary type work." The hearing examiner rejected Ruby Martin's claim for disability. So did the Appeals Council and the federal district court. It remained for the court of appeals, in 1974, to rule Ruby Martin eligible for disability payments. The court specifically rejected end organ damage (heart, brain, kidneys, and eyes) "as an exclusive prerequisite to the establishment of a disability from hypertension." The court stressed instead the language of section 423 (d) (1) (A) of 42 U.S. Court of Appeals (the Social Security Act), which provides that the test for the acceptance of a disability claim is an "inability of the claimant to engage in any substantial, gainful activity by reason of any medically determinable physical or mental impairment which can be expected to last for a continuous period of not less than 12 months."[32] Ironically, no scientists and doctors were present to explain to the Social Security Administration that end organ damage is not an essential prerequisite to disability from hypertension, and that, in fact, as a policy, women and men as young as forty years of age might be truly disabled despite the fact that their bodies may not appear to be ravaged.

These two cases clearly identify the hypertensive as the disadvantaged or the victim. While scientists, doctors, psychologists, and bureaucrats have relatively easy access to resources for the study of diseases like sickle cell and hypertension, the victims, by contrast, are caught in a whirlwind: potential refusal of job opportunities, declining access to insurance policies, extraordinary waits for screening and detection, folkways and mores against certain medication, and a possible denial of disability benefits because of an insistence on the ability to engage in work despite labels of affliction with killer diseases. If the victims dare cry out that they are not being treated as beneficiaries of these health programs, eyebrows are raised. A scientist may sigh and ask why those "ungrateful people don't appreciate what we are doing in their behalf." Disgruntled policymakers and members of the general public resort to their persistent image of the black victim as one whose hand is constantly stuck out, begging for handouts from public coffers. It matters not that

32. Martin v. Secretary of Health, Education, and Welfare, 492 F.2d 905 (4th Cir. 1974).

the victim may have labored for years at some tedious job: he is a disgraceful beggar, not a deserving beneficiary.

III

Early criminologists were fond of theorizing about the constitutional sources of the criminal mind and hypothesizing the existence of the "born criminal."[33] Their theories apparently have not been laughed out of existence. In fact, since 1966 at least, numerous articles on crime and genetics, specifically the XYY chromosome, have appeared in both foreign and national medical/science journals.[34]

Europeans led the competition to determine whether an XYY chromosome augured criminal behavior. Studies were done in England, Scotland, Denmark, and other places to test the proposition that an XYY chromosome pattern could lead to criminal behavior.[35] Very few XYY prison inmates or criminals institutionalized in hospitals were discovered. Those who were tended to be tall (between 175 and 210 cm).

33. See, for example, Cesare Lombroso, *Crime: Its Causes and Remedies* (Boston: Little, Brown & Co., 1911).

34. See, for example, W. H. Price, "Criminal Patients with XYY Chromosome Complement," *Lancet* 1 (March 1966): 565–66 and "Sex Chromosome Abnormalities: How Strong Is the Link with Crime? *Manitoba Medical Review* 48 (January 1968): 26–27; W. H. Price and P. B. Whatmore, "Criminal Behavior and the XYY Male," *Nature* 213 (February 1967): 815; P. A. Jacobs et al., "Chromosome Studies on Men in a Maximum Security Hospital," *Annals of Human Genetics* 31 (May 1968): 339–58; J. Nielsen et al., "Intelligence, EEG, Personality Deviation, and Criminality in Patients with the XYY Syndrome," *British Journal of Psychiatry* 115 (August 1969): 965; R. Veylon, "XYY Chromosome and Criminality: Has Genetics Rediscovered the Born Criminal?" *Presse medicale* 77 (February 1969): 333–35; A. W. Griffiths, "The XYY Anomaly," *Medicine, Science and the Law* 11 (April 1971): 73–76; E. Engel, "The Making of an XYY," *American Journal of Mental Deficiency* 77 (September 1972): 123–27; T. Urdal et al., "Criminality and Length of the Y Chromosome," *Lancet* 1 (April 1974): 626–27.

35. Nine XYY males were discovered and studied in the State Hospital, Carstairs, in Scotland. The scientists concluded: "There is no reason to believe that these patients would have indulged in crime had it not been for their abnormal personalities" (W. H. Price and P. B. Whatmore, "Behaviour Disorders and Pattern of Crime among XYY Males Identified at a Maximum Security Hospital," *British Medical Journal* 1 [March 1967]: 536). Some forty-two patients out of 155 in a Danish institution were singled out for observation, and two "were found to have 47 chromosomes and sex chromosomes XYY or 5% of the 37 patients [whose chromosomes were analyzed] and 1.29% of the total 155 patients" (J. Nielsen, T. Tsuboi, G. Stürup, and David Romano, "XYY Chromosomal Constitution in Criminal Psychopaths," *Lancet* 2 [September 1968]: 576). Some thirty-four prisoners out of 1,200 inmates were selected for study in an Australian prison whose inmates committed crimes ranging from murder to drunkenness to vagrancy. Thirty were found to have "normal chromosome constitution, 3 had 47 chromosomes with an XYY chromosome complement, and 1 was an XYY/XYYY mosaic." All four of these men were labeled psychopaths (Saul Wiener, Grant Sutherland, Allen Bartholomew, and Bryan Hudson, "XYY Males in a Melbourne Prison," *Lancet* 1 [January 1968]:150).

American scientists decided to follow up on the studies of their European counterparts, perhaps seeming to ignore the height considerations present in the European studies—and more critically, appearing to accept the theory that an XYY chromosome pattern could spawn criminal behavior.

Many American scientists have scoffed at the idea of a chromosome connection to crime. For example, Jonathan Beckwith, a microbiologist, and Jonathan King of M.I.T. argue that the XYY syndrome represents a "dangerous myth."[36] On the other hand, "a leading geneticist" remarked during a meeting in 1972: "We can't be sure XYY actually makes someone a criminal, but I wouldn't invite an XYY to dinner."[37] That geneticist might have been deeply disturbed to read a newspaper headline the next day asserting: "XYY Chromosome Person May Murder You at Dinner!"

Let me single out for discussion the Harvard Medical School "Study of Chromosomal Abnormalities in Newborns."[38] This study, commenced in 1968, has been conducted by psychiatrist Stanley Walzer and geneticist Park Gerald. Over 15,000 babies born at the Boston Hospital for Women were examined as of November 1974 to determine whether they possessed an XYY or XXY chromosome syndrome. Female voices, which should have protested, were silent, unaware, or unconcerned about the necessity of organizing around this issue just as vehemently as they have around abortion.

While Gerald and Walzer themselves apparently do not adhere to the notion of an XYY criminal, still they argue that "XYY boys are at risk for developing some rather ill-defined behavior problems," problems such as "impulsivity and difficulty controlling themselves." They further contend that XYY is a "disease" entitling "afflicted" youngsters to medical treatment. As for the XXY chromosome syndrome, Walzer indicates that some of these male youngsters may have high IQs but may manifest "speech and language difficulties" as well as a "significant reading deficit."

Although the National Institute of Health has approved the project for funding for an additional three years and even though the case has been before the Harvard Medical School's Standing Committee on Medical Research, the storm over it has not ceased, especially on the consent issue. In September 1974, Dr. Jay Katz, psychiatrist and professor at the Yale Law School, wrote: "In its present form Dr. Waltzer's [sic] initial approach to the parents is neither straightforward enough nor does it comply with what I consider the requirements of informed consent to

36. Culliton, "Sickle Cell Anemia: The Route from Obscurity to Prominence," pp. 715–16.

37. Ibid., p. 715.

38. Barbara J. Culliton, "Patient's Rights: Harvard Is Site of Battle over X and Y Chromosomes," *Science* 186, no. 4165 (November 1974): 715–17.

imply. Elements of 'fraud, deceit,' and even 'duress and overreaching,' proscribed by the Nuremberg Code, are clearly present." Walzer tries to dismiss the lack of informed consent charge: "We're not doing anything behind anyone's back. Nor have we been disrespectful of our parents. They are grateful for what we're doing." It is ironical that the consent forms are being revised. More shocking, women at one time were asked to give consent for the chromosome study on their babies during labor. Parallels here can be drawn to the sterilization controversy where women, especially poor women being treated in city hospitals, wake up to find that their childbearing capacity has been cut off.[39] In addition to individual voices raised against the Walzer-Gerald study, two groups managed to unleash enough public sentiment to stop the Boston screening. A Boston based group, Science for the People, and a Cambridge (Mass.)–Washington, D.C. organization, the Children's Defense Fund, mobilized to point out the grossly unjust effects which the Walzer-Gerald study could produce for children stigmatized for life due to an XYY or XXY chromosome pattern. As a result, Dr. Walzer announced that he would discontinue screening newborn babies. Nonetheless, he persists in his determination to follow up on forty-five children (having "normal" and "abnormal" chromosome patterns) already screened into the study.[40] Thus, some XYY and XXY American children already are marked.

Imagine a five-year-old beginning public school with a record clearly stamped XYY. Can anyone categorically state that teachers and principals of the school he attends would not expect this five-year-old to be a trouble maker and treat him accordingly? Can anyone categorically state that the XYY male youngsters would not be isolated for experimentation with drugs, experimentation of the kind which has occurred recently in Boston?

There is another point in bringing up the XYY controversy. Admittedly, XYY and XXY have not been identified as racial phenomena. Yet, certain segments of our population have been singled out as prone to crime. These include such racial minorities as blacks, Puerto Ricans, and Mexican Americans—regarded as common perpetrators of crimes against person and property—as well as Italians, who are identified repeatedly with organized crime. It should not be surprising, then, that scientists in Mexico and America chose to study some Mexican and black American prisoners to see whether they possessed the XYY chromosome. In one Mexican prison, ninety-four of the tallest inmates out of a prison population of 941 were selected for the study. Eleven irate prisoners absolutely refused to participate. Five others were excluded because of "faulty technique" in their tests. Four abnormal cases were

39. See, for example, American Civil Liberties Union litigation brought in behalf of Elaine Reddick Trent and Nial Ruth Cox.
40. *New York Times* (June 20, 1975).

found, although none apparently of the XYY variety.[41] Another Mexican study of 236 prisoners turned up not one single XYY chromosome.[42] Some 100 black male prisoners were analyzed by an associate professor of biology at the University of Alabama, Birmingham: not one single prisoner had the XYY chromosome. As Professor Fattig put it: "There was no evidence to suggest XY/XYY mosaicism in any of the individuals studied, nor were any 47, XXY males discovered."[43] While some may have thought that that should be the end of the inquiry, Professor Fattig was not satisfied, feeling that questions remained unanswered: " . . . is it more likely that sociological factors which lead to incarceration of some negro males differ from those of other groups, obscuring the true 'criminal incidence' of the XYY disorder? . . . What is the genetic significance of the suggested association between criminal behavior, tallness, and the XYY syndrome?"[44] What was the professor's suggested route to finding responses to his questions? "The answer to all of these questions probably lies in large-scale cytogenetic screening of newborns or young children accompanied by long-term psychological and sociological follow-up"![45]

Women may feel themselves untouched by such a chromosome study. They are not. One scientist examined 221 Mexican female prisoners, attempting to find chromosome abnormalities. Out of the 221 females, one "was found to have a mosaicism 46,XX/47,XXX, with less than 10% of the abnormal cell line."[46] Not satisfied, the scientist suggested possible additional research: "As an extra X-chromosome is found also in criminal males, and more consistently than an extra Y, it is suggested that the former anomaly is linked more often with criminal activities than the latter."[47] Is it not possible, too, if the movement to pay more attention to female prisoners gathers steam, that theories of female criminality may become rooted in stereotypes about the female metabolism?[48]

41. The researchers explained that there were two Klinefelter's and two mosaics (46, XY/47, XY, C + 46, XY/47, XY, GX) (Carlos Zavala, Q. F. B. Guillermina Mora, and Ruben Lisker, "Estudios cromosomicos en una prison Mexicana," *Revista de investigacion clinica* 22, no. 3 [July-September 1970]: 251–56).

42. Leonora Buentello and Salvador Armendares, "Estudio cromosomico en prisoneros del sexo masculino en ena penitenciaria Mexicana," *Revista de investigacion clinica* 22, no. 3 (July–September 1970): 257–60.

43. W. Donald Fattig, "An XYY Survey in a Negro Prison Population," *Journal of Heredity* 61, no. 1 (February 1970): 10.

44. Ibid.

45. Ibid.

46. Carlos Zavala et al., "Abberaciones gonosomicas en reclusas," *Revista de investigacion clinica* 23, no. 4 (October–December 1971): 303.

47. Ibid.

48. Another association between women and crime was made by a female researcher in a study looking for a possible connection between menstruation and crime. See Katharina Dalton, "Menstruation and Crime," *British Medical Journal* (December 1961), pp. 1752–53.

IV

In 1906 Dr. Robert Bean, a medical doctor from the University of Virginia who had immersed himself in the study of the Negro brain (all in the name of science, he claimed), published some of his results. In general, he suggested the Negro brain was smaller than the Caucasian brain although, he happily concluded, "the brain of the American negro weighs more than the native African, which is no doubt because of the greater amount of white blood in the American negro."[49] As if this were not enough, Dr. Bean argued that the physical size of the brain determined radically differing traits. It is odd how similar his description of the "Negro" is to that of "femininity":

> The Caucasian has the subjective faculties well developed: the negro, the objective. The Caucasian, and more particularly the Anglo-Saxon, is dominant and domineering, and possessed primarily with determination, will power, self-control, self-government, and all the attributes of the subjective self, with a high development of the ethical and esthetic faculties and great reasoning powers. The negro is in direct contrast by reason of a certain lack of these powers, and a great development of the objective qualities. The negro is primarily affectionate, immensely emotional, then sensual, and under provocation, passionate. There is love of outward show, of ostentation, of approbation. He loves melody and a rude kind of poetry and sonorous language. There is undeveloped artistic power and taste—negroes make good artisans and handicraftsmen. They are deficient in judgment, in the formulation of new ideas from existing facts, in devising hypotheses, and in making deductions in general. They are imitative rather than original, inventive, or constructive. There is instability of character incident to lack of self-control, especially in connection with the sexual relation, and there is a lack of orientation, or recognition of position and condition of self and environment, evidenced in various ways, but by a peculiar "bumptiousness," so called by Prof. Blackshear of Texas, this is particularly noticeable.[50]

Not satisfied with that pronouncement, Dr. Bean continued and carefully concluded that the state of the Negro brain could not be altered, even by education: "Having demonstrated that the negro and the Caucasian are widely different in characteristics, due to a deficiency of gray matter and connecting fibers in the negro brain, especially in the frontal lobes, a deficiency that is hereditary and can be altered only by intermarriage, we are forced to conclude that it is useless to try to elevate the negro by education or otherwise, except in the direction of his

49. Robert Bennett Bean, "The Negro Brain," in *The Development of Segregationist Thought*, ed. I. A. Newby (Homewood, Ill.: Dorsey Press, 1968), p. 48.
50. Ibid., pp. 52–53.

natural endowments."[51] In 1910, sociologist Howard Odum pointed to the imitative as opposed to creative analytical intellectual powers of the colored population:

> . . . The mind of the negro is easily sensitive to sound, and words which are sounded in sequence, similar sounding words or words of alliterative sound are retained by the negro child. They are very fond of riddles stated in rhymes and take delight in remembering the answers to them. They learn readily to do things by imitation and become comparatively skilful in a short time. They remember names and faces well. However, there are many negro children who have an almost total lack of mental perception, whose minds are so dense that they can scarcely learn anything. The percentage of such cases increases with age.[52]

The nation, at least publicly, backed away from the theories of men like Bean and Odum. However, memories are short. In the 1960s and 1970s came Jensen, Herrnstein, and Shockley. (Christopher Jencks's name sometimes has been added to the list, but Jencks clearly recognizes that "the importance of genetic differences between races is political rather than scientific.")[53] Jensen hit blacks, as a group. Compared with Caucasians and other minorities, he concluded, they are probably dumber: "As a group, Negroes perform somewhat more poorly on those subjects which tap abstract abilities. . . . In tests of scholastic achievement . . . Negroes score about 1 standard deviation (SD) (15 IQ points) below the average for whites and orientals and considerably less than 1 SD below other disadvantaged minorities tested in the Coleman study—Puerto Rican, Mexican-American, and American Indian. The 1 SD in Negro performance is fairly constant throughout the period from grades 1 through 12."[54] Then Jensen attacked educational efforts in behalf of blacks: "There is an increasing realization among students of the psychology of the disadvantaged that the discrepancy in their average performance cannot be completely or directly attributed to discrimination or inequalities in education.[55]

In a faint attempt to persuade that he was not being racist in his views, he threw out another one of his hypotheses: with respect to the "g or abstract reasoning ability, I would be inclined to rate

51. Ibid., p. 53.

52. Howard Odum, "The Education of Negroes," in *The Development of Segregationist Thought,* pp. 63–64.

53. Christopher Jencks, *Inequality* (New York: Basic Books, 1972), p. 83.

54. Arthur R. Jensen, "How Much Can We Boost IQ and Scholastic Achievement?" *Harvard Educational Review* 39, no. 1 (Winter 1969): 81–82.

55. Ibid., p. 82. For a different perspective on compensatory education programs, see E. W. Gordon and D. A. Wilkerson, "Compensatory Education," *IRCD Bulletin,* vol. 5, no. 1 (1966), and *Compensatory Education for the Disadvantaged* (New York: College Entrance Examination Board, 1966).

Caucasians for the whole somewhat below Orientals, at least those in the United States. A case can be made for conjecture on the basis of the existing evidence, but this is not the appropriate place for it."[56] After Jensen came Herrnstein[57] and Eysenck.[58] Eysenck, a British psychologist, insists that proof of black genetic inferiority can be found in the fact that the West African slave ancestors of black Americans were not intelligent enough to have escaped from slave traders! What can only be called the hysteria of scientific inaccuracy became much more pronounced after Jensen. Those trying to follow the debate had to become only more confused. There was the scholarly and thought-provoking response of Walter Bodmer and Luigi Cavalli-Sforza stressing environmental influences on intellectual performance.[59] In behalf of the libeled black population came the study by Sandra Scarr-Salaptek concluding that "the difference in mean IQ between the races can be affected by giving young black children rearing environments that are more conducive to the development of scholastic aptitudes."[60] Others were quick to attack the learned study of Dr. Scarr-Salaptek by taking digs at her sampling techniques, the insufficiency of her data base, and her classification scheme.[61]

Perhaps the majority female sex can begin to comprehend the anger of blacks over the Jensen insistence on black genetic intellectual inferiority by examining "scientific studies" affecting all females.[62] Ironically,

56. Arthur R. Jensen, "Reducing the Heredity-Environment Uncertainty: A Reply," *Harvard Educational Review* 39 (Summer 1969): 449–83.

57. Richard Herrnstein, "IQ," *Atlantic Monthly* 228 (September 1971): 43–64.

58. See H. J. Eysenck, *Race, Intelligence and Education* (London: Temple Smith, 1971).

59. Walter Bodmer and Luigi Cavalli-Sforza, "Intelligence and Race," *Scientific American* 223, no. 4 (October 1970): 19–29.

60. Sandra Scarr-Salaptek, "Race, Social Class and IQ," *Science* 174, no. 4016 (December 1971): 1294.

61. See "Heritability of IQ by Social Class: Evidence Inconclusive," *Science* (December 1973), pp. 1042–47; L. J. Eaves and J. L. Jinks, "Insignificance of Evidence for Differences in Heritability of IQ between Races and Social Classes," *Nature* 240 (November 1972): 84–88.

62. Some evidence has suggested that females perform better up to a certain age, when tests show a change. "Girls begin to do worse on a few intellectual tasks, such as arithmetic reasoning, and beyond high school the achievement of women now measured in terms of productivity and accomplishment drops off even more rapidly" (Naomi Weisstein, "Psychology Constructs the Female," in *Women in Sexist Society: Studies in Power and Powerlessness,* ed. Vivian Gornick and Barbara K. Moran [New York: New American Library, 1972]). See also Eva P. Lester, Stephanie Dudek, and Roy C. Muir, "Sex Differences in the Performance of School Children," *Canadian Psychiatric Association Journal* 17, no. 4 (1972): 273–78; Eleanor E. Maccoby, "Sex Differences in Intellectual Functioning," in *The Development of Sex Differences* (Stanford, Calif.: Stanford University Press, 1966). Scientists have also pointed to supposedly valid distinctions in cognitive ability in males and females, which do not have their genesis in environmental-sociological factors but in physiological differences between male and female. For a glaring example, see Donald Broverman, Edward L. Klaiber, Yutaka Kobayashi, and William Vogel, "Roles of Activation and Inhibition in Sex Differences in Cognitive Abilities," *Psychological Review* 75, no. 1 (1968): 23–50. For a comment on Broverman, see Eleanor E. Maccoby and Carol N. Jacklin, *The Psychology*

although women are often intellectually denigrated, a rather common belief is that black females are genetically brighter than black males. As Charles Kerr reminds us: "Jensen extends his original argument to look for a genetically based intellectual difference between the sexes in different races. Rather scanty IQ data indicate that American Negro girls have higher IQs than Negro males by an order of magnitude greater than in white samples."[63] These "scanty data" are often translated into biological fact. It is not unusual to hear elementary and secondary school teachers insist that black females are more intelligent than black males. Nor is it rare to see university professors, even those at the graduate and professional level, maintain that black females students are brighter than black males. So successful have these educators been in their pronouncements that even numerous male blacks are convinced that their female counterparts outdistance them in intellectual prowess. Few have listened to or digested Kerr's conclusions: "Interpretation of this difference is confounded by biological and cultural advantages for Negro girls, who are relatively more favoured with regard to infant mortality, scholastic achievement, typically matriarchal households and vocational opportunity."[64] To be sure, Kerr gets off to some stereotypical thinking, but he goes on to conclude: "Jensen shows his Achilles heel by attempting a simplistic explanation in elementary Mendelian terms. He postulates an X-chromosomal dosage effect, with disadvantageous recessive loci on the single male X chromosome accounting for lesser capabilities of that sex. This is an untenable proposition. . . ."[65]

Jensen and others raise the sharp question of whether scientific disputes should be played out in the public arena, all in the name of so-called scientific inquiry and First Amendment freedoms. Should scientists rush to print with unproved theories which, in a background of racist society, are bound to be dissected into choice and delectable bits, displayed and consumed with joy, while the more neutral, unappetizing portions are discarded as annoying interferences with an enjoyable meal? Various publics in this country tend not to be sophisticated or patient enough to read carefully and digest the language of the scholar, debates like that stemming from Dr. Scarr-Salaptek's work, or the less rigorous work of the Jensens. People are eager to consume—especially that which borders on the sensational. They seldom understand or are interested in hypotheses. They search constantly for facts and truth in simple form. As a consequence,

---

of Sex Differences (Stanford, Calif.: Stanford University Press, 1974), pp. 42, 44, 47, 53, 101–2, 122–25.

63. Charles Kerr, "Race, Intelligence and Education—Continued," *Medical Journal of Australia* 1 (January 1973): 200.

64. Ibid., p. 200.

65. Ibid.

hypotheses soon become facts. Pressured and hurried politicians, policy-makers, journalists, and other groups speak and write in easy-to-grasp headline style. So when even Arthur Jensen writes: ". . . all we are left with are various lines of evidence, no one of which is definitive alone, but which, viewed all together, make it a not unreasonable hypothesis that genetic factors are strongly implicated in the average Negro-white intelligence difference. The preponderance of the evidence is, in my opinion, less consistent with a strictly environmental hypothesis than with a genetic hypothesis, which, of course, does not exclude the influence of environment or its interaction with genetic factors,"[66] that scholar's language translates into: "Jensen says Negroes are born dumber than whites." And when Jensen further writes that "the evidence so far suggests the tentative conclusion that the pay-off of preschool and compensatory programs in terms of IQ is small,"[67] that translates into: "compensatory education for blacks is a waste of money."

The fact that Christopher Jencks argues that 45 percent of the "variance in IQ scores is explained by genetic factors" (with a possible 10–20 percent error in either direction), while Jensen insists on an 80 percent figure,[68] has not seemed to trouble America's unthinking people—does not seem to impress on their minds that the psychologists may not really know what they profess to think they know. Some people, no doubt, are also ignorant of Price-Williams's 1961 study of the Tiv people in Nigeria, which revealed that Tiv test results on " 'Piaget-type' tasks were approximately equivalent to that of Swiss children, provided that familiar materials . . . were used."[69] Nor are the unthinking and nonreading aided by the clarifications of the psychologists, geneticists, and other scientists who created the storms in the first place. For those learned men are quick to deny responsibility for erroneous interpretations of their data or misuse of their data for far-reaching policy decisions. When Arthur Jensen came under heavy attack for what some perceived as clear errors in his work,[70] he responded in a closed, elitist manner: "One of my purposes in writing 'How Much Can We Boost IQ and Scholastic Achievement?' was to provoke discussion among qualified persons of some important issues I believe have been relatively neglected in our common concern with improving the education of children called disadvantaged. Therefore it is a source of great satisfaction to see that the Editors have solicited and received extensive discussions of my arti-

66. Jensen, "How Much Can We Boost IQ and Scholastic Achievement?" p. 82.
67. Ibid., p. 108.
68. Jencks (n. 53 above), p. 65.
69. John Radford and Andrew Burton, "Changing Intelligence," in *Race and Intelligence,* ed. Ken Richardson, David Spears, and Martin Richards (Baltimore: Penguin Books, 1972), p. 33.
70. See "How Much Can We Boost IQ and Scholastic Achievement? A Discussion," *Harvard Educational Review* 39, no. 2 (Spring 1969): 273–356; Kerr, "Race, Intelligence and Education—Continued."

cle from several distinguished psychologists and an eminent geneticist—men whose own research in a variety of fields most germane to the contents of my article is widely known and respected."[71] Jensen, then, is prepared to recognize as competent critics of his work the "qualified" or the "distinguished psychologists" as well as the "eminent geneticist." He does not seem equally prepared, though, to accept criticism from educators or people who have a personal stake in the mushrooming of his ideas. There is a fascinating account of how one school teacher (turned temporarily Harvard University student) attempted to critique Professor Richard Herrnstein's highly controversial *Atlantic Monthly* article on IQ.[72] What power of the word could George Purvin, social studies teacher on leave from Manhasset High School in Long Island, New York muster against the 107 giants of the Harvard faculty who raced to defend Richard Herrnstein's right to speak about IQ, no matter how racist might be the implications of his "masterpiece"?[73]

Then, too, Jensen does not really bother himself with those who seized upon the content of his work, alleged errors and all, after it was published in the Congressional Record. What did Jensen care about gleeful and horrified public officials who might quote from his study, in outrageous baritone tones, or school board members bent on resisting the edicts of desegregation who could now justify their stance on the ground that little white children should not be mixed with dumb black children? What did Jensen care if the White House engaged in policy discussions emanating from the substance and flavor of his work?[74]

Jensen callously dismisses a letter from William F. Brazziel of Virginia State College who called attention to a school desegregation suit in Greenville and Caroline County schools. Defense attorneys for the resisters of desegregation contended that "white teachers could not understand the nigra mind." And what did they use as proof of that contention? Jensen's theories.[75] Brazziel wrote: "It will not help one bit for Jensen or the HER [*Harvard Educational Review*] editorial board to protest that they did not intend for Jensen's article to be used in this way. For, in addition to superiority in performing conceptual cluster tricks on task sheets, the hard line segregationist is also vastly superior in his ability to bury qualifying phrases and demurrers and in his ability to

71. Arthur R. Jensen, "Reducing the Heredity-Environment Uncertainty: A Reply," *Harvard Educational Review* 39, no. 3 (Summer 1969): 449.

72. See Herrnstein (n. 57 above).

73. See George Purvin, "Introduction to Herrnstein 101," in *The New Assault on Equality: IQ and Social Stratification*, ed. Alan Gartner, Colin Greer, and Frank Riessman (New York: Perennial Library, 1974), pp. 131–62.

74. See the discussion on Jensen in *Harvard Educational Review* (n. 70 above), and Alexander Thomas and Samuel Sillen, *Racism and Psychiatry* (New York: Brunner/Mazel, Inc., 1972), p. 31.

75. William F. Brazziel, "A Letter from the South," *Harvard Educational Review* 39, no. 2 (Spring 1969): 348–56.

distort and slant facts and batter his uneducated clientele into a complete state of hysteria where race is concerned."[76] Jensen, for one, did not protest. He hid behind the First Amendment: "Brazziel's letter seems to be saying in part that my paper should not have been published in the first place. I would plead for more faith in the wisdom of the First Amendment. To refrain from publishing discussions of research on socially important issues because possibly there will be some readers with whose interpretation or use of the material we may disagree is, in effect, to give those persons the power of censorship over the publication of our own questions, findings, and interpretations. It is only when all the available facts, issues, and questions can be openly examined and discussed by everyone that we can put stock in the maxim that 'the truth will out.' "[77] The short answer to Jensen's appeal to the First Amendment, of course, is that not all speech is constitutionally protected—not fighting words, nor obscenity, nor some forms of libel. Moreover, if Jensen and Herrnstein are to appeal to the First Amendment, must they not accept an obligation to answer all their critics with more than the arrogance of the all-knowing scholar? Should they be afforded the luxury of conversing only with other scientists from the same halls who may well entertain the same theories? Have they not given up that luxury by the popularizing, or at least not stopping the popularization, of their ideas in the mass media? Did Richard Herrnstein have to choose the *Atlantic Monthly* as the mechanism for advocating his theory of a hereditary meritocratic elite? Once having selected that widely read publication as the instrument for the dissemination of his ideas, could he then run from the George Purvins of the world or dismiss his critics as nuisances not up to his intellectual stature, and therefore in no position to criticize his writing?

Nor does Herrnstein bother to protest those who willfully and harmfully cast hypothesis into truth. He says, "It is not the function of scientists to pass moral judgment on their work. . . . after all, nature doesn't give a crap about the character of the scientist."[78] Herrnstein has reportedly also indicated that "he doesn't know what implications his work should have for public policy, but he does not believe it supplies a rationale for curtailing social problems. He points out that adherents of racism and oppression need no scientific rationale for their convictions."[79] Maybe not, but they certainly delight in using as a basis for their rationale and convictions the words of Richard Herrnstein, Bronx High School of Science, Harvard Ph.D. and professor of psychology at Harvard; and of Arthur Jensen, professor of psychology at the

76. Ibid., p. 349.
77. See Jensen, "Reducing the Heredity-Environment Uncertainty: A Reply," pp. 479–80.
78. Constance Holden, "R. J. Herrnstein: The Perils of Expounding Meritocracy," *Science* 181, no. 4094 (July 1973): 38.
79. Ibid.

University of California, Berkeley; and William Shockley, physicist at Stanford University.[80]

The impact of studies like those of Jensen and Herrnstein on the victims has not yet been adequately documented. Who knows how much of a role those studies have played in the dismantling of educational programs for minorities in elementary, secondary, and higher education? Who knows how much those studies have increased an environment of hostility or indifference toward the learning abilities of racial minorities? And if the scorecard is high on the side of injury, can the First Amendment be held up in justification of what well may be irreparable harm to young children? Surely the First Amendment was not meant to have that effect. Surely the First Amendment would ask, at least, that scientists who have thrown the first stone of verbal injury (which can have a permanent effect on the psychological and physical well being of humans) join issue with those who would take away the lethal effects of that stone rather than remaining mute while the effects spread.

The effects, however, are real and lethal. A very provocative study entitled *Children out of School in America* reveals the pernicious practice of misclassifying minority youngsters, especially with regard to intelligence. The case of B.J., a ten-year-old black youngster in New Bedford, Massachusetts, is instructive. One day, B. J. reported to his parents that he had been placed in a "special class," a class for mentally retarded children. Investigation soon revealed that an IQ score of 85 had been recorded for B. J., but further investigation revealed that B. J. had never taken an IQ test. When this gross injustice was uncovered, eight more black youngsters in the same school, which incidentally was only 18 percent black, were found to have been placed in special classes. After testing, B. J. and seven of the other eight youngsters were found to be intelligent enough to be in regular classes.[81] Similar experiences with misclassification have been reported in Philadelphia and California. In fact, as of 1971, some 9,284 youngsters in California had been taken out of classes for the mentally retarded on the basis of new testing. Not surprisingly, perhaps, Mexican-Americans in large numbers had been placed erroneously in mentally retarded classes.[82]

80. As for Shockley, his approach to the subject of IQ presents excellent fuel for the racist who likes to relate little stories to wide-eyed audiences. The racial tinge to Shockley's work allegedly began when "he read of a black acid-throwing teenager, one of seventeen children born to a woman who had an IQ of 55" (Cornish Rogers, "William Shockley! IQ and Equality," *Christian Century* 91, no. 15 [April 1974]: 414–16). Unconcerned about the reactions to his theories, Shockley merely stresses his resentment to social taboos on the subjects of eugenics, dysgenics, and race.

81. Children's Defense Fund, *Children out of School in America* (Cambridge, Mass.: Children's Defense Fund, 1974), pp. 25–26.

82. Ibid., p. 26, n. 11.

V

My concentration thus far on the dangers, misconceptions, and racial biases of the treatment of health, criminality, and intelligence, each as an isolated phenomenon, has been intentional. I wanted to stress the singular harm, in each instance, that psychologists, doctors, and scientists cause in conjunction with policymakers. Viewed as isolated phenomena, the damage done is gigantic. Viewed as a composite phenomenon, the damage is larger and unjustifiable on any rational grounds. However, an even greater danger is apparent. Let us superimpose our discussion of supposed black American genetic inferiority or weaknesses in the domain of health, crime, and intelligence on what is rapidly becoming a harangue about economic scarcity: oil and energy shortages, disappearing food supplies, depression, and suffering. The foundation may be etched for devious, rapid, and extraordinary remedies; as Geoffrey Barraclough reminds us: "Economic strain breeds desperate remedies, and economic strain is building up inexorably, nationally and worldwide."[83] Here in America, the uncertainty of economic security, the anxiety of a perceived, impending, unalterable worldwide doom, are in the wind. The weakness of America's humanity is the nagging thought that doom can only be staved off by hanging on to what one has and ignoring those that have not. The result, despite all the horrified denials, may well be the loss of huge segments of the world's colored population.

When William and Paul Paddock first published their work *Famine—1975!* (in 1967) and suggested therein the possibility of selecting out some nations from existence so that others could survive in times of famine, few paid attention.[84] When Paul R. Ehrlich echoed that idea (called "triage") in his 1968 book, *The Population Bomb,* few still responded.[85] By 1974, however, the ideas of Paddock and Ehrlich were being uttered to policymakers. Thus, Garett Hardin, an ecologist-biologist from the University of California, appearing before a congressional subcommittee, "calmly advised the lawmakers that we would be

83. Geoffrey Barraclough, "The Great World Crisis I," *New York Review of Books* (January 23, 1975).

84. The Paddocks wrote: "Nations in which the population growth trend has already passed the agricultural potential [should be cut off from food aid for the benefit of the rest]. This, combined with inadequate leadership and other divisive factors, makes catastrophic disasters inevitable. . . . To send food to them is to throw sand in the ocean" (Wade Greene, "Triage: Who Shall Be Fed? Who Shall Starve?" *New York Times Magazine* [January 5, 1975]). See also William and Paul Paddock, *Famine—1975! America's Decision: Who Will Survive?* (Boston: Little, Brown & Co., 1967). The Paddocks discuss their definition of triage on p. 206.

85. See Paul R. Ehrlich, *The Population Bomb* (Binghamton, N.Y.: Vail Ballou Press, 1969), pp. 141–54.

doing a favor to starving countries as well as ourselves if we refused to send them any more food."[86] Those were views Hardin had held since 1968; they also are views which now have been reprinted "in no less than 48 different collections" under the title "Tragedy of the Commons."[87] Congressman John Dingell, for example, stated after Hardin spoke: ". . . whether we make a conscious decision of the kind that Hardin has discussed in his papers, the hard fact of the matter is that nature is probably going to make those judgments for us. . . . Triage is going to come upon us whether we like it or not."[88] Even Philip Handler, President of the National Academy of Sciences, at one time "gently" accepted and urged a policy of triage;[89] there is some feeling that Handler may have stepped away from that "gentle" acceptance.

Few ask whether the facts, or the truth, dictate the justice of a triage-type solution. Truth seems to be a luxury that can be ill-afforded in times of economic panic, that cannot be tolerated by nations clinging to the power of their days of glory. Truth finding is skipped over in a hasty search for the scapegoat: some 400–800 million Asians and Africans are faced with starvation in the next few years, and it's their own fault! Never mind that colonialists, or multinational corporations, or agribusinesses usurped arable land for growth of money-making exports![90] Never mind that the bellies of Westerners are puffed outward with excessive food intake! Never mind that Westerners wasted energy for years![91] Never mind that the redistribution of resources, the economic and social development of the poorer peoples of the world, and the ironing out of the inequities of trade could save millions over the next several years![92] Never mind that citizens of the world have a right to adequate nutrition and an equitable share of the world's resources![93] Economic disaster and famine will be and are the fault of Asians, Africans, and black Americans—all of whom have too many children!

86. Greene, p. 11.

87. Ibid.

88. Ibid., p. 44.

89. Ibid.

90. See, for example, Transnational Institute, *World Hunger: Causes and Remedies* (Washington, D.C.: Transnational Institute), and the *World Bank Annual Report* for 1974.

91. Barraclough, p. 22. He says, ". . . millions of dollars were spent by oil companies, utilities, electrical appliance manufacturers, and the automobile industry to persuade the consumer to squander energy, and the government aided and abetted the waste 'through promotional pricing, tax advantages, and other forms of subsidies.' There was nothing necessary or inevitable about these developments. Energy was wasted because, so long as oil was cheap, there was no incentive to save." And, ". . . in 1973 it was the US, with only 6 percent of the world's population, that was consuming one-third of the world's total energy output, at a cost of only 4 percent of its gross national product."

92. See the thought-provoking note, "World Hunger and International Trade: An Analysis and a Proposal for Action," *Yale Law Journal* 84, no. 5 (April 1975): 1046–77.

93. Ibid., pp. 1047, 1067, 1069.

When the day of reckoning comes, then, many of these Asians and Africans especially may not be saved.[94]

Into this atmosphere of panic thinking step the Shockleys of this world who wonder, "Can it be that our humanitarian welfare programs have already selectively emphasized high and irresponsible rates of reproduction to produce a socially relatively unadaptable human strain?"[95] They supply a remedy before an answer to the inquiry can be given: sterilization—of blacks, of women—with bonuses for sterilization.[96] The justification for letting America as a nation stand aside and the colored people of the world die is that injustice raises its head if we permit more babies to be born into poverty and famine. What is meant, in reality, is that our power position is weakened if we share.[97]

The problem, then, is not just Jensen, nor the repercussions of an assault on hypertension and sickle cell anemia in American blacks, nor the suggestibility of born criminals—born *minority* criminals. It is the problem of arrogance of power, erroneous human fears of suffering, and an unwillingness of the greedy to share.[98] Nonetheless, the emergence and rapid escalation of an atmosphere of panic sufficient to tolerate notions of genetic selectivity has been aided and abetted by the Jensens, the Herrnsteins, the Shockleys, and the scientists who explore the health problems of blacks and the genetic strains of the criminal population without a wary eye for the misuse of their theories and findings and without an attempt to introduce safeguards which can minimize the risk of social injustice. If thousands should cry out in the next five years "Let the darker peoples of the world starve unto death, or be sterilized into extinction: they are dumb, unhealthy, poor, an unnecessary burden!" that should not be surprising.[99]

Daniel Callahan of the Institute of Society, Ethics, and the Life Sciences has posed the critical question: "Once we have declared all-out

94. The Paddocks's list of those who cannot be saved if triage is applied includes Haiti, Egypt, and India (Paddock, p. 222).

95. Thomas and Sillen (n. 74 above), pp. 42–43.

96. Rogers (n. 80 above), p. 415.

97. Barraclough, p. 21. He says, ". . . what is driving Western politicians to despair is not the plight of the poor nations but the plight of the wealthy nations, and above all else the dislocation of the economic system which has made the wealthy nations wealthy."

98. In ibid., p. 23, he states that "whatever the rights and wrongs of the argument, one fact is clear and that is that the companies have not suffered. According to Szulc, 'the majors' profit in Middle East operations in early 1974 increased on the average from $.30 to over $1 a barrel' and in the case of Aramco, the biggest of all producing companies, 'from $.80 early in 1973 to $4.50 a barrel in March 1974.' "

99. Given the notion of triage, the writings of the Paddocks et al., and the resistance of national and international governments and agencies to take sufficiently serious steps to redistribute resources and to augment the levels of economic development for the poorer nations of the world, it is not farfetched to conclude that advocacy of death by starvation or death by sterilization may become a reality by the late twentieth century.

war against genetic disease, how can we keep that attitude from spilling over to become an all-out warfare against the afflicted and defective in our midst?" And he adds profoundly: "It is not easy to draw a clear distinction between hating a disease and hating those who have the disease."[100] But, I might add, it is easy to label the comparatively powerless "diseased."

Let me suggest that what is needed in response to Callahan's deep concern, and that of others, is a new scientific responsibility: one which accents and clearly publicizes the difference between theory and fact, one which foresees many of the negative repercussions of scientific findings or the misuse of scientific inquiries by policymakers. Scientists should take the lead in organizing sessions on the implications of their research—the likely response from various segments of the public and the kind of educational campaign needed to avoid the overflow of negative repercussions. Scientists, too, need to reprimand racists, sexists, and others who seize from their findings all that may have a racist or sexist tinge, twisting phrasing to suit their own evil uses. Scientists should not tolerate scientists who use the First Amendment to wave scientific theories and words that maim the hopes and aspirations of other human beings. They should say to the misguided that their perceptions of and response to articulated scientific theories may well propel them into disastrous acts of irresponsibility. This is the new scientific responsibility, one which sheds the shield of indifference for a more humane armor which protects not only the scientist in his investigation but also humanity from the capricious uses of scientific inquiry.

Barnard College

100. See the testimony of Daniel Callahan, U.S. Congress, Senate, Committee on Labor and Public Welfare, *Quality of Health Care—Human Experimentation*, 93d Cong., 1st sess. 1973, p. 484.

Biology and Equality: A Perspective on Sex Differences

Helen H. Lambert

The biological basis of observed differences between the sexes is an explicit or implicit issue in both feminist and antifeminist literature. On both sides, this issue generates more controversy than one might expect from the definition of "biological."[1] The antagonism seems to arise from certain broad assumptions about biological differences; for instance, that they are universal, inevitable, desirable, and that they justify the social inequality of the sexes. This essay will examine these assumptions and then will discuss their possible relevance to feminism.

First, biological sex differences are not necessarily universal, that is, they do not obtain between any male/female pair chosen at random. Some women are taller than some men; the men of some racial groups have no beards. Males and females of the human species do differ, on the average, in many aspects of anatomy, body chemistry, and behavior, and biologists do speak of sexual "dimorphism" (two forms), which suggests a qualitative rather than a quantitative difference. But for many

A much earlier version of this paper was presented at the Barnard College Conference, "The Scholar and the Feminist II," April 12, 1975. I thank Carolyn Elliott, Susan Horn-Moo, Suzanne Lambert, Jane Mansbridge, Nancy Miller, Kathy Ralls, and the late Melissa Richter for comments and/or useful discussions. I also thank the National Science Foundation for financial support, and the Wellesley Center for Research on Women in Higher Education and the Professions for a congenial working environment during 1976–77.

1. "Of or related to biology or to life and living things; belonging to or characteristic of the processes of life" (*Webster's New International Dictionary*, 3d ed., s.v. "biological").

This essay originally appeared in *Signs*, vol. 4, no. 1, Autumn 1978.

physical and behavioral sex differences, we find instead a bimodal distribution along the same quantitative measure, with some overlap between the two sexes. As anyone who has ever played with statistics knows, even a very small difference between two group means can become statistically significant if the sample sizes are large enough.[2] The magnitude of a sex difference is not necessarily correlated with biological causation. Even a perfect correlation between sex and some other characteristic would not prove a biological basis for the latter.[3]

A common assumption that biological differences are inevitable is more troublesome. In a sense, we can avoid even that most biological of sex differences, reproduction, with contraception. Future technology may make it possible (not to say prudent or desirable) to abolish other biological sex differences, though modifiability by social and cultural environment is more often discussed than that by technology. Those who maintain that biological sex differences are inevitable generally have two kinds of evidence. The first is evolutionary. Both physical and behavioral dimorphisms are believed to have been favored in evolution because of their selective advantage in attracting or competing for mates, in producing or caring for offspring, and/or in utilizing resources.[4] That a particular sex difference does exist suggests that it was adaptive for the species, at least in past environments. However, far too often, both data gathering and interpretation take place in the unconsciously assumed framework of that familiar model from ethology and anthropology (also found in the popular literature): the larger, more aggressive male, out defending the territory while the smaller, more submissive female tends the youngsters. In addition, evolution theory (like other functionalist explanations) does not tell us "why" in the sense of ultimate purpose or cause, but only "how" on a large time scale; in this case, how sex differences may have come about on the species level by the mechanism of natural selection. Furthermore, differences which were once adaptive for the human species may no longer be.[5] A final problem with evolu-

2. It would be helpful if the various disciplines which study sex differences used a measure of difference which gave a better idea of the degree of overlap between the sexes. David Tresemer ("Measuring Sex Differences," *Sociological Inquiry* 45 [1975]: 29–32) has suggested several measures of association which could be used. In the meantime, we usually have to inspect the data (if they are included in the report) to determine overlap.

3. Except in the trivial sense that biological sex, by definition, is a necessary (but, one hopes, not sufficient) precondition for the social inequality of the sexes.

4. G. C. Williams, *Sex and Evolution* (Princeton, N.J.: Princeton University Press, 1975); V. C. Wynne-Edwards, *Animal Dispersion in Relation to Social Behaviours* (London: Oliver & Boyd, 1962); B. D. Campbell, ed., *Sexual Selection and the Descent of Man* (Chicago: Aldine Publishing Co., 1972); M. T. Ghiselin, *The Economy of Nature and the Evolution of Sex* (Berkeley: University of California Press, 1974).

5. An interesting theoretical model of the sum of biological and cultural evolution is M. W. Feldman and L. L. Cavalli-Sforza, "Cultural and Biological Evolutionary Process: Selection for a Trait under Complex Transmission," *Theoretical Population Biology* 9 (1976): 238–59.

tionary explanations is that they presume (though not always explicitly) an intrinsic origin for the sex difference. To evolve by natural selection, a sex difference must have had not only a function (which conferred the selective advantage) but also a genetic basis (because natural selection works on genes). This basis is assumed by an evolutionary model even if the mechanisms linking genes and sex difference are totally unknown.

The second kind of evidence for the "biological-is-inevitable" assumption concerns such proximate mechanisms. A considerable body of research tends to indicate that the genetic and hormonal processes that bring about the sexual differentiation of the reproductive system also operate in the development of other sex differences, including some behavioral ones, at least in laboratory animals. Such findings have frequently been generalized by analogy to "explain" observed sex differences in human behavior. In order to evaluate this claim, it is necessary to review the relevant ontogenetic mechanisms.

Briefly, the current paradigm of sexual differentiation in mammals involves both genes and hormones. The primary sex difference is chromosomal—the two sex chromosomes of a human male are an X and a Y; those of a human female are both X's. This difference amounts to only about 2 percent of the total genetic material,[6] because the other forty-four (non-sex) chromosomes present in both sexes do not differ systematically between males and females. The obvious physical differences between the sexes come about chiefly through the selective expression of different parts of the same genetic material in the two sexes. A common mechanism by which this sex-specific gene expression occurs in mammals involves, initially, the male's Y-chromosome, and later, the hormones (mainly androgens) secreted by the male embryo's testes. Like other steroid hormones, androgens actually enter the cells they affect via specific cellular receptors and interact directly with the genetic material.[7] Androgens can thus, apparently, induce selective gene expression in the cells which will form the male sexual apparatus. If androgens are not present at the right time in embryonic life, or if the cellular receptors for them are blocked or defective, the same (genetically XY) cells can form female structures; similarly, the XX cells of a female embryo can form male genitals if exposed to androgens in utero.[8]

6. An X is about 2.5 percent of the total, and a Y is about 1 percent (H. M. Golumb and G. F. Bahr, "Analysis of an Isolated Metaphase Plate by Quantitative Electron Microscopy," *Experimental Cell Research* 68 [1971]: 65–74). Most of the Y's genetic material is probably inactive (H. Cooke, "Repeated Sequence Specific to Human Males," *Nature* 262 [1976]: 182–86). However, there seems to be at least one functional gene on it, which may have a role in early sexual differentiation (for a review see W. K. Silvers and S. S. Wachtel, "H-Y Antigen: Behavior and Function," *Science* 195 [1977]: 956–60).

7. B. W. O'Malley and W. T. Schroader, "The Receptors of Steroid Hormones," *Scientific American* (February 1976), pp. 32–43; J. Gorski and F. Gannon, "Current Models of Steroid Hormone Action: A Critique," *Annual Reviews of Physiology* 38 (1976): 425–50.

8. This embryonic bipotentiality can be described in various ways, but such statements as "Female differentiation occurs by default" (C. Hutt, *Males and Females* [New York:

Such "organizational" effects of androgens probably represent not only a turning-on of that which is appropriate to the male, but also a turning-off of that which is appropriate to the female. Typically the switch is accessible only at a certain critical time during development.[9] The prototypical experiments to demonstrate this androgen-induction mechanism are (1) depriving a genetic male of his own androgens (either by removal of the testes or by blocking androgen action at the cellular level) and observing a female pattern of development, and (2) treating a genetic female with exogenous androgen and obtaining a male pattern. Such experiments have been performed on the embryos of several mammals,[10] and the androgen-induction mechanism seems to be a general one for genital dimorphism.[11] Clinical data show that it holds for human genital sex differences as well.[12] The androgen-induction mechanism is appealing in its simplicity and has been widely and popularly assumed to be responsible for the ontogeny of other mammalian sex differences. However, a survey of the literature[13] reveals that this mech-

---

Penguin Books, 1972], p. 18) or "All mammalian sexual organs are innately female" (M. J. Sherfey, *The Nature and Evolution of Female Sexuality* [New York: Random House, 1972], p. 42) suggest a confusion of biology with ideology. In fact, whether sex differences arise from a difference in genes or from a hormonally induced expression of different genes seems to have little social significance.

9. There is considerable species variation both in critical period for these effects and in the degree of their reversibility. For instance, some genital masculinization can be induced in females by androgen exposure after birth. In humans this is limited to clitoral enlargement, but in the female rat, a "penis bone" (normal male equipment for rats) is formed (A. Glucksman and C. P. Cherry, "Hormonal Induction of an Os Clitoridis in the Neonatal and Adult Rat," *Journal of Anatomy* 112 [1972]: 223–31). In neither case are the internal sex organs masculinized. Similarly, human males can experience breast enlargement due to hormone imbalance—or even excessive marihuana use (J. Harmon and J. Aliapoulos, "Gynecomastia in Marihuana Users," *New England Journal of Medicine* 287 [1972]: 936). In the male rat, even nipple development is normally prevented by his prenatal androgen exposure (K. Kratochwil, "*In Vitro* Analysis of the Hormonal Basis for Sexual Dimorphism in the Embryonic Development of the Mammary Gland," *Journal of Embryology and Experimental Morphology* 25 [1971]: 141–53).

10. This is not the case for all vertebrates. For instance, in birds, the male pattern is the "anhormonal" one while "female" hormones are necessary to bring about the female pattern. It has been suggested that in mammals, where high levels of female hormones are a necessary accompaniment to intrauterine life, the male embryo needed a reverse strategy to differentiate as a male (R. K. Burns, "Role of Hormones in the Differentiation of Sex," in *Sex and Internal Secretions,* ed. W. C. Young [Baltimore: Williams & Wilkins, 1961], pp. 76–158).

11. The testes also produce a nonandrogen substance which induces the regression of the female component of an ambisexual precursor structure, the Mullerian duct (N. N. Josso, "*In Vitro* Synthesis of Mullerian-inhibiting Hormone by Seminiferous Tubules Isolated from Calf Testis," *Endocrinology* 93 [1973]: 829–34).

12. A readable review of the animal experimental and clinical data is J. Money and A. A. Ehrhardt, *Man and Woman, Boy and Girl* (Baltimore: Johns Hopkins Press, 1972).

13. H. Lambert, "Comparative Ontogeny of Mammalian Sexual Dimorphisms" (forthcoming).

anism has, in fact, not yet been demonstrated in the development of many nongenital sex differences in mammals.[14]

Early exposure to androgen[15] has effects on the nervous system. One of the best studied effects is on the neural control of the pituitary gland, which lies just underneath the brain. The pattern of pituitary gonadotrophic-hormone secretion is generally cyclic in females and relatively constant in males, corresponding to the periodic occurrence of ovulation versus the continuous production of sperm. The proximate basis of this sex difference is a differential sensitivity of the hypothalamus of the brain to steroid feedback from the gonads (testes or ovaries),[16] a sensitivity established in the male direction by exposure to

14. It does, though, seem to apply to some differences, e.g., liver metabolism and body weight regulation (see P. De Moor and C. Denef, "The Puberty of the Rat Liver," *Endocrinology* 82 [1968]: 480–92, and 83 [1968]: 791–98; J. Gustafsson and A. Stenberg, "On the Obligatory Role of the Hypophysis in Sexual Differentiation of Hepatic Metabolism in Rats," *Proceedings of the National Academy of Sciences* 73 [1976]: 1462–65; G. N. Wade, "Sex Hormones, Regulatory Behaviors, and Body Weight," *Advances in the Study of Behavior* 6 [1976]: 201–71).

15. The term "androgen" is used consistently here in spite of considerable evidence that, at least in nervous tissue, some effects of androgens are mediated via their local transformation to estrogens (e.g., F. Naftolin et al., "The Formation of Estrogens by Central Neuroendocrine Tissues," *Recent Progress in Hormone Research* 31 [1975]: 295–319). Since estrogens are usually considered "female" hormones, this finding is sometimes pointed out by feminists with an air of triumph. It is worth remembering that all of the steroid sex hormones are closely related biochemically; that the pathways for their manufacture and metabolism in the body intersect at several points (the usual processing sequence, in both sexes, is progestins to androgens to estrogens); and that we normally find small amounts of estrogens in males and of androgens in females. However, the androgen-to-estrogen process (aromatization) does not make the effects described here any less sex-specific, because males have more aromatizing enzymes in their brains and females do not have much androgen to thus be transformed, at this stage of development. We lack a very good explanation for the fact that the normally high levels of maternal and placental estrogens do not masculinize the female fetus. Possible explanations are sex-specificity of nuclear receptors (B. R. Westley and D. F. Salaman, "Role of Oestrogen Receptor in Androgen-induced Sexual Differentiation of the Brain," *Nature* 262 [1976]: 407–8), binding of estrogen by plasma proteins (D. C. Anderson, "Sex-Hormone Binding Globulin," *Clinical Endocrinology* 3 [1974]: 69–96), and a protective effect of higher progesterone levels in female fetuses (B. H. Shapiro et al., "Neonatal Progesterone and Feminine Sexual Development," *Nature* 264 [1976] 795–96; and F. C. Hagemenas and G. W. Kittinger, "Influence of Fetal Sex on Plasma Progesterone Levels," *Endocrinology* 91 [1972]: 253–56).

16. There are also effects of early steroid exposure on the pituitary (C. A. Barraclough and J. L. Turgeon, "Ontogeny of Development of Hypothalamic Regulation of Gonadotropic Hormone Secretion: Effects of Perinatal Steroid Exposure," in *Developmental Biology of Reproduction*, ed. C. L. Markert and J. Papastantinou [New York: Academic Press, 1975]) and other organs (R. J. Gellert, J. Lewis, and P. H. Petra, "Neonatal Treatment with Sex Steroids: Relationship between the Uterotrophic Response and the Estrogen Receptor in Prepubertal Rats," *Endocrinology* 100 [1977]: 520–28). Probably a major aspect of these differentiating processes is the development and regulation of specific cellular receptors, which mediate later (pubertal and adult) responses to hormones in all these tissues.

androgen at a critical time in early development,[17] although the full effect is not manifested until puberty. Sex differences in the biochemistry and neural connections of this region of the brain have been demonstrated in rats.[18] These differences can be modified by early removal of the testes in males or early androgen treatment of females. Such results had led many people to the idea that the brain, like the genital tract, is initially bipotential—capable of developing in a male or female direction according to the hormonal milieu during certain critical stages of development.

In common laboratory animal species, many behaviors show reliable sex differences in relative frequency. These include sexual or reproductive behaviors, such as mounting a mate or building a nest, as well as nonsexual behaviors, such as specific responses to stress or learning certain tasks. It is customary, where there is a significant sex difference in

17. The critical period for this effect is later than that for genital differentiation. In the rat, conveniently for investigators, it is even postnatal. A review of these experiments is R. A. Gorski, "Gonadal Hormones and the Perinatal Development of Endocrine Function," in *Frontiers in Neuroendocrinology*, ed. L. Martini and W. F. Ganong (New York: Oxford University Press, 1971), pp. 237–90. There are also species differences. The cyclic female pattern apparently can occur even after prenatal androgen exposure in Rhesus monkeys and women (R. W. Goy and J. A. Resko, "Gonadal Hormones and Behavior of Normal and Pseudohermaphroditic Non-Human Female Primates," *Recent Progress in Hormone Research* 28 [1972]: 707–33; J. Money and A. A. Ehrhardt, "Gender Dimorphic Behavior and Fetal Sex Hormones," *Recent Progress in Hormone Research* 28 [1972]: 735–63), and the capacity for cyclical response may remain intact even in adult male primates if their own androgens are removed (F. J. Karsch, D. J. Dierschke, and E. Knobil, "Sexual Differentiation of Pituitary Function: Apparent Difference between Primates and Rodents," *Science* 179 [1973]: 484–86). However, there are sex differences in hypothalamic sensitivity in primates and in pituitary hormone content, which develop parallel to the secretion of androgens by the fetal testes (R. A. Steiner, "Sexual Differentiation and Feedback Control of Luteinizing Hormone Secretion in the Rhesus Monkey," *Biology of Reproduction* 15 [1976]: 206–12; and S. L. Kaplan, M. M. Grumbach, and M. L. Aubert, "The Ontogenesis of Pituitary Factors in the Human Fetus," *Recent Progress in Hormone Research* 32 [1976]: 161–243).

18. C. Libertun, P. S. Timiras, and C. L. Kragt, "Sexual Differences in the Hypothalamic Cholinergic System before and after Puberty: Inductory Effect of Testosterone," *Neuroendocrinology* 12 (1973): 73–85; R. B. Clayton et al., "Effects of Testosterone on Brain RNA Metabolism in Neonatal Female Rats," *Nature* 226 (1970): 810–12; J. A. Moguilevsky et al., "Metabolic Evidence of Sexual Differentiation of the Hypothalamus," *Neuroendocrinology* 4 (1970): 264–69; N. Ladosky and L. C. J. Gaziri, "Brain Serotonin and Sexual Differentiation of the Nervous System," *Neuroendocrinology* 6 (1970): 168–74; G. Dorner and J. Staudt, "Structural Changes in the Preoptic Anterior Hypothalamic Area of the Male Rat Following Neonatal Castration and Androgen Treatment," *Neuroendocrinology* 3 (1968): 136–40; G. Raisman and P. M. Field, "Sexual Dimorphism in the Neuropil of the Preoptic Area of the Rat and Its Dependence on Neonatal Androgen," *Brain Research* 54 (1973): 1–29; D. W. Pfaff, "Morphological Changes in the Brains of Adult Male Rats after Neonatal Castration," *Journal of Endocrinology* 35 (1966): 415–16; R. G. Dyer, N. K. MacLeod, and F. Ellendorf, "Electrophysiological Evidence for Sexual Dimorphism and Synaptic Convergence in the Preoptic Anterior Hypothalamic Areas of the Rat," *Proceedings of the Royal Society of London* B 193 (1976): 421–40.

frequency of the behavior, to speak of "sex-typical" frequencies, or of "female-type" and "male-type" behaviors, even though in many cases what is being described is a difference between the means of two over-lapping distributions. For quite a few such behavioral sex differences in animals, the prototypical experiments (early castration of males and early androgen treatment of females) suggest that the androgen-induction mechanism has some function. A genetic male can be feminized (i.e., display a female frequency of the behavior in adulthood) and a genetic female can be masculinized (i.e., show a male frequency) by manipulation of hormones during the perinatal period.

Among the diverse behaviors for which this result has been shown (in rodents, dogs, and monkeys) are urination posture,[19] copulatory posture,[20] juvenile play patterns,[21] saccharin preference,[22] open-field ambulation,[23] maze learning,[24] avoidance learning,[25] object-discrimination reversal tasks,[26] reactivity to and recovery of function after brain lesions,[27] and fighting.[28] Such a variety of sexually differ-entiated behaviors strongly suggests that the effects of early androgen are not limited to the "lower" brain. It seems well demonstrated that, at

19. F. A. Beach, "Hormonal Effects on Socio-sexual Behavior in Dogs," in *Mammalian Reproduction*, ed. H. Gibian and E. J. Plotz (Berlin: Springer-Verlag, 1970).

20. R. A. Gorski, "Neuroendocrine Regulation of Sexual Behavior," *Advances in Psychobiology* 2 (1974): 1–58.

21. R. W. Goy, "Early Hormonal Influences on the Development of Sexual and Sex-related Behavior," in *The Neurosciences: Second Study Program*, ed. F. O. Schmitt (Cambridge, Mass.: M.I.T. Press, 1970).

22. G. N. Wade and I. Zucker, "Taste Preferences of Female Rats: Modification by Neonatal Hormones, Food Deprivation and Prior Experience," *Physiology and Behavior* 4 (1969): 935–43.

23. H. H. Swanson, "Alteration of Sex-typical Behavior of Hamsters in Open-Field and Emergence Tests by Neonatal Administration of Androgen or Estrogen," *Animal Behavior* 15 (1962): 209–16.

24. J. L. M. Dawson, Y. M. Cheung and R. T. S. Lau, "Effects of Neonatal Hormones on Sex-based Cognitive Abilities in the White Rat," *Psychologia* 16 (1973): 17–24, and "Developmental Effects of Neonatal Sex Hormones on Spatial and Activity Skills in the White Rat," *Biological Psychology* 3 (1975): 213–29; J. L. M. Dawson, "Effects of Sex Hormones on Cognitive Style in Rats and Men," *Behavior Genetics* 2 (1972): 21–42.

25. W. W. Beatty and P. A. Beatty, "Effects of Neonatal Testosterone on the Acquisition of an Active Avoidance Response in Genetically Female Rats," *Psychonomic Science* 19 (1970): 315–16.

26. P. S. Godman and R. M. Brown, "The Influence of Neonatal Androgen on the Development of Cortical Function in the Rhesus Monkey," *Neuroscience Abstracts* 1 (1975): 494.

27. M. Dennis, "VMH Lesions and Reactivity to Electric Footshock in the Rat: The Effect of Early Testosterone Level," *Physiology and Behavior* 17 (1976): 645–49; A. G. Phillips and G. Deol, "Neonatal Gonadal Hormone Manipulation and Emotionality Follow-ing Septal Lesions in Weanling Rats," *Brain Research* 60 (1973): 55–64; P. S. Goldman et al., "Sex-dependent Behavioral Effects of Cortical Lesions in the Developing Rhesus Monkey," *Science* 186 (1974): 540–42.

28. D. A. Edwards, "Early Androgen Stimulation and Aggressive Behavior in Male and Female Mice," *Physiology and Behavior* 4 (1969): 333–38, and dozens of others.

least in animals, early androgen exposure normally has differentiating effects on the neural substrates of behavior[29] (for instance, via connections among nerve cells) which render certain actions more readily evoked in one sex than in the other, other things being equal. This is not the same as saying that early hormonal milieu is the sole determinant of these behaviors. Some of these same animal behaviors have been shown to vary with extrinsic influences such as prenatal stress, handling in infancy, daily light cycle, and other experiential factors.[30]

The notion that "innate" factors, such as genes or hormones, influence human behavior is often called (usually pejoratively) "biological determinism." To equate biological with intrinsic, inflexible, or pre-programed is an unfortunate misuse of the term biological. Behavior is itself a biological phenomenon, an interaction between organism and environment. Influences extrinsic to the individual affect biological events within the organism—events by which, in many cases, the structural entities and functional mechanisms within the organism develop. We often say that these internal structures and mechanisms are modified by the external input, but the concept of modification here is tricky. It can be falsely taken to imply an extrinsically induced deviation from some hypothetical, "normal" course of development that would occur in the total absence of extrinsic influences. In fact, such absence would be the abnormal situation. The innately determined structure and function normally develop in interaction with the environment. The visual system provides excellent examples of environmental input essential for the proper maturation of otherwise "prewired" neural circuits.[31] At present science has a very inexact conception of how genes, hormones, or external input affect the biological basis of behavior at the cellular and molecular levels, but it is certain that both intrinsic and extrinsic factors are important.

29. A readable discussion of these effects is B. D. McEwen, "Interaction between Hormones and Nerve Tissue," *Scientific American* (July 1976), pp. 48–58.

30. E.g., A. Jolley and J. H. Adam, "Gestational Stress: Effects on Open-Field Behavior and Heart Rate Reactivity in Rat Offspring," *Biological Psychology* 3 (1975): 231–36; I. Ward, "Prenatal Stress Feminizes and Demasculinizes the Behavior of Males," *Science* 175 (1972): 82–84; L. P. Mos, "Light-rearing Effects on Factors of Mouse Emotionality and Endocrine Organ Weight," *Physiological Psychology* 4 (1976): 503–10; H. H. Swanson, "Interaction of Experience with Adrenal and Sex Hormones on the Behavior of Hamsters in the Open-Field Test," *Animal Behavior* 17 (1968): 148–54.

31. C. Blakemore, "Developmental Factors in the Formation of Feature-extracting Neurons," in *The Neurosciences: Third Study Program*, ed. F. O. Schmitt and F. G. Worden (Cambridge, Mass.: M.I.T. Press, 1974); B. G. Cragg, "The Development of Synapses in Kitten Visual Cortex during Visual Deprivation," *Experimental Neurology* 146 (1975): 445–51; M. Cynader and G. Chernenko, "Abolition of Direction Selectivity in the Visual Cortex of the Cat," *Science* 193 (1976): 504–6; M. S. Banks, R. N. Astin, and R. D. Letson, "Sensitive Period for the Development of Human Binocular Vision," *Science* 190 (1975): 675–77.

That "environmental" factors, especially social ones, are important in the development of human behavioral sex differences is commonly emphasized, but it is less often pointed out that such factors may have biological effects, for instance, on wiring patterns in the brain. In adults, social stimuli have physiological consequences, especially where reproduction is concerned.[32] It seems likely that infants and children are even more sensitive to experiential effects on physiology as well as on behavior,[33] and possibly such effects are not the same for the two sexes.[34] Especially in the case of higher mental functions, a precise separation of the biological bases into those which are intrinsic in origin and those which are not may be an unrealistic goal.

An example of the kind of circumstantial evidence one has to work with in studying the interaction of intrinsic and extrinsic factors in the development of human behavioral sex differences is spatial visualization test performance (hereinafter called spatial ability).[35] Such performance shows one of the most consistent sex differences in the cognitive-

32. M. McClintock "Menstrual Synchrony and Suppression," *Nature* 229 (1971): 244–45; L. E. Kreuz, R. M. Rose, and J. R. Jennings, "Suppression of Plasma Testosterone Levels and Psychological Stress," *Archives of General Psychiatry* 26 (1972): 479–82; A. A. Loeser, "Effect of Emotional Shock on Hormone Release and Endometrial Development," *Lancet* i (1943): 518–27; A. E. Rakoff, "Endocrine Mechanisms in Psychogenic Amenorrhea," in *Endocrinology and Human Behavior,* ed. R. P. Michael (London: Oxford University Press, 1968); R. M. Rose, T. P. Gordon, and I. S. Bernstein, "Behavioral and Environmental Events Influencing Primate Testosterone Levels," *Journal of Human Evolution* 3 (1974): 517–25.

33. H. H. Harlow and M. K. Harlow, "The Affectional Systems," in *Behavior of Non-Human Primates,* ed. A. M. Schrier, H. F. Harlow and F. Stollnitz (New York: Academic Press, 1965), vol. 2; M. R. Rosenzweig, "Effects of Environment on Development of Brain and Behavior," in *The Biopsychology of Development,* ed. E. Tobach, L. R. Aronson and E. Shaw (New York: Academic Press, 1971); R. A. Spitz, "Hospitalism," in *Psychoanalytic Study of the Child,* ed. R. S. Eissler et al. (New York: International Universities Press, 1945), vol. 1.

34. S. Levine, "Differential Response to Early Experience as a Function of Sex Difference," pp. 87–98, and G. P. Sackett, "Sex Differences in Rhesus Monkeys Following Varied Rearing Experiences," pp. 99–122, both in *Sex Differences in Behavior,* ed. R. C. Friedman, R. M. Richart, and R. L. van de Wiele (New York: John Wiley & Sons, 1974).

35. Measures of spatial visualization include the spatial subsections of the Thurstone Primary Mental Abilities Test, the Differential Aptitude Test, and specifically spatial tests such as the Identical Blocks test—all of which involve mental visualization and/or manipulation of objects and configurations, often in three dimensions. The spatial factor has also been shown to be a major component of "field independence" (e.g., Embedded Figures and Rod-Frame tests) as well (J. A. Sherman, "Problem of Sex Differences in Space Perception and Aspects of Intellectual Functioning," *Psychological Review* 74 [1967]: 290–299; R. P. McGilligan and A. G. Barclay, "Sex Differences and Spatial Ability in Witkin's 'Differentiation' Construct," *Journal of Clinical Psychology* 30 [1974]: 528–32; P. E. Vernon, "The Distinctiveness of Field Independence," *Journal of Personality* 40 [1972]: 366–91; J. S. Hyde, E. R. Geiringer, and W. M. Yen, "On the Empirical Relation between Spatial Ability and Sex Differences in Other Aspects of Cognitive Performance," *Multivariate Behavioral Research* 10 [1975]: 289–309).

intellectual sphere.[36] Although the difference is not large, it is of some practical interest because its direction correlates with the poor representation of women in occupations where spatial ability might be useful.[37]

One hypothesis for the origin of this sex difference is genetic. In several studies,[38] the order of correlations among the spatial-test scores of parents and offspring is consistent with a recessive gene for superior spatial ability located on the X-chromosome.[39] However, some recent studies,[40] using similar tests, have not found the order of correlations predicted by the X-linked model. The issue remains unresolved. Even in the studies which did find the predicted order, the magnitude of the observed correlations only explained about half the total variance in individual scores, so the postulated X-linked gene cannot be the only source of variation.[41]

36. Of the sixty or more studies reviewed by E. Maccoby and C. Jacklin (*The Psychology of Sex Differences* [Stanford, Calif.: Stanford University Press, 1974], pp. 91–133), the majority found a significant difference in favor of males after about ten years of age, although females may excel at earlier ages (S. Coates, "Sex Differences in Field Dependence between the Ages of 3 and 6," *Perceptual and Motor Skills* 39 [1974]: 1307–10). The magnitude of the adult sex difference is not always reported, but where it is, the average male score exceeds the average female score by one-half to one standard deviation of either mean score. In nonstatistical terms, this means that only about one-quarter of the females score above the male average.

37. I. M. Smith, *Spatial Ability* (London: University of London, 1964); J. O'Connor, *Structural Visualization* (Boston: Human Engineering Laboratory, 1943).

38. R. D. Bock and D. Kolakowski, "Further Evidence of Sex-linked Major Gene Influence on Human Spatial Visualizing Ability," *American Journal of Human Genetics* 25 (1973): 1–14; R. E. Stafford, "Sex Differences in Spatial Visualization as Evidence of Sex-linked Inheritance," *Perceptual and Motor Skills* 13 (1961): 428; L. C. Harlage, "Sex-linked Inheritance of Spatial Ability," ibid., 31 (1970): 610; W. M. Yen, "X-linked Major Gene Influences on Selected Types of Spatial Performance," *Behavior Genetics* 5 (1975): 281–98. In an elegant experimental design using other genetic markers on the X-chromosome of brothers, Goodenough et al. have recently found evidence of X-linkage of field dependence, although not for other spatial tests (D. R. Goodenough, et al., "A Study of X-Chromosome Linkage with Field Dependence and Spatial Visualization," ibid., 7 [1977]: 373–87).

39. Such a gene should be expressed in any male who has it (since he has no other X-chromosomes). But among females, only those who have a recessive gene on both X's would express the trait. The X-linked model predicts a higher correlation between the scores of mothers and sons (since mothers give their sons an X) than between fathers and sons (since no X passes between them). Mother-son and father-daughter correlations should be equal and mother-daughter correlations intermediate. Similar predictions can be made for the order of sibling correlations.

40. J. C. DeFries et al. "Parent-Offspring Resemblance for Specific Cognitive Abilities in Two Ethnic Groups," *Nature* 261 (1976): 131–33; T. Williams, "Family Resemblances in Abilities: The Wechsler Scales," *Behavior Genetics* 5 (1975): 405–9; T. J. Bouchard and M. G, McGee, "Sex Differences in Human Spatial Ability: Not an X-linked Recessive Gene Effect," *Social Biology* 24 (1977): 332–35.

41. The postulated gene could not, in any case, account for the scarcity of women in fields where spatial ability might be important. Females with two superior but recessive

Since the sex difference in spatial ability is not usually found in young children,[42] sex hormones might be involved in its development, but the available evidence does not support a simple prenatal androgen-induction mechanism. Two clinical syndromes provide human approximations to the two prototypical animal experiments of nonandrogenized genetic males and of androgenized genetic females. In one of these syndromes, genetic males with defective cellular receptors for androgens have feminine external genitals, and a majority of those tested showed the average female performance profile: lower spatial than verbal scores.[43] In the other syndrome, women who were exposed prenatally to excessive steroid hormones[44] have higher spatial scores than the female average, but their verbal scores are also unusually high. However, in both groups, the possible effects of prenatal hormone "imprinting" (or lack thereof) may often be confounded by ambiguous genitals and/or parental concern about the child's gender. A few other findings support a possible connection between sex hormone levels in adult or adolescent life and spatial ability. For instance, it is reported that West African males feminized by *kwashiorkor* (malnutrition) show poor spatial performance compared with their normal male tribesmates.[45] Women born without ovaries perform very poorly on spatial tests, yet their verbal scores are normal or high.[46] Among normal individuals, one study found a negative correlation in adult males between somatic signs

---

genes should be as skillful at these tasks as males with one superior gene. For the recessive-gene frequency of .5 suggested by the X-linked model, half the men would express the trait, while only a quarter of the women [(.5)²] would be double-recessive and, hence, express the trait. It follows that if the gene were an absolute determinant of success in say, engineering or architecture (socialization effects and outright discrimination being nil), one should find about half as many women as men in these fields, and of course we find fewer than that.

42. See n. 36 above.

43. D. N. Masica et al., "IQ, Fetal Sex Hormones and Cognitive Patterns: Studies in the Testicular Feminizing Syndrome of Androgen Insensitivity." *Johns Hopkins Medical Journal* 124 (1969): 34–43. However, this study only involved fifteen patients.

44. They could be exposed to the hormones either from their own malfunctioning adrenal glands (adreno-genital syndrome) or by medical prescription (to avert threatened miscarriage) (V. G. Lewis, J. Money, and R. Epstein, "Concordance of Verbal and Nonverbal Ability in the Adrenogenital Syndrome," *Johns Hopkins Medical Journal* 122 [1968]: 192–95; J. M. Reinisch, "Effects of Prenatal Hormone Exposure on Physical and Psychological Development in Humans and Animals: With a Note on the State of the Field," in *Hormones, Behavior and Psychopathology,* ed. E. J. Sachar [New York: Raven Press, 1976]; S. W. Baker and A. A. Ehrhardt, "Prenatal Androgen, Intelligence, and Cognitive Sex Differences," in Friedman et al. [see n. 34 above]; L. S. McGuire and G. S. Omenn, "Congenital Adrenal Hyperplasia. I. Family Studies," *Behavior Genetics* 5 [1975]: 165–73). The more recent studies have found that the families of these women also have high IQ's.

45. J. L. M. Dawson, "Cultural and Physiological Influences on Spatial-Perceptual Processes in West Africa," *International Journal of Psychology* 2 (1967): 115–28, 171–85.

46. A. Silvert, P. H. Wolff, and J. Lilienthal, "Spatial and Temporal Processing in Patients with Turner's Syndrome," *Behavior Genetics* 7 (1977): 11–21; D. C. Garron, "Intelligence among Persons with Turner's Syndrome," *Behavior Genetics* 7 (1977): 105–27; J.

of androgenicity (e.g., body hair) and good spatial performance,[47] while another found a negative correlation for adolescent males but a positive correlation for adolescent females.[48] These apparently contradictory (and mostly unreplicated) results suggest that if sex hormones directly affect spatial ability, they may have different effects in the two sexes. Of course, genetic and hormonal factors are not mutually exclusive; genetic influence could be mediated via sex hormones and/or the different maturation rates in the two sexes.[49]

A possible neurological mechanism for the development of sex differences in spatial ability is differential lateralization of the cerebral hemispheres in the two sexes.[50] The relative functional specialization of the two hemispheres of the cerebral cortex is well established; in right-handed people, the left hemisphere is usually dominant for language (sequential, analytic processing) while the right hemisphere tends to dominate for spatial (parallel, gestalt) functions. The relationship between cerebral laterality and sex, however, is still unclear. One theory proposes that females are earlier and more laterally specialized than are males for language function,[51] which somehow leaves less "room" for

---

Money, "Turner's Syndrome and Parietal Lobe Functions," *Cortex* 9 (1973): 387–93; J. Money and D. Alexander, "Turner's Syndrome: Further Demonstration of Specific Cognitional Deficiencies," *Journal of Medical Genetics* 3 (1966): 47–48; J. W. Shaffer, "A Specific Cognitive Deficit Observed in Gonadal Aplasia (Turner's Syndrome)," *Journal of Clinical Psychology* 13 (1962): 403–6. These women also have only one X chromosome, which presents some problems for the X-linked recessive gene hypothesis. One would expect such a gene to be expressed as often in single-X females as in males, unless expression of the gene is somehow dependent on ovarian hormones (D. C. Garron, "Sex-linked Recessive Inheritance of Spatial and Numerical Abilities and Turner's Syndrome," *Psychological Review* 77 [1970]: 147–52). Adding to this puzzle is the finding that in children with Turner's syndrome androgens (presumably of adrenal origin) are elevated but in adults they are lower than in normal women (M. G. Forest, A. M. Cathierd, and J. S. Bertrand, "Total and Unbound Testosterone Levels in the Newborn and in Normal and Hypogonadal Children," *Journal of Clinical Endocrinology and Metabolism* 36 [1973]: 1132–42). Of course, the other physical disabilities of women with Turner's syndrome may possibly have prevented them from developing spatial abilities.

47. E. L. Klaiber, D. M. Broverman, and Y. Kobayashi, "The Automatization Cognitive Style, Androgens and MAO," *Psychopharmacologia* 11 (1967): 320–36.

48. A. C. Peterson, "Physical Androgyny and Cognitive Functioning in Adolescence," *Developmental Psychology* 2 (1976): 524–33.

49. There may be X-linked genes affecting maturation rate (S. M. Garn and C. G. Rohmann, "X-linked Inheritance of Developmental Timing in Man," *Nature* 196 [1962]: 695–96) and secondary sex characteristics like body hair (see n. 48).

50. H. Lansdell, "Sex Difference in the Effect of Temporal Lobe Neurosurgery in Design Preference," *Nature* 194 (1962): 852–54, and "Sex Difference in Hemispheric Asymmetries of the Human Brain," *Nature* 203 (1964): 550; D. C. Taylor, "Differential Rates of Cerebral Maturation between Sexes and between Hemispheres," *Lancet* i (1969): 140–42.

51. A. W. H. Buffery and J. A. Gray, "Sex Differences in the Development of Spatial and Linguistic Skills," in *Gender Differences: Their Ontogeny and Significance*, ed. C. Ounstead and D. C. Taylor (Edinburgh: Churchill Livingstone, Ltd., 1972).

bilateral development of spatial skills. An alternative theory attributes male superiority on spatial tests to greater lateralization in males, at least for spatial tasks.[52] Both of these theories assume a developmental etiology of the observed sex differences in spatial ability, and a competition effect between verbal and spatial functions. The two theories differ in whether bilateral representation for spatial processing leads to superior performance or not.[53] It would seem that a choice between these theories could easily be made on the basis of correlations between spatial test performance and the degree of lateralization. However, the several measures of cerebral laterality used (handedness; EEG; assymetrical responses to visual, auditory, or tactile stimuli) do not always agree. Different functions are apparently lateralized to different degrees and at different developmental rates.

Quite a number of studies support the hypothesis of greater specialization of the two hemispheres in adult males than in females, at least for spatial tasks, and of the association of greater right-hemisphere specialization with better spatial performance.[54] In addition, some studies on children indicate marked sex differences in the developmental course of lateralization for a variety of functions; for some of these girls are more and/or earlier lateralized,[55] while for others, boys are.[56] One study of adolescents found that late sexual maturation, in both sexes, was

52. J. Levy, "Lateral Specialization of the Human Brain: Behavioral Manifestations and Possible Evolutionary Basis," in *The Biology of Behavior,* ed. J. A. Kiger (Corvallis: Oregon State University Press, 1972).

53. J. C. Marshall, "Some Problems and Paradoxes Associated with Recent Accounts of Hemispheric Specialization," *Neuropsychologia* 11 (1973): 463–70.

54. D. M. Tucker, "Sex Differences in Hemispheric Specialization for Synthetic Visuospatial Functions," *Neuropsychologia* 14 (1976): 447–54; J. McGlone and A. Kertesz, "Sex Differences in Cerebral Processing of Visuospatial Tasks," *Cortex* 9 (1973): 313–20; D. Kimura, "Spatial Localization in Left and Right Visual Fields," *Canadian Journal of Psychology* 23 (1969): 445–52; W. J. Ray, M. Morell, and A. W. Frediani, "Sex Differences and Lateral Specialization and Hemispheric Functioning," *Neuropsychologia* 14 (1976): 391–94; C. J. Furst, "EEG Alpha Asymmetry and Visuospatial Performance," *Nature* 260 (1976): 254–55; J. Levy and M. Reid, "Variations in Writing Posture and Cerebral Organization," *Science* 194 (1976): 337–39; J. McGlone and W. Davison, "The Relation between Cerebral Speech Laterality and Spatial Ability with Special Reference to Sex and Hand Preference," *Neuropsychologia* 11 (1973): 105–13; D. A. Lake and M. P. Bryden, "Handedness and Sex Differences in Hemispheric Asymmetry," *Brain and Language* 3 (1976): 266–82.

55. L. Ghent, "Developmental Changes in Tactual Thresholds on Dominant and Non-dominant Sides," *Journal of Comparative and Physiological Psychology* 54 (1961): 670–73; P. H. Wolff and I. Hurwitz, "Sex Differences in Finger Tapping: A Developmental Study," *Neuropsychologia* 14 (1976): 35–41; M. B. Denckla, "Development of Motor Coordination in Normal Children," *Developmental Medicine and Child Neurology* 16 (1974): 729–41; D. Kimura, "Speech Lateralization in Young Children as Determined by an Auditory Test," *Journal of Comparative and Physiological Psychology* 56 (1963): 899–901.

56. R. G. Rudel, M. B. Denckla, and E. Spalten, "The Functional Asymmetry of Braille Letter Learning in Normal, Sighted Children," *Neurology* 24 (1974): 733–38; S. F. Witelson, "Sex and the Single Hemisphere: Specialization of the Right Hemisphere for Spatial Processing," *Science* 193 (1976): 425–27.

correlated with higher spatial than verbal test scores and, in one group, with the degree of cerebral lateralization as well.[57] This is of particular interest since girls usually mature earlier than boys and because a sex difference in spatial performance is not usually found until puberty. Although another study found no correlation between pubertal timing and spatial performance,[58] the relation between lateralization and sex differences in developmental pace seems a promising avenue for future research.[59]

If there are behavioral concomitants of sex differences in cerebral lateralization they are probably not limited to spatial ability or even to cognitive tasks. For instance, women show more lateral differentiation of EEG than males do during biofeedback training.[60] Because patterns of cerebral dominance have long been thought by some to be related to the general cognitive abilities, it seems probable that sex differences in laterality will be the popular version of "biology is women's destiny" for the next few years, as sex differences in aggression have been for the last few.[61] However, it must be obvious from the discussion of laterality studies that "lateralization" is not a unitary phenomenon, and the question of whether (not to mention how) greater hemispheric specialization leads to better intellectual performance in general has yet to be resolved.

57. D. P. Waber, "Sex Differences in Cognition: A Function of Maturation Rate?" *Science* 192 (1976): 572–74, and "Sex Differences in Mental Abilities, Hemispheric Lateralization, and Rate of Physical Growth at Adolescence," *Developmental Psychology* 13 (1977): 29–38. However, this study found no overall sex differences in spatial performance or laterality.

58. See n. 48 above.

59. It is perhaps worth noting that some sex differences in cerebral laterality may exist very early in life (J. F. Wada, R. Clarke, and A. Hamm, "Cerebral Hemispheric Assymetry in Humans," *Archives of Neurology* 32 [1975]: 239–46; D. L. Molfese et al., "Cerebral Asymmetry: Changes in Factors Affecting Its Development," *Annals of New York Academy of Sciences* 280 [1976]: 821–33). Whether such differences emerge in infancy or puberty does not really resolve the question of intrinsic versus extrinsic causation, however.

60. R. J. Davidson and G. E. Schwartz, "Patterns of Cerebral Lateralization during Cardiac Biofeedback versus the Self-Regulation of Emotion: Sex Differences," *Psychophysiology* 13 (1976): 62–68; R. J. Davidson et al., "Sex Differences in Patterns of EEG Asymmetry," *Biological Psychology* 4 (1976):119–38.

61. In fact, this has already begun to happen (see K. Lamott, "Why Men and Women Think Differently," *Horizon* 19 [1977]: 41–45). There are also some interesting, if controversial, relationships between culture or social class and laterality, and of course the same intrinsic/extrinsic issues and methodological problems exist for these ("The Effects of Sex, Race and Socioeconomic Class," *Neuropsychologia* 14 [1976]: 363–70; D. Geffner and I. Hochberg, "Ear Laterality Performance of Children from Low and Middle Socioeconomic Levels on a Verbal Dichotic Listening Task," *Cortex* 7 [1971]: 193–203; J. E. Bogen et al., "The Other Side of the Brain. IV. The A/P Ratio," *Bulletin of the Los Angeles Neurological Society* 37 [1972]: 49–61; J. A. Zook and J. H. Dwyer, "Cultural Differences in Hemisphericity: A Critique," *Bulletin of the Los Angeles Neurological Society* 41 [1976]: 87–90; W. D. ten Houten et al., "Discriminating Social Groups by Performance on Two Lateralized Tests," *Bulletin of the Los Angeles Neurological Society* 41 [1976]: 99–108).

Those who believe that sex differences in spatial ability are determined primarily by intrinsic factors argue both from ontogeny and from evolution. The ontogenetic argument notes that a male superiority in solving certain kinds of mazes has been found in the rat,[62] and that this sex difference can be modified by early hormone exposure.[63] There are similar recent studies in monkeys on a delayed-alternation task.[64] The evolutionary argument proposes that spatial ability would have been of great value to our (male) ancestors in territorial behaviors, in the visual disembedding of prey or predators from camouflage, and in the aiming of projectiles, either for defense or for obtaining food.[65]

Often cited in opposition to these biological aspects of sex differences in spatial ability are the different experiences of males and females: with their own bodies, within the family constellation, and in society at large. For instance, it has been suggested that the "hidden" female sex organs make it more difficult for little girls to develop a clearly articulated body concept, which in turn may affect their ability to (cognitively) differentiate self from surrounding, figure from ground.[66] A related idea, drawing on psychoanalytic theory, is that the girl's differentiation of self from nonself might be less acute because both her primary and her eventual identifications are with a parent of the same sex. However, these processes seem to predict an earlier sex difference in spatial ability than is, in fact, observed. Perhaps more consistent with its appearance in late childhood is a sex-role hypothesis. Various studies report a correlation in females between good spatial performance and

62. R. C. Tryon, "Studies in Individual Differences in Maze Ability," *Journal of Comparative Psychology* 12 (1931): 1–22; R. J. Barrett and O. S. Ray, "Behavior in the Open-Field, Lasley III Maze, Shuttle-Box and Sidman Avoidance as a Function of Strain, Sex and Age," *Developmental Psychology* 3 (1970): 73–77.

63. See n. 24 above.

64. See n. 26 above.

65. J. A. Gray and A. W. H. Buffery, "Sex Differences in Emotional and Cognitive Behavior in Mammals including Man: Adaptive and Neural Bases," *Acta Psychologia* 35 (1971): 89–111. Such arguments have also been advanced by J. Money (n. 46 above), pp. 387–93, Masica et al. (n. 43 above) (1969), Dawson et al. (n. 24 above) (1972), and P. A. Hamburg ("The Psychobiology of Sex Differences: An Evolutionary Perspective," in Friedman et al. [see n. 34 above]). One research group even went so far with this idea as to correlate the accuracy of spear throwing in a group of adolescent males with their performance on the Primary Mental Abilities spatial test; the correlation only predicted 14 percent of the variance in throwing performance (D. Kolakowski and R. M. Malina, "Spatial Ability, Throwing Accuracy and Man's Hunting Heritage," *Nature* 251 [1974]: 410–12). If, in fact, there were a gene for superior spatial ability (expressed via hormones and/or laterality), which had a significant selective advantage, one might expect it to have become more widely distributed in the population during all those millennia of hunting and gathering than the 50 percent gene frequency suggested by familial correlations for the postulated X-linked gene.

66. H. A. Witkin et al., *Psychological Differentiation* (New York: John Wiley & Sons, 1962), p. 220.

high "masculinity" scores,[67] identification with the father,[68] preferring to be a boy,[69] and aggressiveness.[70] Spatial ability has also been related to the putatively greater abstraction required to learn the male role,[71] to the greater dependency and passivity of girls,[72] and to the greater opportunities provided by the culture for boys to learn and practice spatial skills.[73]

If, indeed, these sociocultural factors influence sex differences in spatial ability, one might expect to find variation across cultures. Yet, most cross-cultural studies have described the same superior male performance on spatial tasks.[74] A notable exception is two studies on Eskimos.[75] The lack of a sex difference in Eskimos has been variously attributed to the permissiveness of Eskimo childrearing, the relative emancipation of Eskimo women, the unique visual features of the Arctic environment, and/or an isolated gene pool among Eskimos.[76] The few other cross-cultural studies reporting no sex differences in some population groups have attempted to correlate sex differences in spatial performance with level of food accumulation, social stratification, and/or

67. G. M. Vaught, "The Relationship of Role Identification and Ego Strength to Sex Differences in the Rod-and-Frame Test," *Journal of Personality* 33 (1965): 271–83; Smith (n. 37 above), p. 234.

68. J. Bieri, "Parental Identification, Acceptance of Authority, and Within-Sex Differences in Cognitive Behavior," *Journal of Physiological and Comparative Psychology* 60 (1960): 76–79.

69. S. C. Nash, "The Relationship among Sex-Role Stereotyping, Sex-Role Preference, and the Sex Difference in Spatial Visualization," *Sex Roles* 1 (1975): 15–32.

70. L. R. Ferguson and E. E. Maccoby, "Interpersonal Correlates of Differential Abilities," *Child Development* 37 (1966): 549–71.

71. D. B. Lynn, "Curvilinear Relation between Cognitive Functioning and Distance of Child from Parent of the Same Sex," *Psychological Review* 76 (1969): 236–40, and "Sex Role and Parental Identification," *Child Development* 33 (1962): 555–64.

72. E. E. Maccoby, *The Development of Sex Differences* (Stanford, Calif.: Stanford University Press, 1966), pp. 44–47.

73. J. A. Sherman, "Problem of Sex Differences in Space Perception and Aspects of Intellectual Functioning," *Psychological Review* 74 (1967): 290–99. Some direct support for this idea is provided by one study in Kenya (S. B. Nerlove, R. H. Munroe, and R. L. Munroe, "Effect of Environmental Experience on Spatial Ability," *Journal of Social Psychology* 84 [1974]: 3–10).

74. Although field-dependence was the cognitive dimension under study, spatial ability has been shown to be a major factor in the tests used (see n. 35 above) (H. A. Witkin, "Cognitive-Style Approach to Cross-cultural Research," *International Journal of Psychology* 2 [1967]: 233–50; M. B. Parlee and J. Rajagopal, "Sex Differences on the Embedded-Figures Test: A Cross-cultural Comparison of College Students in India and in the United States," *Perceptual and Motor Skills* 39 [1974]: 1311–14; C. G. Pande, "Sex Differences in Field-Dependence: Confirmation with Indian Sample," *Perceptual and Motor Skills* 31 [1970]: 70).

75. J. W. Berry, "Temne and Eskimo Perceptual Skills," *International Journal of Psychology* 1 (1966): 207–29; R. MacArthur, "Sex Differences in Field Dependence for the Eskimo: Replication of Berry's Findings," ibid., 2 (1967): 139–40.

76. J. Kagan and N. Kogan, "Individual Variation in Cognitive Processes," in *Carmichael's Manual of Child Psychology*, ed. P. H. Mussen (New York: John Wiley & Sons, 1970).

Westernization.[77] They have not been able to do so consistently. Even if they had, this would not prove that intrinsic factors do not affect spatial ability but only that the sum of the interaction between intrinsic and extrinsic factors varies with culture and geography. Of course the same can be said for the studies of genetic, hormonal, and laterality factors in spatial ability. The limitation in all cases is the correlational nature of the evidence.

These various biological and sociocultural explanations for sex differences in spatial ability could all be correct. There could be genetic and/or hormonal factors that pre-dispose females to verbal, left-hemisphere development and to reciprocally interact with the different experiences of girls in the family and in the larger society. The composite result may be a biological difference, in the sense of having a basis in neural organization but not in the sense of being determined entirely by intrinsic factors specific to women. Nor is it clear that the usual outcome is inevitable. There is evidence that practice and specific instruction can improve spatial visualization performance in both sexes.[78] It seems likely that complex, multiple causation is the rule for this and many other behavioral sex differences. Social environment multiples and magnifies, in many ways, an average kernal of intrinsic predisposition.

Some people still seem to believe that enough research will eventually yield exact percentages of intrinsic and extrinsic contributions to sex differences. This is unrealistic. Research on the development of sex differences may be useful in the design of effective social-intervention measures to change them, where that is desirable, but it is not likely to give us some sort of quantitative answer, a "biological limit" of socialization.

In many arguments about the biological factors in sex differences, one has the impression that it is not the factual inevitability of the biological that is the problem, but something more valuational, even moralistic. The most extreme position maintains that biological differences as such are desirable and should not be changed, even if countersocialization measures are effective in modifying them. This point of view is more than a rationalization for the belief that biology is inevitable; it places a

77. J. W. Berry and R. C. Annis, "Ecology, Cultural and Psychological Differentiation," *International Journal of Psychology* 9 (1974): 173–93; J. W. Berry, "Ecological and Cultural Factors in Spatial Perceptual Development," *Canadian Journal of Behavioral Science* 3 (1971): 324–36; H. A. Witkin and J. W. Berry, "Cross-cultural Studies of Psychological Differentiation," *Journal of Cross-cultural Psychology* 6 (1974): 4–87.

78. E. H. Brinkmann, "Programmed Instruction as a Technique for Improving Spatial Visualization," *Journal of Applied Psychology* 50 (1966): 179–84; J. A. Sherman, "Field Articulation, Sex, Spatial Visualization, Dependency, Practice, Laterality of the Brain and Birth Order," *Perceptual and Motor Skills* 38 (1974): 1223–35; A. G. Goldstein and J. E. Chance, "Effects of Practice on Sex-related Differences in Performance on Embedded Figures," *Psychonomic Science* 3 (1965): 361–62. It would be interesting to know if such improvement were accompanied by shifts in cerebral laterality.

positive value on the "natural" in either aesthetic (*vive la différence*) or religious (God/Nature knows best) terms. A less mystical, and more prevalent, variant of the naturalistic position views biological differences as desirable, not per se, but by virtue of their complementary relation to present social institutions, which are also assumed to have roots in human biology. While the postulates of this argument may be correct, it (like most functionalist arguments) is circular; it assumes the persistence of the status quo and thus the desirability of socialization reinforcing biological predisposition. In opposition, some feminists find themselves denying the biological component in sex differences and even arguing that sex differences must be abolished by whatever means necessary, in order to achieve social equality of the sexes.[79] To the extent that some feminists take this tack, they are apparently accepting a necessary relationship between the fact of difference, especially biological difference, and an unequal division of social rewards.

An unexamined but crucial assumption is the more general notion that differences among persons justify social inequality. Biological (usually meaning intrinsically determined) differences are often regarded as more reasonable bases for unequal social rewards than are differences that result from variations in the environment, which are held to have more of a claim to compensatory special treatment.[80] There is more consensus on the moral obligation of society as a whole to make restitution, as it were, for differences shown to be caused by imperfections in the social system (e.g., discrimination) than to compensate for differences in natural endowments. In this view it becomes quite important to know whether an observed difference is due to biology or to society.[81] Since most characteristics (cognitive abilities, personality, motivation, energy) that are often considered relevant to social benefits (income, status, responsibility, power) are influenced by both intrinsic and extrinsic factors, this distinction between biological and social causation of difference involves manifold practical difficulties.

One may also wonder what moral justification can be offered for it. It seems obvious that one does not deserve one's genes or hormones any

79. S. Firestone (*The Dialectic of Sex* [New York: William Morrow & Co., 1970]) even regards "ectogenesis" as necessary.

80. Special treatment here is meant to include both remedy of difference (e.g., compensatory training) and compensation for the social results of difference (e.g., preferential hiring or admissions policies). Either represents a social investment contrary to what the usual meritocratic criteria would dictate.

81. Some discussions of this limitation of "equal opportunity" are J. H. Schaar, "Equality of Opportunity and Beyond," in *Nomos IX: Equality,* ed. J. R. Pennock and J. W. Chapman (New York: Atherton Press, 1967); B. A. O. Williams, "The Idea of Equality," in *Philosophy, Politics and Society,* ed. P. Laslett and W. C. Runciman (Oxford: Oxford University Press, 1964); T. Nagel, "Equal Treatment and Compensatory Discrimination," *Philosophy and Public Affairs* 2 (1973): 348–64.

more than one deserves other accidents of birth,[82] such as social class, which may determine how one's intrinsic potential is eventually expressed as a social contribution. Social justice may not require redress of cosmic injustice but does not seem to exclude it by definition.[83] There seems to be no a priori reason to omit from the liabilities that are collectively assumed those such as sex, dealt out by nature's lottery. That women, as a group or individually, bear the economic cost of their female biology in differential retirement benefits[84] and in maternity leaves which are ineligible for disability pay[85] is sometimes condoned because the relevant differences are biological. A more likely explanation is that such policies have, so far, been an expedient (though morally arbitrary) way of distributing these particular costs. But we are not obliged to run society by actuarial principles. We could decide to distribute the cost of old age or reproduction in various other ways. The desirability of alternative methods does not necessarily depend on whether sex differences in longevity or reproductive function are biological. Similarly, the peculiar arrangement whereby many women receive economic reward for their social contributions (in child care, homemaking, and community work) only indirectly, via their husbands' income, is neither morally nor practically required by the fact (if indeed it is a fact) that women are biologically better parents than men.[86] Social policies

82. If it does not seem obvious, read H. Spiegelberg, "A Defense of Human Equality," *Philosophical Review* 53 (1944): 101–24; D. D. Raphael, "Justice and Liberty," *Proceedings of the Aristotelian Society* 51 (1951): 167–96. A somewhat more plausible argument can be made for entitlement to, if not desert of, the social benefits which accrue to one as fruits of one's natural endowments, as long as such benefits are gained in a just (or at least, agreed-upon) social system. But this argument for entitlement, if accepted, seems to be equally applicable to advantages which result from class origin and other circumstances which also, essentially, depend on luck, and hence offers no moral basis for distinguishing biological from social causation of difference.

83. Of course it may be argued that society at large does not deserve to pay for the remedy or recompense of biological differences and/or their social consequences because other members of society are entitled to the full fruits of their receipts from the natural lottery. Clearly this depends on one's conception of the social contract. Contrast, for instance, R. Nozick, *Anarchy, State and Utopia* (New York: Basic Books, 1974) and J. Rawls, *A Theory of Justice* (Cambridge, Mass.: Harvard University Press, 1971).

84. "Sex-based mortality tables . . . should only be permitted to the extent that it can be established that mortality differences are due to the genetic advantage of females" (D. Halperin, "Should Pension Benefits Depend upon the Sex of the Recipient?" *American Association of University Professors Bulletin* 61 [1975]: 43–48).

85. Such a policy "divides potential recipients into two groups—pregnant women and nonpregnant person," and is therefore judged nondiscriminatory, General Electric v. Gilbert, U.S. 74-1589 (1976).

86. A. Rossi, after examining the evidence for a "biologically based potential for heightened maternal investment in the child . . . that exceeds the potential investment by men in fatherhood," concluded that "an egalitarian ideology in fact involves profound difficulties when applied to child-rearing" ("A Biosocial Perspective on Parenting,"

could be devised to more equitably distribute the costs and benefits of parenting,[87] even while retaining the family as an economic and social unit.

Of course, there are utilitarian arguments for making, or trying to make, a distinction between biological and social causation of differences that are relevant to social contribution. If biological differences are regarded as more basic, essential, and harder (or at least more expensive) to change while socially induced differences are viewed as superficial accretions, mere cultural artifacts subject to easy modification, then remedy of socially induced differences may have a better cost/benefit ratio than remedy of biological differences. To the extent that such a remedy results in better matching of talents to job, it may result in increased overall productivity, repaying the social investment of the remedy.[88] But this conclusion depends on the questionable premise of higher social cost for changing biological differences and on the relevance of specific differences to specific spheres of social function. The utilitarian, market considerations generally advanced to justify unequal rewards for different social contributions offer no additional basis for distinguishing the origin of differences in contribution. If meritocracy makes any sense, only performance counts; intrinsic potential has no significance unless it actually results in socially valuable contributions. In fact, in this society, we do accept some collective responsibility for those who are less fortunate in natural endowments, as well as for victims of social injustice, in order to achieve a more nearly equal result in an otherwise meritocratic system. Our operating criteria for redress or recompense of differences (or their social consequences) are usually severity of need and cost of remedy—not the cause of the difference.

Thus the important matters for feminists to address would seem to be not the cause of sex differences, but how much of our collective resources should be devoted to equalizing and in what respects, and how to balance this against other social goals.[89] In brief, arguments about sexual equality often get stuck on whether sex differences are socially or biologically caused. This battleground, which feminists generally have not chosen, is strategically unwise. In many cases the question is unanswerable and arguing about it delays social reform. Moreover, such a

---

Daedalus 106 [1977]: 1–22). But at least some egalitarian ideologies could accommodate nonidentity of social function (i.e., division of labor) were there greater parity in status, power, and income.

87. For some suggestions, see M. J. Bane, *Here to Stay* (New York: Basic Books, 1976), chap. 7.

88. Implementation of such policies also involves the identification and "tracking" of individuals according to presumed intrinsic capacities, procedures not without social risk even using more reliable technology for identification than is presently available.

89. For instance, universal literacy may well have priority over equalizing everyone's spatial visualization, but this is true regardless of the origin of differences in spatial ability—or literacy.

stance, by implication, grants to biological differences a special, but morally unjustified, value status. We would more profitably use our energies devising practical measures to achieve greater social equality in ways which are compatible even with presently observed sex differences, and focusing, not on their causation, but on their possible relevance to specific social functions. If a particular sex difference is incompatible with important aspects of social equality, we should argue for compensatory measures, independent of biological causation. No doubt greater social equality will, in itself, tend to decrease sex differences to a biological minimum, whatever that may prove to be. But equality will be slower in coming if we appear to base our demands on the proposition that sex differences are due wholly to nurture rather than partially to nature.

Department of Biology
Northeastern University

Social and Behavioral
Constructions of Female Sexuality

Patricia Y. Miller and Martha R. Fowlkes

The twentieth century has occasioned a major revolution in Western
conceptions of the sexual. Freud, of course, is most widely credited with
the secularization of human sexuality, recasting superego and libido in
the roles traditionally played by their predecessors, "good" and "evil."
While the Freudian conceptualization is constructed on a solid foundation of Judeo-Christian values, these are manifest only in the kinds of
disguised forms Freud himself, in his dream analysis, trained us to perceive and explore. More critical than the foundation of his thought,
however, is the apparent opposition to theological constructions in the
Freudian edifice. For it is the semblance of scientific neutrality that
Freud brought to the consideration of sexuality rather than the content
of his provocative ideas that anticipated the full-blown emergence of
what Robinson has called "the modernization of sex."[1] The hallmark of
"modernization" is detached neutrality or, at least, the appearance of
neutrality as thinkers and researchers—guided by the precepts of the
scientific method—struggle to reconcile their own culturally linked
biases with the data at hand. A few have been particularly influential.
Their work does not in all cases represent an ideal application of the
tools of science to human conduct but, in most instances, it is the kind of
large-scale, well-funded, controversial research that must necessarily be

This paper represents a joint effort. The order of authors was established by the toss
of a coin.
    1. P. Robinson, *The Modernization of Sex* (New York: Harper & Row, 1976), p. vii.

This essay originally appeared in *Signs*, vol. 5, no. 4, Summer 1980.

assimilated in the work of others and in the public consciousness. It is this research that is the subject of review in this paper.

"The modernization of sex" has advanced under the standard of democracy. Male and female sexuality have warranted equal time, if not equal thought, in the few serious major studies extant. In consequence, female sexuality may have received particular benefits since, compared to male sexuality, it has been substantially more shrouded in misconception. Attending this small flurry of research is a growing acceptance of female sexuality as a topic of discussion among the larger public as well as the publication of numerous popular treatises on, exhortations to, and celebrations of the sexuality of women.[2] While actual knowledge and understanding have not kept pace with the assertion of female sexuality, be it heterosexual or homosexual, the thesis of this paper is that research and thinking are moving, not incrementally but productively nonetheless, toward a more accurate, if fragmented, understanding.

The majority of women enact their sexuality in the world of men, and usually marriage. Yet a search of existing literature reveals scant attention to the sexuality of adult women in the context of their normative social roles and relationships. The behavioral paradigms that dominate much of the work in sex research have led to the abstraction of sexual behavior and response from the social and interactional setting. Ironically, though, the researchers' own social norms tend to make their way into interpretations of the behaviors and responses they have observed or collected. This has the effect of imbuing behavioral data with social meanings, but the meanings are those of researchers, not of the women they study.

Researchers whose orientation is sociological are typically less interested in the normative than the nonnormative. In the study of social class, for example, sociologists have devoted far more time and energy to studying the organization and life-styles of the lower and working classes and the upper class than to those of the middle class. Similarly, in the sociological study of female sexuality, literature abounds on adolescent or premarital sexuality, and studies of extramarital sex occur with some frequency. Literature (albeit of varying quality) on lesbian sexuality and relationships proliferates. Increasing attention is paid to the sexuality of the elderly as well. In all of this, the woman's married heterosexual experience serves as a boundary delineator: it is an outcome, a point of departure, a way station. The most conscientious of sociological discussions, which take care to emphasize the emergent and socially con-

---

2. Obvious examples are: D. Reuben, *Any Woman Can* (New York: Bantam Books, 1972); A. Comfort, ed., *The Joy of Sex* (New York: Crown Publishers, 1972); Boston Women's Health Collective, *Our Bodies, Ourselves* (New York: Simon & Schuster, 1971); D. Martin and P. Lyon, *Lesbian/Woman* (New York: Bantam Books, 1972); S. Abbott and B. Love, *Sappho Was a Right-on Woman* (New York: Stein & Day, 1972).

structed nature of female sexuality, tend to fall down or, perhaps more accurately, to fall away in their approach to the adult heterosexual female. She is assumed to have one role—married—and the meaning and experience of a woman's sexuality are inferred from taken-for-granted normative assumptions about that role. This argument is borne out by the dearth of recent material on sexuality in marriage in so pertinent a publication as the *Journal of Marriage and the Family.*

The most comprehensive and, in their day, controversial studies are Kinsey's of male and female sexual behavior,[3] which broke new ground in depicting the norms of sexual activity, not in terms of what people should do, but in terms of what they, in fact, do. The overriding argument of Kinsey's work is that ideas of what should be ought to be refashioned to be consonant with what people actually—and naturally—do. Yet insofar as what people actually do represents a departure from the expected conventions and mores of social behavior, Kinsey rests his case for tolerance of those departures on the grounds that they are in the service of socially desired norms, particularly those of heterosexual marriage. Thus, for men and women alike, all premarital erotic and sexual activity is seen as facilitating adjustment to marriage. Kinsey is critical of the double standard of male and female sexual behavior that works disproportionately to inhibit women's sexual activity outside of marriage.[4]

However, Kinsey's support for increased female sexual autonomy is rendered suspect by his own double standard. Although Kinsey is concerned that women are able to experience sexual satisfaction as it is manifest in orgasmic outlet, his advocacy of women's premarital sexual activity is prompted as much by his commitment to the satisfaction of male sexual needs and desires as to that of women's. Premarital sexual activity is seen as a useful socializing agent for women in their relationships with men, as it enables women to learn to adjust emotionally to various types of men.[5] Since Kinsey's own data show that coitus (both marital and premarital) is far less frequently a source of orgasmic release for women than for men, he is essentially encouraging women to seek premarital coitus in the service of male physiological release.

A similar bias is revealed in Kinsey's interpretation of his finding that female sexual unresponsiveness is a major factor in women's premarital coital reluctance. A man, of course, may reach orgasm through coitus without doing much at all to arouse or stimulate response in his female partner. Rather than portraying men as the victims of women's

3. A. C. Kinsey, W. B. Pomeroy, and C. E. Martin, *Sexual Behavior in the Human Male* (Philadelphia: W. B. Saunders Co., 1948); A. C. Kinsey et al., *Sexual Behavior in the Human Female* (Philadelphia: W. B. Saunders Co., 1953).

4. Kinsey et al., *Female*, pp. 307–30.

5. Ibid., pp. 259–66.

unresponsiveness, it would be fairer to suggest that women may be victimized by men's self-interest and failure to show the consideration necessary to incorporate female orgasm into the coital experience.

The female orgasm itself is emphasized by Kinsey as a source of satisfaction and reassurance to the male. He states that "orgasm cannot be taken as the sole criterion for determining the degree of satisfaction which a female may derive from sexual activity. . . . Whether or not she herself reaches orgasm, many a female finds satisfaction in knowing that her husband or other sexual partner has enjoyed the contact, and in realizing that she has contributed to the male's pleasure."[6] The assumption that women can do quite nicely without an orgasmic accompaniment to coitus may be correct. But it is supported by no data from the females themselves about the place and meaning of sexual satisfaction in their marriage relationships.

Overall, when compared to males, Kinsey's females show a greater uniformity in their sexual behavior with respect to the impact of selected social variables, and they also exhibit consistently less sexual responsiveness and lower frequencies of orgasm than men. The differences between the sexual responsiveness of men and women are attributed to the greater "psychologic conditionability" or flexibility of men.[7] Although Kinsey does acknowledge the essential likeness of the physiological experience of sexual arousal and orgasm for men and women (in anticipation of the work of Masters and Johnson a decade later), he also views women as having less well-developed sexual interests than men and as more willing than men to subordinate those interests to an overriding commitment to home and family.

What Kinsey overlooked, of course, was the powerful influence of sex role itself as a variable in conditioning sexual response and behavior. Men are expected and socialized to be autonomous in their activities, including their sexual activities. Whereas males are encouraged to give full expression to their sexuality as an indication and demonstration of their masculinity, female sexual response has traditionally been thought to be appropriately derived from relationships with men and their needs. It is unquestionably true that women's sexuality is affected by socialization into the female sex role and the subordinate status attached to it in common ways that both supersede and preclude the potential effect of other social variables (at least, the social variables selected by Kinsey).

Kinsey's determination that women are generally less capable of sexual autonomy by virtue of being less psychologically conditionable itself reinforces a double standard for male and female sexual behavior.

6. Ibid., p. 371.
7. Ibid., p. 684.

It also constitutes a license for the perpetuation of predatory male sexual activity. It is precisely this predatory context which renders a derivative, rather than an autonomous, female sexuality functional. If women are to be sexually autonomous in a fashion similar to men, they must also participate in the social, political, and economic autonomy that has been granted to men. To suggest otherwise is to suggest that women should simply become more readily available to men for sex on the terms that men determine as a consequence of their dominant social roles. This would have the effect, on the one hand, of enhancing women's vulnerability to exploitation by men as sex objects and, on the other, of removing their sexuality from marriage (or potential marriage), thereby weakening the major institutional source of security and social approval accessible to them in a patriarchal society.

In contrast to Kinsey, Masters and Johnson in *Human Sexual Response* addressed the question of women's sexual autonomy solely and specifically in terms of physiological functioning.[8] Their documentation of a specifically orgasmic physiology for the female, including her multiple orgasmic potential, is both testimony and tribute to women's innate sexual capacities. In what we can only hope is a final debunking of the Freudian distinction between the vaginal and clitoral orgasm, Masters and Johnson use the evidence of their laboratory observations to recognize the clitoris: "The clitoris is a unique organ in the total of human anatomy. Its express purpose is to serve both as a receptor and transformer of sensual stimuli. Thus the female has an organ system which is totally limited in physiologic function to initiating or elevating levels of sexual tension. No such organ exists within the anatomic structure of the human male."[9]

Thus Masters and Johnson established unequivocally the physiological ability of women to achieve orgasm and dispelled any doubts that females might in any way be possessed of a lesser sexuality than males. Their findings indicate that women's orgasmic capacities are a given, and that the key variable in the achievement of successful orgasmic outcome for women is the quality of sexual attention they receive. Gone are both Kinsey's nonresponsive woman and the woman who develops her sexual responses in order to be able to fulfill male needs. Ironically, however, Masters and Johnson's own data show clearly that the sexual autonomy of women is manifest not only in their physiological functioning but in the nature of their response to sexual stimulation as well. For women "the maximum physiologic intensity of orgasmic response subjectively reported or objectively recorded has been achieved by self-regulated mechanical or automanipulative techniques. The next highest level of

8. W. H. Masters and V. E. Johnson, *Human Sexual Response* (Boston: Little, Brown & Co., 1966).
9. Ibid., p. 45.

erotic intensity has resulted from partner manipulation, again with established or self-regulated methods, and the lowest intensity of target-organ response was achieved during coition."[10]

It is not difficult to read this statement as the ultimate testimony to the autonomy of female sexuality. The female orgasmic response is demonstrably not correlated with—or, more accurately, is negatively correlated with—conventional heterosexual coitus. However, if the clitoris is no respecter of heterosexual intercourse, Masters and Johnson find that the vagina is. While the clitoris behaves like a wanton child, female vaginal response and activity are portrayed by Masters and Johnson as staunch defenders of the wisdom of copulation and, therefore, presumably, of the conventional heterosexual relationship:

> The vaginal barrel performs a dual role, providing the primary physical means of heterosexual expression for the human female and serving simultaneously as an integral part of her conceptive mechanism. . . . *To appreciate vaginal anatomy and physiology is to comprehend the fundamentals of the human female's primary means of sexual expression* [emphasis added]. In essence, the vaginal barrel responds to effective sexual stimulation for penile penetration. Just as a penile erection is a direct physiologic expression of a psychologic demand to mount, so expansion and lubrication of the vaginal barrel provides direct physiologic indication of an obvious mounting invitation.[11]

Masters and Johnson's conceptualization of the vagina—preparing for and accommodating to copulation and facilitating in multiple and essential ways the conceptive process—is a powerful corrective to the autonomous functioning and responsiveness of the clitoris. Their discussion of the vagina must be read basically as a statement of women's intrinsic heterosexuality and as a rejection of sexual activity that does not involve the vagina. Women's sexuality has again been reined in to conform to traditional heterosexual expectations. To be sure, an autonomous female sexuality is ultimately a social construction rather than simply a physiological reality. However, Masters and Johnson must be faulted for overriding their physiological data in a way that maintains rather than challenges traditional social constructions of women's sexuality.

For all that the majority of women themselves may sympathize with Masters and Johnson's sympathies for the heterosexual alliance, their research tells us much more about how people *do* sex than about sexual relationships as such. While it is helpful to know something of the sexual techniques that are the precondition for bringing women to orgasm, it

10. Ibid., p. 133.
11. Ibid., p. 68.

seems more salient to have some understanding of the kinds of relationships and communications between partners that predispose them
to knowing or wanting to know those techniques.[12] The major variable
influencing the achievement of sexual fulfillment for women as well as
men is, undoubtedly, not merely technique but the quality of the relationship itself. Masters and Johnson offer no recognition that effective
and meaningful sexual activity is not defined solely by the achievement
of orgasm, but may involve a wide range of behaviors and contacts that
are expressive of caring and commitment between partners. Masters and
Johnson have followed Kinsey's behaviorist lead to its logical conclusion.
They have earned for women the right to have equal time and space with
men on the sexual production line.

Gagnon and Simon's book, *Sexual Conduct,* presents a social interactionist perspective on sexual behavior, a perspective which stands as a
welcome and important antidote to the narrowly focused work both of
Kinsey and of Masters and Johnson.[13] For Gagnon and Simon, sexual
actors are neither mere representatives of statistical constituencies nor
demonstrators of the mechanical operation of the sexual machine.
Rather, they view sexual behavior as social behavior, entered into and
endowed with meaning by social actors whose interpersonal relationships are bound together by common and shared understandings
and communications that are the product of their common and shared
social and cultural worlds. Sexual behavior is, above all, learned behavior; sexual conduct is "neither fixed by nature or [*sic*] by the organs
themselves."[14] Social scripts organize our understanding of the participation in sexual activity. A sexual situation exists when the actors involved are responding to a socially constructed definition of what is
sexual with strategies for doing the sexual. "The social meaning given to
the physical acts releases biological events."[15] In other words, it is the
meaning assigned to behaviors by the actors involved—not the behaviors
themselves—that determines whether a situation or activity is sexual or
not.

These ideas would seem to hold much promise for understanding
the many and complex ways in which the sexual takes its cues from and is
embedded in the social. It is reasonable to expect Gagnon and Simon to
offer both insights and information about the meaning and management of sexuality in the context of that most salient and normative of
institutions by which relationships between adult men and women are
arranged, marriage. Indeed, Gagnon and Simon themselves acknowl-

12. It is also possible that individual variations in male or female physiology (surprisingly, unexamined by these researchers) account for variability in orgasmic response.
13. J. H. Gagnon and W. Simon, *Sexual Conduct* (Chicago: Aldine Publishing Co.,
1973).
14. Ibid., p. 9.
15. Ibid., p. 22.

edge that "the management of sexual commitments within a marital relationship characterizes the largest part of postadolescent sexual experience in our society."[16]

Yet their treatment of adult sexuality in marriage is limited in scope and shallow in content. To be sure, they do acknowledge the paucity of data on the topic of marital sexuality. However, it is disappointing that they do not apply their sociological perspective to formulating constructive questions and issues about the ways in which marriage per se and the evolving marriage relationship over time lend particular and characteristic kinds of meaning to the sexual behavior of men and women. Instead we are simply told that sex declines in salience after the formation of the marital unit. Women are assigned considerable responsibility for this decline because of the extra (and presumably distracting) burdens placed on them in their maternal roles. This is surely another way of saying that men's sexual energy and enthusiasm are sustained at the price of their wives' fatigue. In the end, Gagnon and Simon are guilty of the very reductionism that their interactionist perspective was meant to alleviate. Sexual behavior in marriage is viewed as consisting of little more than a variety of available bodily positions. Marriage itself is presented as a kind of social-interactional plateau from which all roads lead downhill.

Reflecting an instrumental male bias, Simon and Gagnon do not consider the possibility that sexual activity in marriage might be more appropriately assessed in qualitative rather than quantitative terms. Whereas adolescent sexual behavior has been described as "behavior in search of meaning,"[17] in marriage or other adult committed relationships, it may well be that, where sexuality is concerned, meaning is in search of behavior. The more complex and comprehensive and settled the intimacy between partners, the wider the array of available behaviors—of which sex is only one—with which to express that intimacy. In this way eroticism may actually come to carry more rather than less meaning as the relationship endures and sexual activity comes to have a focused and shared meaning in the affective repertoire of the couple.

Gagnon and Simon's inability to see beyond the form of marriage to the processes and interactions contained within it is the consequence of their having superimposed a symbolic-interactionist perspective on a developmental model of sexual behavior. The linear-historical emphasis of developmental analysis eventually and inevitably cancels out the symbolic-interactionist emphasis on the situational construction of behavior. Marriage is seen less as an entry into adult social and sexual roles than as the end product of childhood and adolescent socialization. In-

16. Ibid., p. 82.
17. P. Y. Miller and W. Simon, "The Development of Sexuality in Adolescence," in *Handbook of Adolescence*, ed. Joseph Adelson (New York: John Wiley & Sons, 1980).

deed, Simon and Gagnon actually characterize marriage as "post adolescent sexual development." In this framework, adult heterosexual relationships are the dependent variables flowing from earlier developmental processes. The potential of marriage to function as an independent variable acting on adult life and as a powerful socializing influence in its own right is virtually overlooked. Despite the deficiencies in their analysis of adult heterosexual relationships, Gagnon and Simon must be credited with bringing the study of sex fully into the sociological domain and with an ambitious attempt to establish a paradigm for thinking about sexual behavior as social behavior. Recently they have both independently shown signs of shedding the cumbersome baggage of developmental theory in moving toward a more effective application of symbolic-interactionist theories to sexual socialization and careers.[18]

Certainly other sociologists who have addressed themselves to the topic of human sexuality, and female sexuality in particular, do not begin to approach Gagnon and Simon's level of theoretical sophistication. Laws and Schwartz, for example, purport to offer a feminist-interactionist perspective on female sexuality.[19] Yet their use of the "scripting" concept does not convey the richness and complexity of the interaction process, but consists more of stereotypic descriptions of the influence of social norms on women's sexual identity and sexual behavior. Their claim to present a dynamic view of sexuality is belied by their static "stage model" of sexual development. While they describe what the alternatives to normative sexual transactions and life-styles are for women, they offer no analysis of how, in terms of social process, women might find their way into those. And they maintain a questionable separation between women's reproductive functioning and women's sexuality. Female reproductive processes and potential are seen as incidental to or contingent upon sexuality rather than as interactive with a woman's sexual self-concept and responsiveness.

Love, Sex and Sex Roles, by Safilios-Rothschild, is best described as a sociologically informed ideological essay.[20] Taking nearly 150 pages to make the point that so long as marriage is organized on the basis of traditionally unequal sex roles for men and women, marriage will not be a partnership of equals, her work reads like a feminist marriage manual. We can hardly quarrel with Safilios-Rothschild's observations about the ways in which social distance between men and women generally is transferred into the marriage relationship. However, the same or similar

18. See the discussion of Simon's (and Miller's) as well as Gagnon's recent work in a review by R. W. Libby of *Human Sexuality: A Comparative and Developmental Perspective*, ed. H. A. Kathchadourian (Berkeley and Los Angeles: University of California Press, 1979) in the *Journal of Sex Research* (November 1979), pp. 325–26.

19. J. L. Laws and P. Schwartz, *Sexual Scripts: The Social Construction of Female Sexuality* (Hinsdale, Ill.: Dryden Press, 1977).

20. C. Safilios-Rothschild, *Love, Sex, and Sex Roles* (Englewood Cliffs, N.J.: Prentice-Hall, Inc., 1977).

observations were made with considerably more depth and eloquence by Simone de Beauvoir over twenty-five years ago.[21] It is also unfortunately true, as Hacker noted in her classic essay, that equality has never been shown to be a necessary precondition for love.[22]

In the case of behaviorists such as Kinsey and Masters and Johnson, the imposition of unexamined social values and stereotypes on their data is understandable. The failure of sociologists to correct those stereotypes is less understandable. Equally disappointing is the sociological failure to advance the level of thinking about the analysis of female sexuality to treat the common and distinctive features of women's sexual experience against the backdrop of the many and varied forms the female life course may take. Although marriage may be the modal adult relationship for women, there are no grounds for assuming homogeneity among marriages or the sexual relationships contained within them. The long and venerable tradition in sociology wherein the impact of social-class and status-group subcultures on all manner of life-styles and values is examined does not, for the most part, extend to the study of sexual life-styles and values. The few exceptions are studies within a given class;[23] comparative, cross-class studies are notably lacking. We also know very little about how the particular stresses and rhythms of work and leisure associated with various kinds of occupational commitments (on the part of both men and women) affect the importance and timing of marital sex. And it is certainly well worth asking whether or how patterns of change in women's political and economic roles and bargaining power are reflected in changes in their sexual roles and bargaining power.

Our culturally cherished notion of the sexually settled, mature, and nurturing adult woman may account for the persistent sociological bias that, once a woman is anchored in the socially approved heterosexual world of marriage, her sexuality speaks for itself and is not a subject for investigation. Yet there is every reason to believe that a woman's sexuality is emergent and takes on new meanings and qualities in the context of the emergent marriage relationship itself. Then, too, the proliferation and popularization of research on women's sexuality and orgasmic potential has led simultaneously to unprecedented expectations and self-consciousness with respect to marital sexual performance and satisfaction. We may well wonder for how many women the potential for enhanced orgasmic experience is outweighed by the pressure to achieve orgasm in the service of their men's sense of sexual adequacy. The overlap of the sexual and the maternal in women's experience has been

21. Simone de Beauvoir, *The Second Sex* (New York: Alfred A. Knopf, Inc., 1953).
22. Helen M. Hacker, "Women as a Minority Group," *Social Forces* (October 1951), pp. 60–69.
23. See, for example, M. Komarovsky, *Blue Collar Marriage* (New York: Vintage Books, 1962); L. Rubin, *Worlds of Pain* (New York: Basic Books, 1976).

almost entirely neglected as an area for study. Both Rossi and Rich have mentioned the highly charged sexual feelings that are often involved in giving birth to and nursing a child.[24] The mothering role itself draws women into new sets of routines, responsibilities, and relationships, which may affect patterns of sexual expression and interest in the marriage. Furthermore, at the risk of pointing out the obvious, many women do not marry, marry in young middle age, or leave their first or subsequent marriages to move into new living situations or relationships. Each one of these arrangements must surely involve a variety of sexual transactions and different kinds of meanings attached to the expression of sexuality in them.

Finally, of course, a substantial minority will experiment with lesbian activity, and for a minority of these such experimentation will occasion the initial sociosexual expression of what is to become a lesbian career.[25] The major (and minor) fixtures of sex research have produced a noteworthy, albeit controversial, record of the experience of these women. Kinsey et al. is unequivocally the major transitional work marking a reconceptualization of scholarly thought on lesbianism in the twentieth century. Insisting on an "objective" posture bereft of the kind of moral biases that distinguish psychoanalytic writing, Kinsey denied the equivalence of act and actor on empirical grounds: "It would clarify our thinking if the terms could be dropped completely out of our vocabulary, for then socio-sexual behavior could be described as activity between a female and a male, or between two females, or between two males and this would constitute a more objective record of the fact."[26] Kinsey countered the etiologists' conception of lesbian activity—as an unfortunate outcome predetermined by the vagaries of development—with the zoologists' recognition of interspecific continuities and their implications. Following the dictum of parsimony, he posited two factors as the primary causes of lesbianism: the universal physiological capacity for same-sex response and the contingencies of opportunity to engage in lesbian relations. The latter are influenced by the "conditioning effects" of the experience itself and, more critically, the positive and negative social evaluations of lesbian activity encountered by the individual. Ever the naturalist, Kinsey argued that "it is . . . difficult to explain why each and every individual is not involved in every type of sexual activity."[27] Conceptualizing *any* exclusive preference as a

24. A. S. Rossi, "A Biosocial Perspective in Parenting," *Daedalus* 106 (Spring 1977): 1–31; A. Rich, *Of Woman Born* (New York: Bantam Books, 1977).

25. Sexual behavior does not necessarily imply anything about sexual identity. There is no common formula by which individuals uniformly weigh and sum the history of sexual activity and feeling to arrive at a satisfactory identification of themselves or others. Our separate treatment of heterosexual and homosexual forms here is an expedient dictated by our sense that somewhat different intellectual issues are relevant in the two cases.

26. Kinsey et al., *Female*, p. 447.

27. Ibid., p. 451.

perversion of the natural, Kinsey extended and inverted the psychoanalytic construction of innate bisexuality.

Structural variables, of course, dominated all of the work produced by Kinsey and his associates. These are not particularly powerful determinants of women's sexual experience, although they are somewhat stronger predictors of lesbianism. Not surprisingly, education and religiosity are modestly associated with the prevalence of lesbian experience. Interpreting these associations, Kinsey argued that "moral restraints" on heterosexuality serve to encourage lesbianism. Thus, lesbian interests are fostered where devout women, deterred from premarital heterosexual expression by church precepts, turn to lesbianism and, in consequence, abandon the ritual observance of their faith.[28] Similarly, the greater lesbian experience of college graduates follows from the disapproval of premarital heterosexual contacts by parents and college administrators. The emergence of heterosexual commitments among such women is delayed in the interest of education; no longer viable contenders in the marriage market, they subsequently drift into "convenient" lesbian liaisons.[29] Observing that same-sex relations occur more often among single and previously married women, Kinsey concludes that heterosexuality must function to immunize women from such relations. He was apparently unable to recognize the role of lesbian interests in diverting or extruding women from marriage.

Kinsey's stance on lesbianism was certainly more neutral than the moral posturings of the developmentalists. But that he devalued lesbian interests is undeniable. Withholding even the possibility of autonomy from these, he invariably constructed lesbianism as derivative of heterosexual experience and its social meanings. Lesbian encounters emerged because the dominant institutions charged with the moral development of the young too effectively discouraged heterosexual relations, while the failure to achieve marriage pushed women toward the mysteries of lesbianism. And so it goes. Isolating heterosexual prudery from the panoply of conflicting social pressures that females experience with respect to their sexuality, Kinsey promulgated the usual distortions that follow from single-variable explanations of any behavior. More critically, Kinsey failed to understand that women who have chosen to live outside of the restrictions of a virtually monolithic gender role may experience lesbianism as not only a viable alternative but as the one viable alternative that provides cohesive institutional forms. And, of course, the sense of autonomy necessary for entertaining alternatives would be higher in just those populations—the better educated, the irreligious, and the unmarried—where the prevalence of lesbian experience is greatest. Lesbian experience may flow from the social meanings attached to hetero-

28. Ibid., p. 465.
29. Ibid., p. 460.

sexuality, but it undoubtedly flows from the social realities of heterosexuality as well.

The larger contradiction in Kinsey's thought—where he alternatively posed as the champion of sexual tolerance and as the defender of the superiority of heterosexuality for women—perhaps stems in part from his own conviction that female sexuality, unlike male sexuality, lacks any integrity of its own. In this, he confused the nature of sexuality with the overriding mandates of sex roles—an understandable lapse for a zoologist. But his commitment to heterosexuality is also rooted in social and moral convictions. Kinsey the zoologist told us that lesbianism was an attractive, adult alternative; Kinsey the moralist moved discussion to "moral restraints" and magical phalluses. Ideological warfare is, after all, a zero-sum game.

The ideological battle over the valuation of sexual alternatives has recently moved to a curious ground. With the publication of their book, *Homosexuality in Perspective,* Masters and Johnson argue that "more" is actually "less." The "more" in this case refers to their observation that "committed homosexual couples generally become more involved subjectively in the sexual interaction than married heterosexual couples."[30] This recalls, of course, Kinsey's findings of higher orgasmic achievement among committed lesbians compared to married women.[31] While Kinsey traced the apparent superiority of lesbian relations to the technical advantage accruing to the same-sex partner who necessarily shares similar psychological and physiological structures, Masters and Johnson cite the "free flow of both verbal and non-verbal communication."[32]

Masters and Johnson seize the day, imaginatively attributing such communication to "necessity"—a necessity derived from the "long-term disadvantages" of homosexuality. These are of two types: (1) Two basic techniques—partner stimulation and fellatio/cunnilingus—are commonly employed by homosexuals. Because there are *only* two, "these techniques of *necessity* must be constantly varied and refined to the utmost to avoid the loss of stimulative effectiveness through long-term familiarity."[33] (2) One or the other of these techniques is employed by each partner, usually in turn. The mutual stimulation that obviously follows from coitus is lacking in this sequential form of organization, and the level of sexual arousal is accordingly depressed. Masters and Johnson concede that where readily available modes of mutual stimulation are used, the second factor ceases to detract from the viability of the homosexual union. But in the general case this obvious expedient is overlooked, and instead, elaborated forms of communication emerge to

---

30. W. H. Masters and V. Johnson, *Homosexuality in Perspective* (Boston: Little, Brown & Co.), p. 212.

31. Kinsey et al., *Female,* p. 477.

32. Masters and Johnson, *Homosexuality,* p. 213.

33. Ibid., p. 214.

counter the inherent disadvantage, ultimately leading to the higher states of sexual arousal observed in homosexuals compared to heterosexuals.

This convoluted line of argument is of interest because it belies the gloss of neutrality Masters and Johnson bring to the study of homosexuality, and also because it contradicts informed thought with respect to coitus. Masters and Johnson themselves have repeatedly insisted that, whatever its tension-producing characteristics, coitus is a significantly less effective technique for achieving sexual release. (Note that with respect to their enthusiasm for coitus, Masters and Johnson are essentially arguing that "less" is "more.") Moreover, coitus does not commonly culminate in simultaneous orgasm but, rather, orgasm occurs *in seriatim*. Thus, overwhelmingly, coitus shares with other sexual techniques this long-term disadvantage.

Holding aside its authors' ideological biases, *Homsexuality in Perspective* reports the results of a project that violates the central canon of research methodology. No data are marshaled in response to their research question: "Is there a fundamental difference in sexual physiology if the respondents are homosexually rather than heterosexually oriented?"[34] This is a study, not of physiology, but of orgasm, presumably as an indicator of physiology. Masters and Johnson leave no stone unturned in their exhaustive demonstration of the comparability in orgasmic achievement among heterosexual and homosexual men and women. But all of the study subjects have been carefully screened for "a history of facility to respond at orgasmic levels to the sexual excitation in masturbation, partner manipulation and fellatio. . . ."[35] So theirs, in fact, is a study demonstrating that homosexual and heterosexual women and men who have secure histories of orgasmic achievement do, indeed, relentlessly achieve orgasm in the laboratory.

Etiology, of course, continues to lurk about on the scientific agenda. The notion that some factor—genetic, hormonal, physiological, or psychological—predisposes certain individuals to homosexuality is ever with us. Masters and Johnson believe their study effectively removes physiological factors from the list.

Their own preference is for a version of the learning model that assumes meaning flows from behavior. The individual who achieves competency in the techniques of sexual pleasuring will, when motivated, achieve comfort from them as well. Masters and Johnson obtain a qualified success in the application of their learning-based treatment programs to men and women suffering from homosexual dysfunction and "homosexual dissatisfaction," respectively. The political nature of the sources of "homosexual dissatisfaction" is undeniable; but for Masters

34. Ibid., p. 124.
35. Ibid., p. 7.

and Johnson, politics are at a considerable remove from the practitioner's commitment to service. In a spirit of sexual democracy, they devalue a homosexual orientation but they are utterly prepared to defend the individual's right to effective homosexual (or heterosexual) sociosexual functioning.

For Masters and Johnson, as for Kinsey, the sex act is *the* problem. It is a problem for the remainder of contemporary research on lesbianism as well, but here it is a problem because it is not a problem. In recent scholarly work, there is widespread agreement that the sex act itself is not a fruitful area for study, that research interest is more productively focused elsewhere. Ironically, the present state of affairs is undoubtedly a consequence of successful attempts by Kinsey's successors at the Institute for Sex Research to extend interest beyond the historic preoccupation with etiology to the exclusion of other concerns.

The impetus for this tradition is found in a modest article written by Simon and Gagnon for their reader, *Sexual Deviance,* and subsequently elaborated in *Sexual Conduct.*[36] Based on eight interviews and a solid grounding in the Kinsey research, Simon and Gagnon bring an instant of clarity to the study of lesbianism with their observation, "the female homosexual follows conventional feminine patterns in developing her commitment to sexuality and in conducting not only her sexual career but her nonsexual career as well."[37] Echoing the Kinsey motif, Simon and Gagnon argue that lesbian sexuality, like female sexuality in general, is less "autonomous," that is to say, less impersonal than male. Lesbians have fewer sexual partners than males and, among lesbians (as among other women), "the body is not seen . . . as an instrument of self-pleasure."[38]

But the sexual is less interesting to these researchers than modes of social adaptation. They view the ties that bind the lesbian to dominant institutions—family, work, and church—as preponderantly problematic and exacerbated by her sexual orientation and its social meanings. Moreover, there is little solace to be found in the private sector. Friendships with men and women (including other lesbians) are fraught with ambivalence; acceptance of self never can be really complete.[39] Simon and Gagnon recognize that their very small sample provides a poor basis for generalization. Predominantly working class in origin, most of the women they interviewed were no older than twenty-five. The kinds of problems that describe their lives—problems in resolving relationships of self to family, work, friends—are not exceptional for young, single

36. W. Simon and J. H. Gagnon, "The Lesbians: A Preliminary Overview," in *Sexual Deviance,* ed. J. H. Gagnon and W. Simon (New York: Harper & Row, 1967), pp. 247–82; and Gagnon and Simon, *Sexual Conduct,* pp. 176–216.
37. Gagnon and Simon, *Sexual Conduct,* p. 178.
38. Ibid., p. 182.
39. Ibid., p. 213.

women attempting to organize viable adulthoods. There are, of course, many ways of constructing a lesbian career. Social class and life-cycle stage constrain options, as do other factors. The kinds of sexual careers individuals forge not only reflect nonsexual sources of influence but afford varying degrees of potential, in themselves, for facilitating or disrupting the individual's interactions in other, nonsexual spheres of life. Simon and Gagnon understand this level of complexity, but the historical moment, biases in their sample, and biases in their mode of inquiry obscure it from them. They leave us with an image of the lesbian as a tortured creature with few sources of consolation. Occasionally they struggle to bring some balance to their image of overriding estrangement and loss, but the sexual and social rewards of being lesbian ultimately elude them.

Despite the authors' candid admission of the limitations in their work, this provisional study by Simon and Gagnon has substantially anticipated subsequent inquiry with respect to lesbians. A focus on continuity and modes of social adaptation is evident in several studies that have been published in recent years.[40] Principal among these is *Homosexualities* by Bell and Weinberg, researchers from Kinsey's old shop, the Institute for Sex Research.[41] At the behest of the funding agency, this study, the largest of its kind, is primarily a study of homosexual men. But a substantial number of lesbians were interviewed as part of the data collection effort. The usual comparisons with a control group of heterosexual women are reported—religion, marital histories, contacts with the mental health establishment, etc. The growing recognition that lesbianism is not monolithic gains expression in the typology Bell and Weinberg construct, and it is this that distinguishes the work. They designate women as "close-" and "open-coupled," "functionals," "dysfunctionals," and "asexuals," based on the sexual and social adjustment of those who are and are not in couples.

Unfortunately the typology, which frames the major thrust of their analysis, generally is of little utility in discriminating the experience of their lesbian respondents.[42] But this is not the case where mental health experience is concerned. The close-coupled, who constitute the modal category, were happier and more self-accepting than the lesbian sample as a whole. They were also less likely to report loneliness, depression, tension, or paranoia. Moreover, the mental health profile of the close-coupled women was indistinguishable from that of the heterosexual con-

40. See, for example, C. Wolff, *Love between Women* (New York: Harper Colophon, 1971); J. S. Chafetz et al., *Who's Queer?* (Sarasota, Fla.: Omni Press, 1976); D. Wolf, *The Lesbian Community* (Berkeley and Los Angeles: University of California Press, 1979).

41. A. P. Bell and M. S. Weinberg, *Homosexualities* (New York: Simon & Schuster, 1978).

42. Seventeen of the 280 tests (6.1 percent) reported for the five types of lesbians in tables 14.1–20.11 are statistically significant.

trols.[43] In contrast, the adjustments of the asexual lesbians were more precarious; these women were significantly more likely than lesbians in general to report loneliness and histories of suicide attempts. The mental health experience of the remaining groups was generally comparable to that of the sample as a whole.[44] Thus, where lesbian forms parallel those valued by the larger society and lesbian women are able to satisfy their sexual and social needs within those forms, substantial mental health benefits accrue. Or, alternatively, the well-adjusted lesbian with a sturdy sense of self may be predisposed to adopt socially valued forms and to find solace in them.

A critical question remains: What are the social and psychological experiences and attributes that move the individual toward one or the other career mode? A partial answer is suggested in an elegant little study designed by Peplau and her associates.[45] Well-reasoned and carefully executed, the study focuses on the relationship between values and the content of lesbian life-styles. Peplau identifies two, somewhat antithetical, constellations of values that appear to organize lesbian commitments. An emphasis on "dyadic attachment" is associated with both romanticism and traditional sex-role values. Lesbians commited to such values manifest them in behavior; their relationships with lovers are more likely to have a domestic base where loving and liking are central. The relationships these women have made are of longer duration, and their endurance in the face of obstacles is anticipated. The second constellation of values, "personal autonomy," is more characteristic of politically radical lesbians. Negatively associated with romanticism and traditional sex-role values, an emphasis on personal autonomy portends less durable sexual relationships and less contact between sexual partners. Consistent with the commitment to autonomy are the more frequent lapses in sexual exclusivity these women report.

Peplau's work should provide a model for further research in the area. The conceptualization evident in this work responds to the well-recognized complexity of human sexual behavior. "Lesbian" is a term that masks immense variability. Different kinds of women come to lesbianism for different kinds of reasons. Personal values are expressed in different kinds of relationships and structural arrangements which, in turn, have consequences of sexual conduct both within and outside of the primary relationship.

In our review of recent work on lesbianism, we have identified a glaring omission which is probably a consequence of the movement away from a central focus on the sexual per se. The slippage between lesbian

43. Numerous studies examine the mental health of those with devalued sexualities; see *Journal of Clinical Psychiatry*, vol. 39 (July 1978), for a nice review article by Hart et al.
44. Bell and Weinberg, tables 21.1–21.28.
45. L. A. Peplau et al., "Loving Women: Attachment and Autonomy in Lesbian Relationships," *Journal of Social Issues* 34 (Summer 1978): 7–27.

behavior and identity transformation has been considered,[46] and the ways women drift into lesbian behavior were described by Barnhart in her study of the Portland community.[47] But the absence of a comprehensive, sensitive study of the coming-out process as problematic for women—where both the "called" and the "chosen" are simultaneously examined or, better yet, discriminated—continues to constitute a major gap in the literature.

Clearly, there has yet to emerge anything that resembles a dominant paradigm organizing the study and interpretation of female sexuality. Paradigms themselves, of course, are emergent, socially constructed phenomena and, as such, require time to coalesce. However partial or narrowly focused the insights and information contained in the various works we have reviewed, each study has, in its way, made a significant contribution in reinforcing the legitimacy of the study of sexuality and in dispelling the Freudian myths that have obscured the understanding of female sexual behavior, thereby constraining its expression. Although we have been particularly critical of the models advanced by the behaviorists, their empirical work most dranatically breaks with traditional Freudian thought, making possible the development of more comprehensive, sociological alternatives to Freudian modes of thinking, inquiry, and explanation. It may be true, as Freud contends, that the sexual sits at the center of personality. But it sits there, nonetheless, as "the changer and the changed," both shaping and being shaped in its dynamic relationship to the self and the society.

Department of Sociology
Smith College (Miller)

Office of the Dean of the College
Smith College (Fowlkes)

46. P. Blumstein and P. Schwartz, "Bisexuality in Women," *Archives of Sexual Behaviors* (April 1976), pp 171–81.
47. E. Barnhart, "Friends and Lovers in a Lesbian Counterculture Community," in *Old Family, New Family,* ed. N. Glazer-Malbin (New York: Van Nostrand Reinhold Co., 1975).

Body, Bias, and Behavior:
A Comparative Analysis of Reasoning
in Two Areas of Biological Science

Helen Longino and Ruth Doell

Introduction

Our intention in this essay is to bring to light the variety in the ways masculine bias can express itself in the content and processes of scientific research. The discussion focuses on the two areas of evolutionary studies and endocrinological research into behavioral sex differences. Although both have attempted to construe the relation between sex and gender, the forms of these disciplines differ from one another in significant respects. Examining them together should lead to a broader, more subtle understanding of how allegedly extrascientific considerations shape scientific inquiry.

While feminists have succeeded in alerting us to the existence of sexually prejudicial aspects of contemporary research that have implications for our understanding of sex differences, their critiques are dulled by a lack of adequate methodological analysis.[1] In her review of several collections of essays on sociobiology and hereditarianism, Donna Haraway remarks on the inconsistency of adopting a Kuhnian analysis of observation as theory- or paradigm-determined on the one hand, and asserting the incontrovertible existence of any fact on the other.[2] In the

We thank the referees who read the original manuscript for their comments and criticisms. This essay was prepared with the assistance of National Science Foundation grant no. OSS 8018095. The views expressed are those of the authors and do not necessarily reflect the views of the National Science Foundation.

1. See, e.g., Ruth Hubbard, Mary Henifin, and Barbara Fried, eds. *Women Look at Biology Looking at Women* (Cambridge, Mass.: Schenkman Publishing Co., 1979); Ethel Tobach and Betty Rosoff, eds., *Genes and Gender I* (New York: Gordian Press, 1977); Ruth Hubbard and Marian Lowe, eds., *Genes and Gender II* (New York: Gordian Press, 1979); M. Kay Martin and Barbara Voorhies, *The Female of the Species* (New York: Columbia University Press, 1975); Evelyn Reed, *Sexism and Science* (New York: Pathfinder Press, 1978); and the special issue entitled "Women, Science, and Society" of *Signs: Journal of Women in Culture and Society,* vol. 4, no. 1 (Autumn 1978).

2. Donna Haraway, "In the Beginning Was the Word: The Genesis of Biological Theory," *Signs* 6, no. 3 (Spring 1981): 469–82, esp. 478.

This essay originally appeared in *Signs,* vol. 9, no. 2, Winter 1983.

introductory and concluding sections of the feminist anthologies to which she refers,[3] we find the authors explaining sexist science both as bad science (it asks "scientifically meaningless" questions and confuses correlation with causation) and as science as usual ("a product of the human imagination created from theory-laden facts," whose "every theory is a self-fulfilling prophecy."[4] Now, if sexist science is bad science and reaches the conclusions it does because it uses poor methodology, this implies there is a good or better methodology that will steer us away from biased conclusions. On the other hand, if sexist science is science as usual, then the best methodology in the world will not prevent us from attaining those conclusions unless we change paradigms. Only by developing a more comprehensive understanding of the operation of male bias in science, as distinct from its existence, can we move beyond these two perspectives in our search for remedies.

In this question of bias it is important to distinguish between psychological, cultural, and logical issues. The tendency of individuals to portray themselves in a favorable light by preferring certain explanations or theories over others is a psychological characteristic of those individuals, as is the tendency to resist alternative theories. What counts as a "favorable light," and the reasons those individuals link certain explanations or theories with their self-esteem, are culturally conditioned. Thus, in sixteenth-century Europe individual resistance to the heliocentric account of the relations between the earth and other heavenly bodies had much to do with the medieval idea that human uniqueness was signified by the earth's location at the center of a universe created by God. Natural philosophers complained that the new astronomy yielded absurd results in the old physics, but someone like Galileo, who (whether influenced by Pythagorean sun worship or rebellious for other reasons) did not subscribe to the dominant views of his culture, could step outside the terms of the prevailing geocentric account and simply develop a new physics. In the nineteenth century, resistance to Darwinian evolutionary theory was again a response to seeing human uniqueness threatened, this time by kinship with the non-rational and unenlightenable apes.

So, today, individual attachment to theories proclaiming the centrality of male development to the development of the human species, or to theories attributing the allegedly differential distribution of social-behavioral traits between men and women to the differential distribution of certain hormones between the sexes, has much to do with the androcentric and patriarchal beliefs of our culture. Since at least the time

3. Hubbard, Henifin, and Fried, eds.; Hubbard and Lowe, eds. Our criticism of these authors' treatment of methodological issues is not meant to deny their contribution to the perception of sexually prejudicial aspects of science.

4. Hubbard and Lowe, eds., pp. 23–24, 144, 149; Ruth Hubbard, "Have Only Men Evolved?" in Hubbard, Henifin, and Fried, eds., p. 9.

of Aristotle in the West, men have thought it important to justify their social dominance by appealing to ostensibly natural differences between males and females.[5] As feminists we can identify the influence of patriarchal culture when we look at these theories and their proponents. Understanding *that* these theories incorporate a male bias and understanding *how* this can be so are, however, two distinct enterprises.

One way to approach the second problem is by examining the logical aspects of theory construction, particularly the determination of what counts as evidence and how such evidence is related to the hypotheses it is called on to support. These methodological categories can then be used as a probe to analyze the structure of inquiry in evolutionary and endocrinological studies. While we concentrate on the biology of evolution and behavior, the analytic procedure we employ can be applied to other fields of inquiry as well.

A comprehensive understanding of bias would require, in addition, historical and sociological analysis of the institutions in which science is produced. Our analysis exposes the points of vulnerability in the logical structure of sciences to so-called external influences, such as culture, individual psychology, and institutional pressures. We shall argue that masculine bias expresses itself differently at different points in the chain of scientific reasoning (e.g., in description of data and in inference from data), and that such differences require correspondingly different responses from feminists. Feminists do not have to choose between correcting bad science or rejecting the entire scientific enterprise. The structure of scientific knowledge and the operation of bias are much more complex than either of these responses suggests.

Facts, Evidence, and Hypotheses

In our everyday world, we are surrounded by facts: singular facts (this ruby is red); general facts (all rubies are red); simple facts (the stove is hot); and complex facts (the hot stove burned my hand). Description of these facts is limited by the capacities of our sense organs and nervous systems as well as the contours of the language we use to express our perceptions. There is always much more going on around us than enters our awareness, not only because some of it occurs outside our sensory range or behind our backs, but also because in giving coherence to our experience we necessarily select certain facts and ignore others. The choice of facts to be explained by scientific means is a function of the reality constructed by this process of selection. What counts as fact—as reality—will thus vary according to culture, institutional perspective, and so on, making this process of selection one point of vulnerability to external influences.

5. See Caroline Whitbeck, "Theories of Sex Difference," in *Women and Philosophy,* ed. Carol Gould and Marx Wartofsky (New York: G. P. Putnam's Sons, 1976), pp. 54–80.

Even the facts that enter our awareness are susceptible to a variety of descriptions. Accounts may be more or less concrete ("a rough-textured, grey, heavy cube" vs. "a building stone"); more or less value-laden ("she picked up the wallet" vs. "she stole the wallet"); and focused on different aspects ("grey" vs. "hard" vs. "cubical"). A good portion of the history of epistemology and philosophy of science consists in the search for some privileged level of description. We are persuaded by arguments that such a search is futile.[6] But the possibility for multiple descriptions of a single reality means that, despite the ideals of scientific description, any given presentation of data may use terms that reflect social and cultural biases when other less value-laden or differently valued terms might do as well. This is another point of vulnerability to external factors.[7]

An even smaller proportion of the change, flow, and movement in the world that enters our awareness functions as evidence. The category "facts" and the category "evidence" are not only not coextensive; they have their being in quite different ways. The structure of the facts we actually or potentially know is a function of our perceptual and intellectual structures. Evidence is constituted of facts taken in relation to something else—beliefs, hypotheses, theories. To speak of evidence is not to speak of bare facts or data awaiting an explanation. It is, instead, to confer on those facts an epistemic relevance to a belief, hypothesis, or theory. To say that this fact (*F*) is evidence for this hypothesis (*H*) is to take *F* as a sign of *H*, or, to use logical terminology, to claim that *F*'s being the case is a consequence of *H*'s being true.

Statements describing facts that are taken as evidence for hypotheses can be more or less direct consequences of the statements expressing those hypotheses. For example, the singular sentence, "This swan is white," is a fairly direct consequence of the generalization, "All swans are white." In contrast, a statement describing discontinuities in the emission spectrum of hydrogen can be considered a consequence of a statement attributing different energy levels to the electron orbits of a hydrogen atom only in conjunction with a number of further assumptions that, for instance, assert a link between macroscopic phenomena like emission spectra and microscopic phenomena like atomic structure.[8] We will use the spatial term "distance" to convey the logical notion of being more or

6. John Austin, *Sense and Sensibilia* (London: Oxford University Press, 1962); Peter Achinstein, *Concepts of Science* (Baltimore: Johns Hopkins University Press, 1968), pp. 157–78.

7. For other approaches to these issues, see Paul Feyerabend, *Against Method* (London: Verso, 1975), pp. 55–119; Sandra Harding, "Masculine Experience and the Norms of Social Inquiry," in *Philosophy of Science Association 1980,* ed. Peter Asquith and Ronald Giere (East Lansing, Mich.: Philosophy of Science Association, 1981), 2:305–24.

8. Helen Longino, "Evidence and Hypothesis," *Philosophy of Science* 46, no. 1 (March 1979): 35–56.

less directly consequential. The less a description of fact is a direct consequence of the hypothesis for which it is taken to be evidence, the more distant that hypothesis is from its evidence. This distance that must be bridged between evidence and hypothesis provides yet another point of vulnerability to external influences.

Distinguishing between facts and evidence implies that which facts acquire scientific legitimacy will be a function of the theories under consideration. This in turn is determined by the explanatory needs of the scientific community, which are a function of specific social, institutional, and political goals. The concepts of evidence, and of the distance between a hypothesis and the evidence supporting it, are our primary analytical tools in the methodological examination of bias that follows. This approach facilitates comparisons within and between the areas investigated and helps make the operation of bias visible in scientific reasoning as well as in data collection and preparation.

The Role of Evidence

Both evolutionary and endocrinological studies have as part of their purpose the elucidation of human nature. Evolutionary studies are concerned with the description of human descent: what happened—the temporal sequence of changes constituting the evolution of humans from an ancestral species—and how it happened—the mechanisms of evolution. Endocrinology attempts to articulate general laws that describe how hormones influence or control anatomical development, physiology, behavior, and cognition. In the former case, researchers use the principles of the general synthetic theory of evolution to develop a historical reconstruction that can clarify what is human and what is natural about human nature. In the latter case, no history is sought; rather, universals about the natural, in the form of causal generalizations, are developed on the basis of contemporary observations, often made in experimental settings.

Both areas of inquiry take place within established research programs, which address particular kinds of questions and abide by particular conventions as to how to go about answering those questions. We will discuss a parallel series of issues for both kinds of research: what questions are asked; what kinds of data are available, relevant, and appealed to as evidence for different types of questions; what hypotheses are offered as answers to those questions; what the distance between evidence and hypothesis is in each category; and finally, how these distances are traversed. Systematically assembling and analyzing this material will make it possible to see some of the variety in the ways masculine bias functions in science.

Evolutionary Studies

The questions.—The main questions addressed in the search for human origins are standardly grouped into two categories: anatomical and social (or behavioral) evolution. Some features considered central to human development are captured by neither category, being individual behaviors (and so not anatomical) that likely facilitated, but do not of themselves involve, social behaviors. In addition to behavior, students of evolution are interested in capacities and dispositions manifested by behavior, such as intelligence and sociability. Finally, there is a set of questions concerning relations between the two types of evolution.[9]

Anatomical questions direct themselves to the nature and sequence of structural changes that differentiate humans from other primates. These include changes in the size and structure of the bones, teeth, muscles, and brain. Questions about social evolution concern interactions among these developing creatures, including such issues as the size of social groups, the emergence of social structures, the role of dominance and aggression within these structures, the development of cooperation and communication, the nature of relations between mothers and infants or between adults of the same and different sexes, the character of sexuality, and the roles of gathering and hunting in early hominid and human societies. Upright posture, bipedalism, dietary habits, and tool use are additional major subjects of inquiries into the evolution of the human features of individual behavior. Questions addressing the relations among these developments ask which anatomical developments affected which behavioral developments and vice versa.

Data.—Our general theories about evolution (which include the claim that species currently in existence evolved from earlier species rather than came into existence *de novo*) tell us which data to use as evidence for particular theories of human descent. The data base prescribed for theory regarding anatomical evolution is spare: fossils— primarily bits of ancient bone, teeth, some partial and disassembled skeletons, and a few footprints—constitute the foundation for our reconstructions. There are relatively few fossil remains of the earliest hominids—so few that the discovery of a tooth can throw accepted truths into dispute again. Twentieth-century developments in the physical and chemical sciences have, however, given us additional direction in how to read the fossils we do have by allowing us to date material and thus place the bones in evolutionary sequence.

There is more room for controversy regarding data pertaining to the evolution of human behavior. Individual, or noninteractive, physical behaviors such as mode of locomotion and diet seem to pose the fewest problems. The development of bipedalism can be read from fossil foot-

---

9. Clifford Jolly, ed., *Early Hominids of Africa* (London: Gerald Duckworth & Co., 1978), presents a good review of principles and methodology in human evolution.

prints and skeletal remains. Dietary habits are inferred from such phenomena as the size, shape, wear, and thickness of enamel on fossil teeth. Some claims about tool use are based on the presence of what appear to be functional objects with hominid remains. The more elusive feature of developing intellectual capacity is documented through the study of the size of fossil craniums and markings indicative of brain structure left in their interiors.

The greatest area of contention concerns data relevant to the evolution of social, interactive behavior in its relation to the development of human anatomy. Here the material appealed to includes not only the fossils used in reconstructions of individual capacities and behaviors and the estimated size and quantity of remains at hominid sites but also observations of modern human hunting and gathering societies and of modern ape and monkey societies. Since there is considerable variation among human as well as nonhuman primate groups, the relevance of the observed behavior of any one of these societies to the reconstruction of the behavior of early hominids is constantly in question. Although developments in immunology and biochemistry suggest a very close relationship between humans and chimpanzees, the description of the social behavior of any species is fraught with uncertainty, anthropomorphism, and ethnocentrism. The behavior of contemporary apes, which represent an evolved rather than an original species, is, in any case, a questionable model for the behavior of our hominid ancestors.[10]

Hypotheses.—In recent years, scientific reconstructions of human descent have centered around two focal images: man the hunter and woman the gatherer.[11] Both lines of explanation attempt an integrated story of anatomical and behavioral development intended to answer the question posed by the theory of evolution: how were developments that we deem central to an emerging human species favored by the processes of selection in the particular environments where hominid remains have been found? The two approaches differ in their assessment of the relative contributions of males and females to the evolution of the species.

10. This point is stressed by Martin and Voorhies (n. 1 above), pp. 109–10; and by R. van Gelder, "The Voice of the Missing Link," in Jolly, ed., pp. 431–49.

11. The classic source for the man-the-hunter view is Sherwood Washburn and C. S. Lancaster, "The Evolution of Hunting," in *Man the Hunter,* ed. Richard Lee and Irven DeVore (Chicago: Aldine Publishing Co., 1968), pp. 293–303. See also William Laughlin, "Hunting: An Integrating Biobehavior System and Its Evolutionary Importance," in Lee and DeVore, eds., pp. 304–20. For woman the gatherer, see Nancy Tanner and Adrienne Zilhman, "Women in Evolution. Part I. Innovation and Selection in Human Origins," *Signs* 1, no. 3 (Spring 1976): 585–608; Adrienne Zihlman, "Women in Evolution. Part II. Subsistence and Social Organization among Early Hominids," *Signs* 4, no. 1 (Autumn 1978): 4–20; Nancy Tanner, *On Becoming Human* (Cambridge: Cambridge University Press, 1981); Adrienne Zihlman, "Women as Shapers of the Human Adaptation," in *Woman the Gatherer,* ed. Frances Dahlberg (New Haven, Conn.: Yale University Press, 1981), pp. 75–120.

The androcentric man-the-hunter perspective assigns a major role to the changing behavior of males, while the gynecentric woman-the-gatherer perspective assigns that role to the changing behavior of females.

Both accounts consider the development of tool use by early hominids a critical behavioral change. As an aid to survival it favored the development of the bipedalism and upright posture necessary to wield tools effectively, and hence the anatomical changes that made new postures possible. From the androcentric perspective the development of tool use is also seen as a major factor contributing to changes in dentition that featured a reduction in the size of the (male) canines. With tool use defensive threats and displays of aggression could be accomplished by brandishing and throwing objects rather than baring or using the canines. Once smaller canines were no longer an evolutionary liability, selective pressures for reduced canine size, such as diets requiring more effective molar functioning, were free to operate. The androcentric account attributes the development of tool use itself to male hunting behavior.

By contrast, the gynecentric story explains the development of tool use as a function of female behavior, viewing it as a response to the greater nutritional stress experienced by females during pregnancy, and later in the course of feeding their young through lactation and with foods gathered from the surrounding savannah. Whereas most man-the-hunter theorists focus on stone tools, woman-the-gatherer theorists see tool use developing much earlier and with organic materials such as sticks and reeds. They portray females as innovators who contributed more than males to the development of such allegedly human characteristics as greater intelligence and flexibility. Women are said to have invented the use of tools to defend against predators while gathering and to have fashioned objects to serve in digging, carrying, and food preparation. The gynecentric view explains the change in male dentition by depicting female sexual choice as an effective selection mechanism: males with less prominent canines, more sociable and less prone to aggressive displays or behavior, were more desirable partners for females than their more dentally endowed fellows.

Distance between evidence and hypotheses.—As a model of evolutionary reasoning against which to measure other inferences, let us take Mary Leakey's discovery of the footprints preserved in compacted volcanic ash at Laetoli.[12] She uncovered three distinct sets of prints, two of which display "the raised arch, rounded heel, pronounced ball, and forward pointing big toe necessary for walking erect." The distance between the steps, and the pressure shown to have been exerted along the foot, are signs of a striding gait. Only an upright, fully bipedal creature could

12. Mary D. Leakey, "Footprints in the Ashes of Time," *National Geographic* (April 1979), pp. 446–57, esp. 452; Mary Leakey and Richard Hay, "Pliocene Footprints in the Laetolil Beds at Laetoli, Northern Tanzania," *Nature* 278 (March 1979): 317–32, esp. 317.

have left these prints; tracks of a quadruped, knuckle walker, or incompletely upright bipedal creature would differ in significant ways from the ones at Laetoli. Standard dating techniques assign these remains an age of 3.59–3.77 million years. Because bipedalism is the criterion for hominid status, this find allows anthropologists to assert with certainty that hominids developed as early as 3.59 million years ago. This inference is further enhanced, but not logically strengthened, by the nearby presence of australopithecine fossils dating to the same period. Presumably members of the hominid population whose skeletal remains were found left their footprints.

The distance between a set of footprints and claims about the origin of an anatomical and locomotive characteristic of an entire species seems large, but in this case the gap is closed by highly reliable generalizations. Observational knowledge of anatomy and physiology enables us to say that the physical features of the prints rule out the possibility that they were produced by a nonbipedal creature. Measurements of the impact produced by an upright body with a striding gait on certain types of surfaces provide direct support for giving the prints the status of evidence for bipedalism. Flecks of biotite in the stratum where the footprints were discovered made it possible to determine the prints' age by means of potassium-argon dating tests. Even when only one of the tests available for determining the age of prehistoric remains is applicable, as at Laetoli, the mutual consistency of different dating methods in other cases makes that test as reliable as the theory on which they all depend.

It is useful to consider two sorts of inferences in evolutionary studies that fall short of this close fit between data and hypotheses. The first is similar in content to Leakey's line of reasoning in the Laetoli case. Until recently many anthropologists believed that a creature called *Ramapithecus* might have been ancestral to the hominid *Australopithecus*. The basis for this hypothesis was a reconstruction of *Ramapithecus*'s jaw and dentition from isolated jaw and teeth remains, which were intermediate between clearly apelike and clearly hominid characteristics. Because the fossil evidence was incomplete with respect to *Ramapithecus*'s transitional status, it did not allow any inferences regarding the rest of *Ramapithecus*'s anatomical structure. The problem here was not a matter of finding assumptions to bridge the distance between evidence and hypothesis, but of expanding the data. As it happens, the later discovery of more complete cranial remains suggests that *Ramapithecus* is *not* the transitional species from ape to human.[13]

While the inconclusiveness of inferences regarding *Ramapithecus* can be traced to a lack of available data, inferences about the behavior of

13. David Pilbeam et al., "New Hominoid Primates from the Siwaliks of Pakistan and Their Bearing on Hominoid Evolution," *Nature* 270 (December 1977): 689–95; David Pilbeam, "New Hominoid Skull Material from the Miocene of Pakistan," *Nature* 295 (January 1982): 232–34.

subsequent hominid species raise a different set of methodological issues. Drawing conclusions about the uses and users of early tools, based on the association of distinctively shaped stones with fossil skeletal remains of *Australopithecus* and *Homo erectus,* raises several problems. First of all, there is the inference from structural features of the stones to their artifactual or (minimally) implemental character. Their roughly similar size, suitability for manual grasping, and (in the case of chipped stones) limited number of chip patterns make it highly unlikely that their concentration in these sites is fortuitous.[14] The odds of finding such uniform concentrations are low enough that the appearance of intentional selection or manufacture, in association with the presence of these stones at sites inhabited by a creature capable of making and using them, counts as evidence that they are indeed crude tools. The background assumption here seems to be the commonsense notion that, in the absence of countervailing factors, what seems to be is a good indication of what is.[15]

Generalizing about the uses and users of these tools is another matter. The stones could have been used to kill animals, scrape pelts, section corpses, dig up roots, break open seed pods, or hammer and soften tough roots and leaves to prepare them for consumption. In attempting to identify specific uses (which then serve as the basis for more elaborate accounts of hominid behavior), anthropologists frequently rely on analogies with contemporary populations of hunters and gatherers. But in contrast to the distinguishing features of anatomy shared by all humans, the behavior and social organization of these peoples is so various that, depending on the society one chooses, very different pictures of *Australopithecus* and *Homo erectus* emerge. If female gathering behavior is taken to be the crucial behavioral adaptation, the stones are evidence that women began to develop stone tools in addition to the organic tools already in use for gathering and preparing edible vegetation. If male hunting behavior is taken to be the crucial adaptation, then the stones are evidence of the male invention of tools for use in the hunting and preparation of animals. Unlike the footprints at Laetoli, the stones are not unequivocal signs of any specific tool-using behavior, nor are we in a position of waiting to discover more data as with the *Ramapithecus* remains. It is rather a matter of choosing a male-centered or female-centered framework of interpretation and assigning evidential relevance to data on the basis of that framework's assumptions.

One great contribution of the female-centered framework lies in its

14. Sherwood Washburn, "The Evolution of Man," *Scientific American* 239, no. 3 (September 1978): 194–208, esp. 201.

15. A more specific version of this principle is offered by L. G. Freeman, "A Theoretical Framework for Interpreting Archeological Materials," in Lee and DeVore, eds. (n. 11 above), pp. 262–67, esp. p. 265: "When [patterned occurrences of the elements the prehistorian studies] are derived from undisturbed contexts they indicate that patterned human behavior was responsible for their existence."

demonstration of how dependent the man-the-hunter story is on culturally embedded, sexist assumptions. At this stage the woman-the-gatherer framework offers the more comprehensive and coherent theory, but this may be due to its elaboration after and partially in response to man-the-hunter theories. Although a determined partisan of the latter could, no doubt, improve on the man-the-hunter story, we suspect that a less gender-biased theory will eventually supersede both currently contending accounts. The issue here, however, is whether there is direct evidence for either of the interpretive frameworks within which the data, in this case chipped stones, acquire status as evidential support for hypotheses regarding the dietary and social behavior of early hominids. Not only do we not now have such evidence; we cannot have it. What the study of contemporary hunting and gathering societies should teach us is that, short of stepping into a time machine, any speculation regarding the behavior and social organization of early humans remains just that.[16] This leaves framework choice subject to influences such as the speculator's preconceived and culturally determined ideas of what human beings are. The distance between evidence and hypothesis cannot be closed by anatomical and physiological knowledge, by principles from the theory of evolution, or by commonsensical assumptions. It remains an invitation to further theorizing or, as some would have it, storytelling.

Endocrinological Studies of Sex Differences

The questions.—Hormones regulate a variety of physiological functions. The role of sex hormones, the estrogens and androgens, in the development and expression of sexually differentiated traits and functioning constitutes a small but intensively researched portion of the entirety of hormonal effects. Questions that have been studied regarding the relation of sex hormones to sexual differentiation can be grouped into three general categories corresponding to the three areas in which sexual differences are believed to be manifest: effects on anatomy and physiology, effects on temperament and behavior, and effects on cognition. Within these areas are further distinctions concerning the timing and mechanism of hormonal activity which refer to whether a particular effect is due to fetal exposure that affects the organism's development, or to adolescent and adult exposure, which may have an activating or a permissive effect.[17]

16. Warnings to this effect are scattered throughout Lee and DeVore, eds. See especially the three papers by Freeman; Lewis Binford, "Methodological Considerations of the Archeological Use of Ethnographic Data"; and Sally Binford, "Ethnographic Data and Understanding the Pleistocene."

17. The analysis in the sections that follow is based on material derived from Gordon Bermant and Julian Davidson, *Biological Bases of Sexual Behavior* (New York: Harper &

The effects of androgens and estrogens on anatomical and physiological differentiation have been studied in relation to their role in the development of primary and secondary sex characteristics—reproductive organs, along with such traits as hair, voice, and body size—as well as their role in regulating postpubertal physiological functioning, including sperm production, cyclicity, and acyclicity. As research regarding the brain's relation to behavior and physiology has become more sophisticated, the role of hormones in the development and organization of the brain has also become an object of inquiry. Studies of hormonal effects on behavior have focused on sexual behavior such as copulatory positioning and the frequency and timing of sexual activity, in addition to nonsexual but seemingly gender-linked behavior and behavioral dispositions like fighting, aggression, dominance, submission, nurturance, grooming, and activity level in play. Questions about cognition address the possible influence of hormones in bringing about the well-known if not well-understood differences in verbal and spatial abilities between boys and girls. For reasons of space we shall limit our discussion to research in anatomy and physiology and in behavior.

Data.—Although there is a large amount of observational and experimental data available to serve as evidence for hypotheses regarding the relation of sex hormones to sexually differentiated characteristics, it is not highly consistent, nor is it all of the same quality. Information relevant to questions regarding anatomical and physical differentiation includes, first of all, observations of male and female body types and the correlation of these with higher and lower average levels of androgens and estrogens circulating in the body. Abnormalities in sex-linked anatomical and physiological characteristics have been correlated with deficiencies or excesses in hormonal levels, for example, the effects of castration on hair distribution and voice. In addition to data on humans, there are numerous animal studies determining the physiological effects of deliberate manipulation of hormone levels both perinatally and postnatally.

Animal experiments have also been performed to determine the effects of hormone levels on sexual behavior, such as frequency of mounting, frequency of assuming the female mating posture, and increased or decreased female receptivity. One of the most extensively studied effects of hormonal activity on nonsexual behavior involves the

Row, 1974); Basil Eleftheriou and Richard Sprott, eds., *Hormonal Correlates of Behavior* (New York: Plenum Press, 1975); Eleanor Maccoby and Carol Jacklin, *The Psychology of Sex Differences* (Stanford, Calif.: Stanford University Press, 1974); John Money and Anke Ehrhardt, *Man and Woman, Boy and Girl* (Baltimore: Johns Hopkins University Press, 1972); Kenneth Moyer, ed., *The Physiology of Aggression* (New York: Raven Press, 1976); Susan Baker, "Biological Influences on Sex and Gender," *Signs* 6, no. 1 (Autumn 1980): 80–96; and the review articles on the biological bases of sex differences in *Science* 211 (March 20, 1981): 1265–1324.

relation of testosterone levels to frequency of fighting behavior in a variety of strictly controlled laboratory situations. In addition to the animal studies, there have been a number of attempts to correlate hormonal output with human behavioral differences. Commonly accepted stereotypes of sex-linked behaviors and their presumed correlation with different hormonal levels often provide the starting point and underlying context for more serious scientific explorations, despite the fact that the unrigorous and presumptive character of such stereotypes precludes their acceptance as genuine data. A more reliable source of information is found in controlled observations of the behavior of individuals with hormonal irregularities. Among the groups studied are young women with CAH (congenital adrenocortical hyperplasia, a condition leading to the excess production of androgens during development, also referred to as AGS [adrenogenital syndrome]), young women exposed *in utero* to progestins, and male pseudohermaphrodites (genetic males with a female appearance until puberty, at which time they become virilized).

Hypotheses.—Of the hypotheses articulated by researchers in this field, we shall restrict ourselves to only a few representative samples focused on sex differences in humans in the categories of inquiry we have distinguished.

The influence of sex hormones on the development of anatomically and physiologically sex-differentiated traits is generally acknowledged, and the mechanisms of development of the male and female reproductive systems are fairly well understood. Thus, it is widely accepted that during the third and fourth months of fetal life the bipotential fetus will develop the internal and then the external organs of the male reproductive system if exposed to androgen. Without such exposure, the fetus will develop female reproductive organs. The mechanisms of central nervous system development, while increasingly studied, are not yet as well understood. It has been hypothesized that androgen receptors play a primary role in sexual differentiation of the human brain, an assertion that rests on the assumption of sexually differentiated modes of brain organization.[18]

Hypotheses regarding the influence of sex hormones on behavior trace their impact either to their perinatal organizing effects, or direct activating or permissive effects. In the arena of sexual behavior, for example, several (largely unsuccessful) attempts have been made to attribute homosexuality to endocrine imbalances.[19] The area of nonsexual

---

18. Robert Goy and Bruce McEwen, *Sexual Differentiation of the Brain* (Cambridge, Mass.: MIT Press, 1980), p. 79.

19. Robert Goy and David Goldfoot, "Neuroendocrinology: Animal Models and Problems of Human Sexuality," in *New Directions in Sex Research,* ed. Eli Rubinstein and Richard Green (New York: Plenum Press, 1975), pp. 83–98. For criticism of such views, see Julian Davidson, "Biological Models of Sex: Their Scope and Limitations," in *Human Sexuality,* ed. Herant Katchadourian (Berkeley and Los Angeles: University of California Press, 1979), pp. 134–49.

behavior has also seen a proliferation of hypotheses. Steven Goldberg, along with other anthropologists, has argued that the social dominance of males is a function of hormonally determined behavior.[20] Such theorists credit aggression with the capacity to determine one's position in hierarchical social structures and then attribute aggressive behavior to the level of testosterone circulating in the organism. Even such thorough and nonpatriarchal scholars as Eleanor Maccoby and Carol Jacklin endorse the claims that males on the whole exhibit higher levels of aggressive behavior than females and that aggressive behavior is a function of perinatal and circulating testosterone levels.[21] However, Maccoby and Jacklin are much more tentative about linking aggression with such phenomena as leadership and competitiveness.[22] Regarding other possible effects, Anke Ehrhardt has argued that gender-role or sex-dimorphic behavior in humans, including "physical energy expenditure," "play rehearsal of parenting and adult behavior," and "social aggression," is influenced by perinatal exposure to sex hormones.[23]

Distance between evidence and hypotheses.—As a model of reasoning in endocrinology, we can take the studies of hormonal influence on differentiation of the genitalia in humans. The current view that testosterone secreted by the fetal testis is required for normal male sex organ development and that female differentiation is independent of fetal gonadal hormone secretion is substantiated by observations in humans and by experimental data on a variety of other mammalian species. Among the human observations, most significant are those of persons affected by various hormonal abnormalities. Genetic males who lack intracellular androgen receptors and are thus unable to utilize testosterone exhibit the female pattern of development of external genitalia. Genetic females exposed *in utero* to excess androgen, either as a result of progestin treatment of their mothers during pregnancy or due to their own adrenal abnormality, exhibit partial masculine development, including enlargement of the clitoris and incomplete fusion of the labia. These observations support the hypothesis that no particular hormonal secretion from the fetal gonad is required for female development, whereas exposure of the primordial tissues to testosterone or one of its metabolites at the appropriate time is both necessary and sufficient for masculine development of the sex organs. This inference is further corroborated by experimental data in a variety of mammalian species whose reproductive anatomy and physiology are analogous to those of humans.

20. Stephen Goldberg, *The Inevitability of Patriarchy* (New York: William Morrow & Co., 1973).

21. Maccoby and Jacklin, pp. 243–47.

22. Ibid., pp. 263–65, 274, 368–71.

23. Anke Ehrhardt and Heino Meyer-Bahlburg, "Effects of Prenatal Sex Hormones on Gender-related Behavior," *Science* 211 (March 20, 1981): 1312–18.

For instance, castration of male fetuses *in utero* invariably results in their developing a female appearance.

In contrast to the security of the hypothesis regarding sex organ development are issues regarding the biochemical pathways testosterone follows in producing its physiological effects. Because the exact mechanism of hormonal action at the cellular level is only partially understood, it is not yet certain how testosterone or one of its metabolites acts in the cell nucleus. In this respect, this issue in endocrinology is analogous to questions regarding *Ramapithecus* in evolutionary studies: lack of certainty will be allayed by more information and further analysis.

The relation between data and hypotheses becomes much more complex in attempts to link hormonal levels with behavior. The inferential steps in Ehrhardt's work on young women with CAH provide an interesting illustration of this complexity. Unlike some of the authors exploring this topic, Ehrhardt is directly engaged in aspects of the research that forms the basis of her thinking. In addition, since she is concerned with the relation between prenatal hormone exposure and later behavior, there is no question of hormone levels being an effect of behavior rather than vice versa. From the point of view of hereditarian theories of gender, Ehrhardt's work, if sound, would indicate a mechanism that mediates between the genotype and its behavioral expression. All these factors confer on her work a pivotal significance.[24]

The data Ehrhardt brings to her line of reasoning include both observations of humans and experimentation with rats. The human observations follow girls affected by CAH, using their female siblings as controls. She documents the fact of the girls' prenatal exposure to greater than normal quantities of androgens and evaluates observations of their behavior as children and adolescents. It is important to remember that these girls were born with genitalia that were surgically altered in later life and that they require lifelong cortisone treatment. The majority were said to exhibit "tomboyism," operationally characterized as preference for active outdoor play, preference for male over female playmates, greater interest in a career than in housewifery, as well as less interest in small infants and less play rehearsal of motherhood roles than that exhibited by unaffected females. One problem with these behavioral observations concerns Ehrhardt's method of data collection. Because these observations were obtained from the girls themselves and from parents and teachers who knew of the girls' abnormal physiological condition, it is difficult to know how much the reports are influenced by observers' expectations.

24. For this representation of Ehrhardt's views we rely primarily on Ehrhardt and Meyer-Bahlburg. Ehrhardt's earlier publications on this subject have been critically discussed by Elizabeth Adkins, "Genes, Hormones, Sex and Gender," in *Sociobiology: Beyond Nature/Nurture?* ed. George Barlow and James Silverberg (Washington, D.C.: American Association for the Advancement of Science, 1980), pp. 385–415.

Leaving this problem aside, let us proceed with the reconstruction of Ehrhardt's line of reasoning. She advances the hypothesis that human gender-role behavior, that is, behavior considered appropriate to one gender or the other but not to both, is influenced by prenatal exposure to sex hormones. This hypothesis is a generalization of the specific explanation offered for CAH women, namely, that engaging in a degree of gender-role behavior thought inappropriate to their chromosomal and anatomical sex is a function of their prenatal exposure to excessive amounts of androgen. It is significant that Ehrhardt's treatment of the CAH women does not consider the observational data (available in some quantity) indicating the effect of early environmental factors in shaping alleged gender-role behaviors.

What justifies the attribution of the CAH girls' behavior to physiological rather than environmental factors? To begin to answer this question, Ehrhardt appeals to research on other mammalian species that seems to show the hormonal determination of certain behaviors. The premise that physiological and behavioral phenomena are continuous throughout mammalian species allows her to assign the allegedly sex-inappropriate activities of CAH women the status of evidence for the hormonal determination of behavior. When she cites recent research on rodent brains and behavior to support her interpretation of the human studies, she is assuming that the rodent and human situations are similar enough that demonstration of a causal mechanism in one species is adequate to support the inference from correlation to causation in the other. There are several recognized difficulties with this assumption. Obviously the human brain is much more complex than the rodent brain. Second, experiments with rodents all involve single factor analysis, while human situations, including that of the CAH girls, are always interactive. Finally, some of the rodent experiments are equivocal in their support of the hormonal determination of rodent behavior.

In addition to these problems in extrapolating from the results of nonhuman animal experiments to humans, alternative explanations are not ruled out by the data Ehrhardt presents. More sociologically and culturally oriented studies have depicted the kind of behavior exhibited by the CAH girls as an outcome of social and environmental factors.[25] Such studies supply a framework within which the girls' behavior can be seen as evidence for certain early environmental influences. Equally plausible is the hypothesis that the girls' behavior is a deliberate response to their situation as they perceive it. Because the alleged tomboyism of CAH girls is not unique to them but shared by many young women

25. Margaret Mead, *Sex and Temperament in Three Primitive Societies* (New York: William Morrow & Co., 1935), is still an excellent source for this point of view. See also Beatrice Whiting and Carolyn Pope Edwards, "A Cross-cultural Analysis of Sex Differences in the Behavior of Children Aged Three to Eleven," *Journal of Social Psychology* 91 (1973): 171–88.

without demonstrated hormonal irregularities, the difference between the CAH girls and the control group is as likely a function of environment or self-determination as a direct product of their hormonal states. Support for the hormonal explanation in the CAH case must include arguments ruling out such alternative explanations. To date such arguments have not been provided.[26]

The considerable distance between evidence and hypotheses regarding the hormonal determination of behavioral sex differences contrasts sharply with the close fit between the two in the case of anatomical sexual differentiation. The human reproductive system is not significantly different from that of nonhuman mammals and the mechanism of anatomical differentiation in the latter is clearly known. While the course of development in the human embryo has not been observed directly as it has been in other species, the hypothesis of hormonal determination in humans can be seen as a causal generalization from instances where hormonal and anatomical abnormalities are correlated in humans in accordance with Mill's rules of agreement and difference in causal reasoning.[27] Because human behavioral dispositions cannot be exclusively associated with prenatal hormonal levels and receptors in the same way as anatomical conditions, the argument regarding gender-dimorphic behavior fails Mill's agreement test and falls back on animal modeling to give the data relevance as evidence in order to bridge the gap between data and hypotheses. Animal modeling, we have argued, precludes any generalization from the situation of the CAH women because it fails to support the specific inference of causation in that case. This leaves the choice of a physiological or an environmental explanation for behavior (or some alternative to the nature/nurture dichotomy), like the choice of framework in evolutionary studies, subject to the preconceived ideas and values of the researcher.

Understanding Male Bias

We consider in this section the implications of our analysis for understanding the expression of male bias in the development of theory. While there are obvious interconnections between the types of research

26. Ehrhardt (see Ehrhardt and Meyer-Bahlburg) notes that the only influence acknowledged by the parents was encouragement of "feminine" rather than "masculine" behavior. Self-reporting is not, however, the most reliable source of information in a sensitive area like child rearing. In addition, effective parental influence is rarely overt or conscious.

27. The rule of agreement states that if F is present whenever E is present and absent whenever E is absent, then F is likely to be a cause of E. According to the rule of difference, if F is present when E is absent, then F is unlikely to be a cause of E. Both presuppose the temporal priority of F to E. See John Stuart Mill, *A System of Logic*, 8th ed. (London: Longmans, Green & Co., 1949), pp. 253–59.

we have discussed, there are also some significant discontinuities and differences.

We noted at the outset that the aims of evolutionary and endocrinological studies are quite distinct. Evolutionary studies attempt to reconstruct prehistory by recovering particulars and relating them in order to describe the development of a particular species, *Homo sapiens*. On this basis, generalizations concerning the interrelation of various aspects of human existence become possible, but their production is not an immediate objective. In contrast, the goal of neuroendocrinological research is to discover the hormonal substrates of certain behaviors by developing causal or quasi-causal generalizations relating the two. To the extent that evolutionary studies are believed to reveal certain behaviors or behavioral dispositions as expressions of human nature and neuroendocrinological studies to reveal hormonal determinants of those behaviors, the otherwise quite disparate aims of these fields intersect.

At a certain historical phase in both lines of inquiry, we find researchers attempting to achieve precisely this kind of synthesis. Evolutionary studies undertaken within a certain framework have been held to demonstrate that the sexual division of labor observable in some contemporary human societies has deep roots in the evolution of the species. Some contend that man-the-hunter stories of males going off together to hunt large animals while females stayed home to nurture their young prefigure contemporary Western middle-class social life in which men engage in public and women in domestic affairs.[28] If these broadly described behaviors or behavioral tendencies could be correlated with the more particularized behaviors and behavioral dispositions studied by neuroendocrinology, a picture of biologically determined human universals would emerge. Evolutionary studies would provide the universals—gender and sex roles that have remained fundamentally constant throughout the history of the species—while neuroendocrinology provided the biological determination—the dependence of these particular behaviors or behavioral dispositions on prenatal hormone distribution. We have employed a logical analysis focused on the character and role of evidence in these areas of inquiry to show that neither claim need be accepted. Their conjunction obviously can fare no better.

It is instructive to note not only the ways these inquiries intersect but also their distinguishing features, particularly in their expression of masculine bias.[29] In evolutionary studies assigning key significance to

28: Cf. Edward Wilson, *On Human Nature* (Cambridge, Mass.: Harvard University Press, 1978; New York: Bantam Press, 1979), p. 95.

29. We follow convention in distinguishing two forms of male bias. "Androcentrism" applies to the perception of social life from a male point of view with a consequent failure accurately to perceive or describe the activity of women. "Sexism" is reserved for statements, attitudes, and theories that presuppose, assert, or imply the inferiority of women,

man the hunter, androcentric bias is expressed directly in the framework within which data are interpreted: chipped stones are taken as unequivocal evidence of male hunting only in a framework that sees male behavior as central not only to the evolution of the species but to the survival of any group of its members. In current neuroendocrinological studies, because there is no comparably explicit androcentric framework for the interpretation of data, the choice of a physiological framework is not directly related to androcentric bias. Feminists, however, have identified sexist bias in the endocrinologists' search for physiological rather than environmental explanations. One reason for attributing masculine bias to this preference is the potential, noted above, for linking physiological explanations with the androcentric evolutionary account to produce a picture of a biologically determined human nature. This possibility has raised concern that some will see in a biologically determined human nature which includes behavioral sex differences sufficient justification for maintaining social and legal inequalities between the sexes. On a personal level, many fear that men (individually or en masse) will use such a view to buttress their resistance to change. Yet these political interests are served only if one assumes that allegedly masculine characteristics are preferable or superior to allegedly feminine characteristics, that the allegedly physiological basis of these attributes makes them immutable, and that such differences provide adequate grounds for female subordination.[30] The popularity of these assumptions does not mean they can withstand critical scrutiny. Nevertheless, their prevalence explains why feminists are alarmed by attempts to provide physiological explanations for behavior.

Certainly some proponents of the physiological view are influenced by sexist motivations, either their own or those of the research directors, review committees, journal editors, and referees who create the climate in which research is produced and received. But is the physiological project itself sexist? With respect to the methodological categories of analysis connected with evidence, we can look at both the description of data presented as evidence and the assumptions mediating inferences from data to hypotheses.

Physiological explanations are clearly sexist in their description of assumed gender-dimorphic behavior. Using a term like tomboyism to describe the behavior of CAH girls reflects an initial acceptance of social prescriptions for sex-appropriate behavior.[31] This body of research is

---

the legitimacy of their subordination, or the legitimacy of sex-based prescriptions of social roles and behaviors.

30. A similar treatment of the legal/social concern appears in Helen Lambert, "Biology and Equality: A Perspective on Sex Differences," *Signs* 4, no. 1 (Autumn 1978): 114–17.

31. Barbara Fried, "Boys Will Be Boys Will Be Boys," in Hubbard, Henifin, and Fried, eds. (n. 1 above), p. 37.

also androcentric and ethnocentric in its assumption that behavioral differences apparent in the investigator's culture represent human universals. However, these are problems of description and presentation; choosing a less value-laden term than "tomboy" might allow for the description of genuine differences, if they exist, that distinguish the behavior of CAH girls from that of their siblings. Cross-cultural study and a more sophisticated vocabulary for the description and classification of behavior might help to avoid the barbarisms of ethnocentrism. Thus, it is at least theoretically possible that the description of data could be revised to minimize the biases of the investigators. We would then have a catalog of human behavior, dispositions, and behavioral differences that might or might not correspond to the socially salient distinctions of sex, race, and ethnicity. Perhaps we would also find physiological correlates for some of these differences. If this is indeed possible, then the masculine bias present in much behavioral description can be considered a function of inadequate analytic and descriptive tools and therefore incidental to the general project of developing a physiological account of behavior and behavioral sex differences. Ironically, then, a feminist critique has the potential to improve and refine this area of inquiry.

Sexism does not seem intrinsic to the interpretation of data as evidence for physiological causal hypotheses. In our discussion of the distance between data and hypotheses and of the assumptions required to close that distance, we did, however, note that the assumption of cross-species uniformity and the adequacy of animal modeling is highly questionable in its application to behavior. What explains its persistence, if not the role it plays in perpetuating sexism? Historical and sociological analysis is required for a full answer to this question. We would simply remind readers that animal modeling is an important aspect of physiological psychology, the branch of science that seeks a physiological explanation for as much of the subject matter of psychology (including cognition, motivation, and behavior) as possible. The scientific attention given to animal modeling can only be understood and successfully criticized if its part in the accomplishments and aspirations of established research programs is taken into account.[32]

Conclusion

The distances between data and descriptive language on the one hand, and between data described and hypotheses on the other, leave

32. In this connection see Donna Haraway, "The Biological Enterprise: Sex, Mind and Profit from Human Engineering to Sociobiology," *Radical History Review* 20 (1979): 206–37; Stephen Rose and Hilary Rose, eds., *The Radicalization of Science: Ideology of/in the Natural Sciences* (London: Macmillan, 1976).

room for several types of androcentrism and sexism to operate. The man-the-hunter genre of evolutionary studies reveals androcentrism at work directly determining the explanatory hypotheses for which data can function as evidence. Hormonal studies display androcentrism in their description of data and sexism as a possible (but not necessary) motive behind their preference for a system of interpretation that rests on unreliable assumptions about animal modeling.

What constitutes an appropriate feminist response to masculine bias in science? Clearly this depends on the way bias is expressed in a given scientific context. Feminist anthropologists have developed alternative accounts of human evolution that replace androcentric with gynecentric assumptions while remaining within the methodological constraints of their disciplines. This strategy may not provide the final word in evolutionary theorizing, but it does reveal the epistemologically arbitrary nature of those androcentric assumptions and point the way to less restrictive understandings of human possibilities. As Donna Haraway has remarked regarding their work: "The open future rests on a new past."[33] Thus, one response is to adopt assumptions that are deliberately gynecentric or unbiased with respect to gender and see what happens.

In the case of the androcentric description of data, discerning masculine bias is only a first step. Questions remain regarding the phenomena shorn of tendentious description. Does androcentric language create or simply misdescribe its object? Some feminist critics have suggested that the entire category "sex differences" is a fabrication supported by sexism and by analytic tendencies in science that emphasize distinctions over similarities.[34] More modestly, it can be argued that the concept "tomboy" identifies but mystifies a slight difference in behavior among young women. An alternative perspective might invent a name for young women who are not tomboys and seek the determinants of their peculiar behavior. Scrutinizing the language used in the description of data can lead either to its disappearance as an object of inquiry or to the reformulation of the questions we ask about it.

When the issue concerns unreliable but not explicitly androcentric or sexist assumptions that are nevertheless suspected of being sexist in motivation, it is important not only to expose their unreliability but also to search for additional determinants. Such determinants may be embedded in the research programs that grant these assumptions legitimacy, or they may be motivated by discriminatory intent other than sexism. Hereditarianism and various forms of biological determinism have been at the service of race and class supremacy as well as male domination. Because particular assumptions motivated by sexism are

33. Donna Haraway, "Animal Sociology and a Natural Economy of the Body Politic. Part II. The Past Is the Contested Zone: Human Nature and Theories of Production and Reproduction in Primate Behavior Studies," *Signs* 4, no. 1 (Autumn 1978): 37–60, esp. 59.

34. Hubbard and Lowe, eds. (n. 1 above), p. 27.

likely to be reinforced by additional types of bias in other contexts, they will not be dislodged by exposing their relation to sexism alone. Assumptions embedded in institutionalized research programs offer a different challenge. Sometimes the critic will be able to show that their use in a given context is inappropriate. Other times she may have to be willing to take on the research project of an entire discipline.

As our methodological critique has shown, the variety of ways masculine bias expresses itself in science calls for—and permits—a variety of tactical responses. It is not necessary for us to turn our backs on science as a whole or to condemn it as an enterprise. In a number of ways, the logical structure of science itself provides opportunities for the expression of the creative and self-conscious sensibility that has characterized recent feminist attempts to transform the sciences.

Department of Philosophy
Mills College (Longino)

Department of Biology
San Francisco State University (Doell)

The Variability Hypothesis: The History of a Biological Model of Sex Differences in Intelligence

Stephanie A. Shields

From the time of the Greek philosophers to contemporary work in sociobiology, a prevailing theme in the scientific study of sex differences has been the assumption that biological structural properties have ineluctable behavioral consequences. The belief that "biology is destiny" was not invented by the Freudian theorists, nor is it exclusive to that model of human behavior. The precise biological mechanisms proposed to produce behavioral differences vary as a function of the cultural and scientific zeitgeist. Darwin's explanation of secondary sex characteristics, for example, is not equivalent to sociobiology's explanation of the same phenomena. Though the scientific form of the explanation may change radically from one generation to the next, there is traditionally a remarkable correspondence between prevailing scientific notions regarding gender differences and sociocultural definitions of gender-appropriate behavior. Furthermore, beliefs about specific behavioral sex differences presumed to emerge from biological sex differences show great stability over time. Emotionality, nurturance, and dependence—qualities ascribed to femaleness by psychodynamic theorists as well as sociobiologists—are qualities that were assigned to female nature in the earliest scientific studies of the psychology of gender. Until this century, male intellectual superiority was regarded by scientists as an established

---

EDITORS' NOTE: Ignorance of classical literature proved in time to be a remediable and thus ineffective barrier to women's higher education. The variability hypothesis, which came to prominence just as women began to enter universities in significant numbers, provided another hurdle: it suggested that men's natural variability in mental powers made them more likely candidates than women for scholarly work.

This essay originally appeared in *Signs*, vol. 7, no. 4, Summer 1982.

fact. Investigatory efforts were directed only at explaining the mechanisms accounting for gender differences in ability, not at establishing whether ability differences actually did exist.

Around the turn of the century, many scientists began to assert a set of beliefs, known eventually as the "variability hypothesis," according to which males are more likely than females to vary from the norm in both physical and mental traits.[1] For some scientists, greater male variability meant that males are more likely than females to exhibit "abnormal" characteristics; for others, it meant that males are spread out over a broader range of variation or are more likely to deviate widely from the mean.[2] It was assumed that males' greater variability, however defined, acts symmetrically: a higher incidence of negatively valued abnormalities (such as physical or mental defects) among males is compensated for by a higher incidence of positively valued ones (such as genius), and a greater concentration of males at the low extreme of a range (e.g., of intelligence test scores or measurements of cranial capacity) is compensated for by a similarly greater concentration of males at the high extremes. Because variation was identified as the driving force of evolution, greater male variability was assumed to produce greater male contribution to evolutionary progress. Any gender differences in variability that were discovered were assumed to be inherent; their possible environmental sources were largely ignored.[3]

In the decades to follow, the variability hypothesis was somewhat revised. The sweeping assertion that males are in general more variable than females gave way to the more qualified assertion that males are more variable on measures of mental ability. Variability became more precisely defined via quantitative expressions of the dispersion of scores in a distribution, though greater variability was still frequently equated with a broader range and greater concentration at both extremes of the range. Statements about the biological sources and social implications of gender differences in variability became more tentative. With these modifications, the hypothesis continued to be advanced and tested, and it is still discussed in the literature today as an unresolved issue.

The variability hypothesis and more recent biologically based theories of sex differences[4] rest on similar assumptions. They share not only an intellectual heritage but also an emphasis on the equivalence

1. Havelock Ellis, *Man and Woman: A Study of Human Secondary Sexual Characters* (London: Walter Scott, 1894) (hereafter cited as *Man and Woman*).

2. Examples of the variety of definitions of variation can be found in Karl Pearson, *The Chances of Death and Other Studies in Evolution* (London: Arnold, 1897), vol. 1; see also Ellis, *Man and Woman*.

3. For example, see Ellis, *Man and Woman*, and Lewis M. Terman, *The Intelligence of School Children* (Boston: Houghton Mifflin Co., 1919).

4. Helen H. Lambert, "Biology and Equality: A Perspective on Sex Differences," *Signs: Journal of Women in Culture and Society* 4, no. 1 (Autumn 1978): 97–117; Marian Lowe, "Sociobiology and Sex Differences," ibid., pp. 118–25.

between biological and psychological models and a tendency to expound on the social application of scientific propositions. The story of opposition to the variability hypothesis in the first decades of the twentieth century also has contemporary parallels. Just as feminist scholars today have questioned the social interpretations made by sociobiologists, so the variability hypothesis was challenged by an earlier generation of woman scientists. Furthermore, the history of the hypothesis was complicated by personal rivalries and other accidental factors that often had a greater influence on the reception of the hypothesis than did the value of the supportive evidence. That is, people and social issues dictated acceptance or rejection; facts had little to do with either. The intermingling of cultural beliefs about sex differences with the conduct of scientific inquiry that is particularly noticeable in discussions of the variability hypothesis makes a study of it important to our understanding of the sociology of science. This article describes the origin and development of the variability hypothesis as it was applied to the study of social and psychological sex differences, with special attention to the changes in its form over time, the social and scientific factors that fostered its acceptance, and possible parallels between the variability hypothesis and contemporary theories of sex differences.

Origins of the Variability Hypothesis: The Influence of Evolutionary Theory

Even the very earliest formulations of evolutionary theory stressed the importance of variation. Interest in the relationship between gender and the tendency to vary, however, was minimal until Charles Darwin proposed that greater male variability could help explain the occurrence of flamboyant secondary sex characteristics in the males of some species. Such characteristics, he maintained, originated in the males' "stronger passions," were strengthened by use, and then were more likely to be transmitted to male than to female offspring. The tendency toward greater variation was not responsible for secondary sex characteristics but made their development more likely.[5] Darwin first tentatively suggested this idea in 1868 in *Variation in Animals and Plants under Domestication.*[6] He was less hesitant to suggest it in the first edition of *The Descent of Man* three years later,[7] and by the time the second edition was

5. Charles Darwin, *The Descent of Man and Selection in Relation to Sex* (London: John Murray, 1871), pp. 272, 275; and ibid., 2d ed. (New York: D. Appleton, 1897), pp. 221, 223 (hereafter cited as *Descent*). Because Darwin changed his ideas about sexual selection between the first and second editions, I have included citations from both editions to show how his basic ideas about sex differences were retained.

6. Charles Darwin, *The Variation of Animals and Plants under Domestication*, 2 vols. (New York: D. Appleton, 1900), 1:36–59.

7. Darwin, *Descent*, 1st ed., pp. 275, 321.

published, he stated it as a matter of fact: "The cause of the greater general variability in the male sex, than in the female, is unknown."[8]

Darwin's writing did not reflect a concern with the social implications of sex differences in variability. He was traditional in his views about human sex differences in abilities, sharing with many other scientists the belief that males are physically more highly evolved than females—"It is the male which has been chiefly modified, since the races diverged from their common and primeval source"—and intellectually superior to females—"Man is more courageous, pugnacious, and energetic than woman, and has a more inventive genius."[9] But he explained these sex differences without reference to variability: He believed that male superiority had been originally produced by both sexual selection and natural selection and was maintained, in general, by a tendency of characteristics acquired in adulthood to be transmitted only to offspring of the same sex. That is, Darwin proposed that the qualities which adult men acquire in their more rigorous "struggle for life" (e.g., perseverance, courage, powers of reason) are more likely to be passed on to their male than their female offspring.[10] He has also noted that it was very fortunate for females that most inherited characteristics are transmitted equally to offspring of both sexes, as "otherwise it is probable that man would have become as superior in mental endowment to woman, as the peacock is in ornamental plumage to the peahen."[11]

Although Darwin himself devoted little attention to sex differences in variability, other scientists took up his suggestion as a biological mandate and developed theories about its social significance. Theorists who, unlike Darwin, described evolution as progressive, directed, and saltatory were especially likely to magnify the effects of variation on the course of human evolution.[12] In their view, variation produced racial progress and thus had a positive value. Greatness, whether of an individual or a society, was achieved through deviation from the norm.

For these reasons, the question of sex differences in variability took on political significance and stimulated much research on the extent of differences in variation. By the 1890s several studies had been conducted to demonstrate that variability was indeed more characteristic of males. Biological studies focused on various physical anomalies, such as

8. Ibid., 2d ed., p. 224.

9. Ibid., 1st ed., pp. 321, 316, 326–30, and 2d ed., pp. 560, 557.

10. Ibid., 1st ed., pp. 327–29, and 2d ed., pp. 565–66.

11. Ibid., 1st ed., p. 329, and 2d ed., p. 565. The remedy for female inferiority lay in education at the appropriate time of life: "In order that woman should reach the same standard as man, she ought, when nearly an adult, to be trained to energy and perseverance, and to have her reason and imagination trained to the highest point; and then she would probably transmit these qualities chiefly to her adult daughters" (ibid., 1st ed., p. 329, and 2d ed., p. 565).

12. Conflicting nineteenth-century views on the nature of evolution are discussed by David L. Hull, "Sociobiology: Another New Synthesis," in *Sociobiology: Beyond Nature/Nurture?* ed. George W. Barlow and James Silverberg (Boulder, Colo.: Westview Press, 1980), pp. 77–96.

incidence of polydactyly, or the extremes in the range of normal variation, such as the incidence of extremely large or small cranial capacity. The biological evidence overwhelmingly favored males as the more variable sex.[13] A few mental scientists of the 1890s also considered the problem, reasoning that if women varied less than men physically, their mental traits should similarly show a narrower distribution.[14] One of these researchers, G. Stanley Hall, who has been credited with founding the study of psychological development, was a strong social Darwinist. In one study, he tested children's familiarity with a wide variety of objects and ideas at entry into grammar school. He concluded that "the easy and widely diffused concepts are commonest among girls, the harder and more special or exceptional ones are commonest among boys."[15] Girls' minds excel at the common and pedestrian, whereas boys' ideas are wider ranging. Because girls are more like one another than are boys, they are, as a group, closer to the "ordinary."

Hall's great leaps of logic were not unique to him. Cautiousness in extrapolating from the data was not a characteristic of nineteenth-century science. Liberal generalizations regarding sex differences in ability and temperament were made by scientists of all theoretical persuasions. While such differences had been accepted for centuries, the framework for interpreting them was new. Sex differences that earlier scientists and philosophers had described as part of a divinely determined order of things were now explained scientifically as the agents of evolutionary purpose. Reification of the concept of variation did nothing to change scientific ideas regarding the content of sex differences, but it had a profound effect on ideas regarding the function of those differences.

Two theories developed in the 1890s were significant in the history of the variability hypothesis, even though neither used variability as a major construct. The first of these was put forward by Patrick Geddes and J. A. Thomson, who based their metabolic theory of sex differences on the belief that males and females exhibit opposing types of chemical processes in every cell of the body.[16] They reasoned that this difference arises from the nature of reproductive cells: ova are large and inert; spermatozoa are small and active. The female is passive, submissive, and biologically economical; the male, active, enterprising, and biologically profligate. These differences in biological economy, the male being "catabolic" and the female "anabolic," produce parallel differences in

13. A large number of references are cited in Darwin, *Descent*, 2d ed., pp. 221–27, and Ellis, *Man and Woman*, pp. 358–72.
14. See G. S. Hall, "The Contents of Children's Minds on Entering School," *Pedagogical Seminary* 1 (June 1891): 139–73; Joseph Jastrow, "A Study in Mental Statistics," *New Review* (London) 5 (December 1891): 559–68, and "Community and Association of Ideas: A Statistical Study," *Psychological Review* 1, no. 2 (1894): 152–58.
15. Hall, p. 143.
16. Patrick Geddes and J. Arthur Thomson, *The Evolution of Sex* (New York: Scribner & Welford, 1890), esp. chap. 19.

intellectual and temperamental traits. Man is progressive and imaginative; woman is conservative and receptive. Geddes and Thómson's theory also proposed sex differences in variability: "That men should have greater cerebral variability and therefore more originality, while women have greater stability and therefore more 'common sense,' are facts both consistent with the general theory of sex and verifiable in common experience."[17]

A second theory which dealt with variability tangentially was that of the eminent criminologist Cesare Lombroso. He described the "criminal type" as a distinct anomalous physical variation from normality, believing that where one finds physical anomalies, mental anomalies are also present. The malformations of brain, skull, and face indicative of criminality were less frequent among women primarily because women represented the "more primitive type of a species."[18] A similar idea had been expressed earlier by Darwin, Herbert Spencer, and several of their contemporaries. Spencer, for example, believed that the stages in the individual's development recapitulated those which had occurred in evolution (i.e., ontogeny recapitulates phylogeny). He maintained that the female's reproductive physiology arrests her development at an earlier stage than the male's. As a consequence, woman physically and intellectually resembles a man whose individual evolution is incomplete.[19] Lombroso, unlike Spencer or Darwin, linked women's less evolved nature to their lesser variability. The female's characteristics were "better organized and fixed in the woman through the action of time and long heredity,"[20] thereby being less subject to variation. The common bond between the modern woman and her primitive ancestor was moral weakness, the "natural form of retrogression in women being prostitution and not crime. The primitive woman was impure rather than criminal."[21]

Havelock Ellis and the Variability Hypothesis

Most of the literature on variability in the 1890s, like studies that considered only physical variation, either discussed variability without

17. Ibid., p. 271. By 1914, they had changed their views on variability. They still advocated their metabolic theory of sex differences but considered the idea of greater male variability a "dangerous" generalization (ibid. [1914; reprint ed., London: Williams & Norgate, 1927], p. 209). They also revised their thinking on the mental differences between the sexes. Their own previous assertion that "men have greater cerebral variability" they characterized as "guesses at truth without scientific precision" whose "day is over" (ibid.).

18. Cesare Lombroso and G. Ferrero, *The Female Offender* (New York: D. Appleton, 1899), p. 109.

19. See Herbert Spencer, *The Study of Sociology* (New York: D. Appleton, 1891).

20. Lombroso and Ferrero, p. 109.

21. Ibid., p. 152.

reference to social significance or expounded on the social significance of variability but in reference to some other more primary idea.[22] Havelock Ellis, the influential sexologist, was the first to gather the scientific literature on sex differences in variability in order to study the issue directly and in depth. In his first major publication on sex, *Man and Woman,* Ellis devoted an entire chapter to "The Variational Tendency of Man." He began with the emphatic statement, "Both the physical and the mental characters of men show wider limits of variation than do the physical and mental characters of women."[23] Like Darwin, Ellis defined the tendency to vary in terms of deviation from the norm. Abnormalities were the focus of study, and counts of their frequency within a population constituted the collection of evidence in favor of the hypothesis. After reviewing data that indicated a greater incidence of physical abnormalities in males, he proposed that the same type of sex difference could be observed in mental traits but to an even more marked degree. There were more males than females in homes for the mentally deficient, and "genius" was "in nearly every department . . . undeniably, of more frequent occurrence among men than among women."[24] Since both genius and idiocy should be regarded as "an organic congenital abnormality," both should be more common among men "by virtue of the same general [biological] tendency."[25] Ellis described the supposed biological mechanisms responsible for sex differences in variational tendency no further, instead emphasizing the adaptiveness of such differences. The variability hypothesis served as an ideal vehicle for his conception of male-female complementarity.

According to Ellis, the social consequences of sex differences in variability are very beneficial, each sex finding in the other what it lacks itself. "The progressive and divergent energies of men call out and satisfy the twin instincts of women to accept and follow a leader, and to expend tenderness on a reckless and erring child, instincts often intermingled in delicious confusion. And in women men find beings who have not wandered so far as they have from the typical life of earth's creatures; women are for men the human embodiments of the restful responsiveness of Nature."[26] Ellis's views of the complementarity of the sexes reflected a common scientific theme of that period. Darwin, Galton, and Spencer had discussed sex differences only in terms of female inferiority. Galton, for example, believed that women's abilities were in

22. For appraisals of the literature a decade later, see discussions in Cyril Burt and Robert C. Moore, "The Mental Differences between the Sexes: IV," *Journal of Experimental Pedagogy* 1, no. 5 (1912): 355–88; and Helen Thompson Woolley, "A Review of the Recent Literature on the Psychology of Sex," *Psychological Bulletin* 7, no. 10 (1910): 335–42.

23. Ellis, *Man and Woman,* p. 358.

24. Ibid., p. 366.

25. Ibid.

26. Ibid., p. 371.

all respects inferior to men's and that characteristically female traits were defects with no adaptive purpose.[27] By the end of the century, however, the trend was toward describing both male and female traits as adaptive in that they complement each other. Complementarity models, of course, continued to include beliefs about female inferiority, but in disguised form.

Ellis's book received little critical attention when it first appeared,[28] but over the years eight editions were published. An altercation between Ellis and Karl Pearson regarding the variability hypothesis was at least partially responsible for bringing the book and the hypothesis to the attention of the greater scientific community. In an essay critique of Ellis's assertions, Pearson faulted the assumptions and evidence on which Ellis had based his hypothesis.[29] Pearson's arguments centered on Ellis's abuse of statistical methods, but the vigor of his criticism probably had its basis in a personal enmity which had developed between the two men several years before. Pearson was contemptuous of Ellis's fascination with the sexual philosophy of James Hinton. Very likely, he was also distressed by Ellis's influence over the South African writer Olive Schreiner.[30]

In his critique of Ellis, Pearson argued that an inherent sex difference in variational tendency could only exist if inheritance of traits was sex specific, a possibility he was unwilling to admit. He further interpreted evolutionary theory as suggesting that "the more intense the struggle the less is the variability, the more nearly are individuals forced to approach the type fittest to their surroundings, if they are to survive."[31] He reasoned that in primitive societies one would expect little difference in variation between the sexes because "the struggle for existence is more nearly identical to both," but that "man in civilised communities has a harder battle for life than women" and so should be *less* variable.[32] His interpretation of evolutionary theory thus led to a prediction that was the exact opposite of Ellis's.

Pearson stated that Ellis had not only reasoned incorrectly but had based his assertion on clearly erroneous data. Pearson felt that for valid comparisons of variability, data must be limited to observations on the range of normal variation of characteristics common to both sexes. He argued that the study of sex differences in variation, therefore, could not rightfully include comparisons of traits known to be secondary sex

27. Francis Galton, *Inquiries into the Human Faculty and Its Development* (London: J. M. Dent, 1907), pp. 21, 39.

28. Phyllis Grosskurth, *Havelock Ellis: A Biography* (New York: Alfred A. Knopf, Inc., 1980), p. 170; see also Vincent Brome, *Havelock Ellis: A Philosopher of Sex* (London: Routledge & Kegan Paul, 1979), p. 86.

29. Pearson, 1:256–377, esp. pp. 256–62, 372–77.

30. Grosskurth, pp. 103–4; Brome, pp. 58 ff.; and Arthur Calder-Marshall, *The Sage of Sex* (New York: G. P. Putnam's Sons, 1959), pp. 97–98.

31. Pearson, 1:258.

32. Ibid., p. 259.

characteristics, like color blindness, or believed probable as secondary sex characteristics, like "greater variety of intellectual productivity in man." Nor should the sexes be compared on the incidence of abnormalities.[33] Ellis, of course, had committed both errors. Pearson thus dismissed the entire corpus of data that Ellis had amassed: "The whole trend of investigations concerning the relative variability of men and women up to the present seems to be erroneous."[34]

Pearson then presented his own data to show that it was the female who was more variable than the male. Confining his measurements to the range of physical variation in normal individuals, he assembled anthropometric data from various sources, from Neolithic skeletons to modern French peasants. He based his contrasts on recently developed statistical techniques and found a slight tendency toward greater female variability. Ellis's variability hypothesis was therefore a "quite unproven principle."[35]

According to one of his biographers, Phyllis Grosskurth, Ellis immediately sent Pearson a long letter of self-defense, placatory and somewhat apologetic in tone. He thanked Pearson for the criticisms which would allow him to present his arguments "more clearly & precisely than before."[36] He did not, however, yield his position regarding greater male variability. Pearson appears not to have responded to the letter. By withdrawing his volume from the Contemporary Science Series of which Ellis was the editor, Pearson added a further insult. Ellis may have felt compelled to respond in print to such a public insult, but he did not publish his rebuttal until five years later, in 1903. In a letter to the *Popular Science Monthly,* Ellis stated that he wished his response "to avoid everything that is *merely* controversial, and to make the article a connected statement, complete in itself, of an argument for which there still seems much to be said."[37]

Although it had taken Ellis some time to frame his reply to Pearson, his anger had not cooled. With righteous indignation Ellis enumerated the defects that made it "sufficiently clear that the inquiry he [Pearson] has carried out, however valuable it may be in other respects, has no decisive bearing on the question he undertook to answer, and can have no very damaging effect on the writers [*sic*] he attacks."[38] Pearson had failed to define "variation," but his greater sin was the inclusion of characteristics which could have been influenced by environmental con-

33. Ibid., pp. 260, 261.
34. Ibid., p. 261.
35. Ibid., p. 377.
36. Havelock Ellis to Karl Pearson, October 25, 1897, Pearson Papers, University College, cited in Grosskurth, p. 171.
37. Havelock Ellis to James McKeen Cattell, September 29, 1901, François Lafitte Collection, Birmingham, England, cited in Grosskurth, p. 172. Grosskurth also notes that after this interchange Ellis always referred to Pearson as "my enemy."
38. Havelock Ellis, "Variation in Man and Woman," *Popular Science Monthly* 62 (1903): 237–53.

ditions and were not pure manifestations of heredity. Ellis expressed admiration for Pearson's statistical methods but vigorously asserted that they were useless for evaluating sex differences in variability. He defended his own study of the incidence of physical and mental abnormalities on the grounds that "it is precisely the anomalies which furnish us with the most reliable evidence."[39] Pearson's attempt at repudiating the variability hypothesis was an embarrassing failure: "If, again, he had refrained altogether from attempting to interpret his own data—a task for which, it is obvious, he was singularly unprepared—and had put them forth simply as a study in natural selection—which is what they really are—his position would, again, have been altogether justifiable. But as the matter stands he has enmeshed himself in a tangle of misapprehensions, confusions, and errors from which it must be very difficult to extricate him."[40]

Pearson's only response to this attack appeared as an arrogant footnote in an article unrelated to variability. Ellis's criticism needed no reply, "for the author does not appear to understand what weight is to be given to scientific evidence as compared with vague generalities."[41]

While Pearson did not pursue the topic further, the variability hypothesis continued to play a central role in Ellis's theory of sex differences. In the eighth revised edition of *Man and Woman* (1934), Ellis was still advocating the incidence of deviation from normality as the proper measure of variability because "the true biological variation is an individual anomaly."[42] There was no reason to examine variation falling within the normal range because to consider "such trifling and ever-shifting variabilities" was to throw a "veil of confusion over the great essential facts."[43] Ellis felt that the "veil of confusion" that had hindered worthwhile investigation of human variation was the identification of biological with mathematical variability, a problem for which he blamed Karl Pearson.

One deficiency in Ellis's arguments that few scientists even thought to challenge was the lack of adequate measures of mental abilities. Although physical characteristics and sensorimotor capacities had been quantified for many years, direct measures of mental abilities through test performance were developed only after the turn of the century. Prior to the development of mental tests, the extremes of variation in intelligence cited by Ellis could be measured only in terms of statistics on the institutionalization of the feebleminded and statistics on social eminence for the brilliant. Genius was considered an innate quality which

39. Ibid., p. 245.
40. Ibid., p. 252.
41. Karl Pearson and Alice Lee, "On the Laws of Inheritance in Man. I. Inheritance of Physical Characters," *Biometrika* 2, no. 4 (1903): 357–462, esp. 372.
42. Havelock Ellis, *Man and Woman: A Study of Secondary and Tertiary Sexual Characteristics*, 8th rev. ed. (London: William Heinemann, 1934), p. 420.
43. Ibid.

would naturally be manifested if it were possessed; thus superior social position was a clear indication of an individual's high mental caliber. There were obvious differences in the frequency with which males and females attained eminence, and the variability hypothesis provided a scientific explanation for this social fact. In 1888, for example, Lombroso explained women's lack of eminence in terms of a tendency to conservatism: "Women have often stood in the way of progressive movements. Like children, they are notoriously misoneistic; they preserve ancient habits and customs and religions."[44] Ellis, in 1904, noted that when women did achieve eminence, it was primarily "on the strength of achievements which would not have allowed a man to play a similarly large part,"[45] a delicate reference to women's opportunities to attain fame through the role of wife or mistress. Women were more or less confined to achieving fame vicariously "on account of the greater rarity of intellectual ability" in them.[46] That the individual's social fate could be dictated as easily by environmental circumstances as by hereditary impulsion was not even considered.[47] The basic facts of the matter were obvious: genius was manifested in social eminence; eminence was almost exclusively a male domain. Therefore, a greater incidence of genius among men than women did not even need to be proved.

Most of the surveys of eminent men which appeared around the turn of the century were modeled after Francis Galton's study of hereditary genius.[48] Lists of eminent persons were compiled and the occupations and families of these individuals examined to detect a pattern in the transmission of genius. The growing popularity of the variability hypothesis is reflected in the explanations given for women's low representation among the eminent. For instance, Ellis's influence is evident in an analysis of historical eminence by James McKeen Cattell, a prominent psychologist interested in the measurement of mental abilities. Cattell compiled a list of the thousand most eminent persons of history on the basis of the number and length of references made to them in historical texts. There were very few women on his list, and he believed that their rarity was best explained by lesser variability: "The distribution of women is represented by a narrower bell-shaped curve."[49]

Only rarely was the suggestion made that nurture as well as nature

44. Cesare Lombroso, *The Man of Genius* (London: Walter Scott, 1891), pp. 138–39.
45. Havelock Ellis, *A Study of British Genius* (London: Hurst & Blackett, 1904), p. 9.
46. Ibid.
47. Joseph Thomas Cunningham was a notable exception. He felt that the "greater size and bodily and mental power of man" could be explained "as directly due to differences in habits" (*Sexual Dimorphism in the Animal Kingdom: A Theory of the Evolution of Secondary Sexual Characters* [London: Adam & Charles Black, 1900], p. 46).
48. Francis Galton, *Hereditary Genius: An Inquiry into Its Laws and Consequences* (1869; 2d ed., New York: Horizon Press, 1952).
49. James McKeen Cattell, "A Statistical Study of Eminent Men," *Popular Science Monthly* 62 (1903): 359–77, esp. 375. Cattell was editor of *Popular Science Monthly* when Ellis published his rebuttal to Pearson in 1903.

might account for the comparative lack of eminent women. Ten years after Cattell published his study of eminent men, Cora Castle, his student, used the same method to identify the most eminent women of history. She largely accepted the belief that lack of talent kept women from achieving as widely or as significantly as men but questioned whether environment might not also be responsible for sex differences in achievement: "Has innate inferiority been the reason for the small number of eminent women, or has civilization never yet allowed them an opportunity to develop their innate powers and possibilities?"[50] The tentativeness of her question reflects the power held by genetic theories. Few scientists questioned whether nurture might be as important as nature in determining the course of development.

Influence of the Testing Movement

With the development of mental tests, the variability hypothesis began to serve not only as an explanation for woman's inferior social position but as a justification for it. Galton's tests had been based on the assumption that measurement of sensory and motor capacities could provide an estimate of intellectual functioning. This approach to mental measurement fell out of favor when it became apparent that there was no correlation between performance on the tests and actual school achievement: strength of grip could not predict the student's grades in mathematics. The work of Alfred Binet in France suggested that questions tapping reasoning skills and general knowledge provided a more direct means of measuring intelligence. The impetus for large-scale test development came during World War I, when psychologists volunteered their services in testing the ability of military recruits. During the war years, testing became a major industry for American psychology. Several million recruits had been tested by the end of the war, and by 1919 some 4 million schoolchildren had been tested.[51]

Edward Thorndike, a student of Cattell's who became an educational psychologist at Columbia Teachers College and a leader of the testing movement in the United States, did much to further acceptance of the variability hypothesis. His contemporaries often cited him as an expert on sex differences in variability.[52] He is discussed in detail here

50. Cora Castle, *A Statistical Study of Eminent Women,* Columbia University Contributions in Philosophy and Psychology, vol. 22, no. 27 (New York: Columbia University, 1913), pp. vii, 1–90.

51. Lee J. Chronbach, "Five Decades of Public Controversy over Mental Testing," *American Psychologist* 30, no. 1 (1975): 1–14.

52. See, e.g., Burt and Moore (n. 22 above); Edward Andrews Lincoln, *Sex Differences in the Growth of American Schoolchildren* (Baltimore: Warwick & York, 1927); and Carl Spearman, *The Abilities of Man* (New York: Macmillan Publishing Co., 1927).

because he was prominent in psychology and education and was the major proponent of the variability hypothesis within American higher education. His early work came at a particularly crucial time, because with the development of more sophisticated mental tests the investigation of sex differences in ability had become more "scientific." When scientists spoke of sex differences, they spoke with authority.

Thorndike based his case for the variability hypothesis on a re-evaluation of studies not directly concerned with the problem and only much later (in the mid-1920s) on his own research. One of the experiments that Thorndike reanalyzed was a 1901 dissertation in which Clark Wissler examined the correlation between mental ability as measured by college grades and by performance on tests like those of Cattell.[53] Wissler had found no striking evidence of sex differences except for women's "inferiority" in head size and strength of grip; he made no mention of variability except in conjunction with sex differences in eyesight. Thorndike, however, in his reappraisal of these experiments, pieced together evidence which appeared to support the existence of greater male intellectual variability and consequent male intellectual superiority.[54]

Like most of his contemporaries, Thorndike believed that heredity was the primary source of individual differences in ability, and that according to one's innate intellectual capacity one could choose or reject, exploit or be exploited by the elements in the environment. The gifted individual should therefore seek out an environment compatible with and nurturant of his or her gifts.[55] According to Thorndike, measurement of general sex differences could only "prove that the sexes are closely alike and that sex can account for only a very small fraction of human mental differences."[56] Nevertheless, an examination of the extremes of the distribution showed that the importance of sex differences in variability was undeniable: "This one fundamental difference in variability is more important than all the differences between the average male and female capacities. . . . a slight excess of male variability would mean that of the hundred most gifted individuals in this country not two would be women, and of the thousand most gifted, not one in twenty."[57] Differences in variability, he believed, indicated that education should

53. Clark Wissler, *The Correlation of Mental and Physical Tests, Psychological Review Monograph*, vol. 3, no. 6, suppl. no. 16 (Princeton, N.J.: Psychological Review Co., 1913).

54. Staff of the Division of Psychology of the Institute of Educational Research, Teachers College, Columbia University, "Sex Differences in Status and Gain in Intelligence Test Scores from Thirteen to Eighteen," and "On the Variability of Boys and Girls from Thirteen to Eighteen," *Pedagogical Seminary* 33, no. 2 (1926): 167–81, 182–84.

55. See Edward Lee Thorndike, *Individuality* (Boston: Houghton Mifflin Co., 1911).

56. Edward Lee Thorndike, *Educational Psychology*, 3d ed. (New York: Teachers College, Columbia University, 1914), 3:185.

57. Edward Lee Thorndike, "Sex in Education," *Bookman* 23 (April 1906): 211–14, esp. 213.

channel women into fields where only a moderate level of ability was necessary: "Not only the probability and the desirability of marriage and the training of children as an essential feature of woman's career, but also the restriction of women to the mediocre grades of ability and achievement should be reckoned with by our educational systems. The education of women for . . . professions . . . where a very few gifted individuals are what society requires, is far less needed than for such professions as nursing, teaching, medicine, or architecture, where the average level is essential."[58]

Thorndike believed that sex differences in variability were accompanied by sex differences in instinct. The male aggressive instincts favored intellectual striving, aggression, and "joy at activity in mental matters"; females, in contrast, were gifted with a maternal instinct compatible with average intellectual ability and manifested in women's "unreasoning tendencies to pet, coddle, and 'do for' others."[59]

Although he believed in instinct, Thorndike was suspicious of tests that purported to measure nonintellectual variables. He believed that they were particularly misleading with regard to sex differences because they applied a single standard to both sexes when "the same degree of emotionality might be called emotional in the case of a man and not emotional in the case of a woman, or vice versa."[60] Thorndike's belief in sex differences in variability was so strong that after evaluating one particularly flawed study of personality traits he asserted the results to be nevertheless worthwhile because "the greater variability found for males . . . is a sign of the trustworthiness of the data. . . . So the amounts of difference are worthy of acceptance until a more adequate study is made."[61] Thorndike had elevated the variability hypothesis to the status of scientific law.

The contradictory pattern in Thorndike's work, by which he first stated the triviality of sex differences, then named them in detail, and finally expounded on their social implications, was repeated by other scientists as well. Hugo Munsterberg, another prominent psychologist, posited the greater intelligence of males despite his stated conviction that it was foolish to assert the mental superiority of one sex over the other. He believed that men were the more intellectually variable sex, women the more emotional sex. The sex difference was most evident in mental abilities. Men showed "the greatest development of intellectual, emotional, and volitional powers in the case of scientific or artistic or political or religious genius and the greatest criminal depravity."[62] The female,

58. Ibid., p. 213.
59. Thorndike, *Educational Psychology,* 3d ed., p. 203.
60. Ibid., p. 199.
61. Ibid., p. 198.
62. Hugo Munsterberg, *Psychology: General and Applied* (New York: D. Appleton, 1914), p. 232.

being nearer the race type, possessed less flamboyant qualities, some of which were assets: "The average female mind is patient, loyal, reliable, economic, skillful, full of sympathy and full of imagination."[63] Her defects, however, were her undoing. "It [the average female mind] is capricious, oversuggestible, often inclined to exaggeration, is disinclined to abstract thought, unfit for mathematical reasoning, impulsive, overemotional."[64] The combination of greater affectability with lesser variability resulted in female mediocrity: "The creative work of women is fair and may represent highly estimable qualities and values, and certainly does not stand below the average of men's, but nowhere reaches the highest mark."[65]

Challenges from Feminist Scholars

Although biologists had long been debating the role of environmental versus genetic factors in producing certain traits, the psychologists and social scientists discussing variability regularly disregarded environmental factors as possible determinants of intellectual performance. In established scientific circles, discussion of the variability hypothesis focused primarily on the appropriate measure of variation, not on the legitimacy of the concept. The challenge to the variability hypothesis, and to the biologically deterministic explanations for gender differences in achievement which accompanied it, was initiated by an upstart element of the scientific community—the new generation of women attending graduate school. Women had not been admitted to graduate school until the 1880s, but by 1910 they accounted for 11 percent of all Ph.D.'s granted.[66] The establishment of institutions such as the University of Chicago, where prohibitions against sex discrimination had been written into the 1892 charter, increased the number of doctoral degrees granted to women. Challenges to the variability hypothesis arose from certain institutions where some of the less orthodox scientific theories of the time were gaining new precedence. At Chicago, for example, John Dewey's functionalism dominated other methods because he had handpicked the faculty in philosophy and psychology. Several women thus trained in the social sciences actively challenged biologically

63. Ibid.

64. Ibid., p. 233.

65. Hugo Munsterberg, "The German Woman," *Atlantic Monthly* 109, no. 4 (1912): 457–67, esp. 464.

66. Thomas Woody, *A History of Women's Education in the United States*, 2 vols. (New York: Science Press, 1929), 2:337. See also Margaret W. Rossiter, "Women Scientists in America before 1920," *American Scientist* 62 (May–June 1974): 312–23, and Rosalind L. Rosenberg, "The Dissent from Darwin, 1890–1930" (Ph.D. diss., Stanford University, 1974).

based theories of sex differences. Although the absolute number of women students was small and the proportion of them who actually contested established views even smaller, the work of these relatively young and unconventional scholars was taken surprisingly seriously by the scientific establishment. It was not that their challenges were accepted; there is no record that any established scientist advocating the variability hypothesis changed his views because of this work. However, the research that these women completed did jar the complacent acceptance of strict determinism. Their work was acknowledged, even though it did not change the course of scientific thinking. Not all the women wrote as feminists, nor did they agree on the specific limitations of determinism. Yet one consistent theme in their work was an advocacy of the need to acknowledge the environment as a major force shaping the development of the individual.

Mary Whiton Calkins, one of the first women psychologists in the United States and quite eminent in the field, did not identify herself as a feminist.[67] But as an instructor at Wellesley College and later as its president, she could not remain apart from debates concerning female intellect. Thus when Joseph Jastrow, a colleague at the University of Wisconsin, after measuring students' associations to simple words in order to describe commonality of associations, reported that "there is less variety among women than men" and that females exhibit "an attention to the immediate surroundings, to the finished product, to the ornamental, the individual, the concrete; while the masculine preference is for the more remote, the constructive, the useful, the general, the abstract,"[68] Calkins quickly challenged his findings: she conducted a study that failed to replicate his results. A series of exchanges between Jastrow and Calkins ensued in one of the psychological journals.[69] In the course of the debate Havelock Ellis himself came to Jastrow's defense.[70] Calkins finally suggested that, although statistical comparisons of men and women can detect differences in their interests, to make distinctions between masculine and feminine intellect per se seemed "futile and impossible, be-

67. Calkins completed all requirements for the Ph.D., including the dissertation, but was denied the degree by Harvard because she was female. She went on to have an extremely distinguished career as a psychologist, including election as the first woman president of the American Psychological Association (see Laurel Furumoto, "Mary Whiton Calkins [1893–1930], Fourteenth President of the American Psychological Association," *Journal of History of the Behavioral Sciences* 15, no. 4 [1979]: 346–56).

68. Jastrow, "Mental Statistics," pp. 563, 564–65.

69. Cordelia Nevers, "Dr. Jastrow on Community of Ideas of Men and Women," *Psychological Review* 2, no. 4 (1895): 363–67; Joseph Jastrow, "Community of Ideas of Men and Women," ibid. 3, no. 1 (1896): 68–71; Mary W. Calkins, "Community of Ideas of Men and Women," ibid., 3, no. 4 (1896): 426–30; and Joseph Jastrow, "Reply to Calkins," ibid., pp. 430–31.

70. Havelock Ellis to James Cattell, cited in Jastrow, "Community of Ideas," p. 70.

cause of our entire inability to eliminate the effects of environment."[71] The sorts of studies advocated by Jastrow seemed to her to tap nothing other than the "cultivated interests" of men and women. For his part, Jastrow maintained that his results were nevertheless credible and that "inasmuch as there is other evidence of greater uniformity amongst women than amongst men," he would still "tend toward the conclusions first suggested."[72] The final chapter in the debate came when a new student at the University of Chicago, Amy Tanner, pointed out the lack of comparability between the Jastrow and Calkins studies, maintaining that "they fail equally because they do not consider the effect of habit."[73] She summarized her assessment of studies of sex differences with a vehemence unusual for the pages of the staid *Psychological Review:* "The real tendencies of women can not be known until they are free to choose, any more than those of a tied-up dog can be."[74]

One of the stars of the psychology graduate program at Chicago, Helen Bradford Thompson (Woolley), took on prevailing biological views of sex differences in an ambitious dissertation project examining sex differences in intellectual capacities, motor skills, sensory abilities, and even emotional processes.[75] Her sample of twenty-five men and twenty-five women was carefully chosen to include subjects who were similar in age and of comparable backgrounds. She found sex differences in the performance of a number of individual tasks, but the differences were slight. Sensory thresholds tended to be lower in women, sensory discriminations better in men. She observed few differences in intellectual abilities, although women exhibited superior memory and made more rapid associations while men were better at

71. Calkins, "Community of Ideas," p. 430. Calkins's arguments had no effect on Jastrow. In 1915 he was still promoting sex differences in variability: "Psychologically the greater variational tendency of men, as likewise the greater conservative tendency of women, radiates to every distinctive aspect of their contrasted natures and expressions" (*Character and Temperament* [New York: D. Appleton, 1915], p. 568). Some years later, Catherine Cox Miles and Lewis M. Terman reviewed research on the problem studied by Jastrow and concluded that where sex differences in the association of ideas were evident, "It has not been proved that these are traceable to primary innate differences" ("Sex Differences in the Association of Ideas," *American Journal of Psychology* 41, no. 2 [1929]: 165–206, esp. 205). "The cumulative evidence indicates that the groups of women who have been compared with the groups of men show a slightly more introverted, subjective, evaluating type of response. This would be expected from any group, without regard to sex, that had been subject to the somewhat more personalized and limited life experience of these women as compared with the men" (ibid., p. 205).

72. Calkins, "Community of Ideas," p. 430.

73. Amy Tanner, "The Community of Ideas of Men and Women," *Psychological Review* 3, no. 5 (1896): 548–50, esp. 550.

74. Ibid.

75. Helen B. Thompson (Woolley), *The Mental Traits of Sex* (Chicago: University of Chicago Press, 1903).

puzzle solving. She found no difference in emotionality. What set Woolley's study apart from other attempts to document sex differences was her awareness that simply selecting a group of males and a group of females for comparison without regard to other group differences like educational level and social background confounded any and all attempts to describe observed differences as the exclusive function of gender. Her discussion of the implications of her findings focused on the inconsistencies and tenuous assumptions on which contemporary biological theories were based. Geddes and Thomson's metabolic theory specifically seemed "to rest on somewhat far-fetched analogies only. . . . Women are said to represent concentration, patience, and stability in emotional life. One might logically conclude that prolonged concentration of attention and unbiased generalization would be their intellectual characteristics. But these are the very characteristics assigned to men."[76] Woolley also criticized Ellis's writings on gender differences in variability. Citing Pearson's objections to the inappropriate use of statistical methods and the equation of anomaly with variability, she argued that even if males were found to be intellectually more variable, one could not infer "that originality and inventiveness are characteristic of the male mind as a whole, in opposition to the female mind, as a whole."[77] Finding biological theories lacking in their ability to explain observed sex differences, Woolley suggested that an environmental interpretation was at least as persuasive. She emphasized the point that a variety of external factors impel and compel the child to assume a behavioral repertoire socially appropriate to gender. Boys and girls are reared differently, not only with respect to physical exercise and coordination but also with respect to ingenuity and inventiveness. "Girls are taught obedience, dependence, and deference. They are made to feel that too much independence of opinion or action is a drawback to them—not becoming or womanly. A boy is made to feel that his success in life, his place in the world, will depend upon his ability to go ahead with his chosen occupation on his own responsibility, and to accomplish something new and valuable. No such social spur is applied to girls."[78] Differences could not be considered simple manifestations of biologically driven instincts; if they were, "it would not be necessary to spend so much time and effort in making boys and girls follow the lines of conduct proper to their sex."[79]

Woolley's research on gender differences was widely read and gave support to the growing awareness that average differences in ability

76. Ibid., p. 173.
77. Ibid., p. 176.
78. Ibid., p. 179.
79. Ibid., p. 181.

between the sexes tended to be quite small.[80] Her critique of the variability hypothesis itself did not have an equivalent impact, however. Gender differences in variability remained one of the major popular theories for explaining sex differences in achievement.

The most systematic critique of the variability hypothesis appeared in a series of studies and articles by Leta Stetter Hollingworth, first a student and later a colleague of Edward Thorndike's at Columbia Teachers College. Her interest in the topic probably stemmed from her contact with Thorndike early in her graduate career when she took an education class from him (1911–12). Thorndike was at that time the country's foremost proponent of the variability hypothesis and was convinced of the social importance of this sex difference: "In particular, if men differ in intelligence and energy by wider extremes than do women, eminence in and leadership of the world's affairs of whatever sort will inevitably belong oftener to men. They will oftener deserve it."[81] Hollingworth disputed such claims, pointing out, like Woolley, that they were most often made without the benefit of verifiable and verifying data.

Hollingworth approached the question of variability from several different research angles. She first examined the literature on sex differences that had already been published and found no evidence of sex differences in variability.[82] Then, in an effort to isolate inherent variations from those that could be influenced by the environment, she undertook a study of neonatal length and weight. After amassing data on over 2,000 infants, she and her associate Helen Montague found either no sex differences in variability on these measures or slightly greater variability among the female infants.[83] After finding no difference in physical measurements, she directed her attention to intellectual variation. Her critiques of the variability hypothesis were the most comprehensive published since Pearson's early work. But while Pearson had focused his attention primarily on the quantification of variation, Hollingworth also found fault with some of the conceptual assumptions upon which the hypothesis rested.

Hollingworth pointed out that disagreements over the appropriate definition of variability had not been resolved over the years. The lack of definition was only one problem in a massive knot of problems. Pro-

80. For example, see Burt and Moore (n. 22 above), and William J. Thomas, "The Mind of Woman and the Lower Races," *American Journal of Sociology* 12, no. 4 (1907): 435–69.

81. Edward Lee Thorndike, *Educational Psychology,* 2d ed. (New York: Teachers College, Columbia University, 1910), p. 35.

82. Leta S. Hollingworth, "Variability as Related to Sex Differences in Achievement," *American Journal of Sociology* 19, no. 4 (1914): 510–30 (hereafter cited as "Variability").

83. Helen Montague and Leta S. Hollingworth, "The Comparative Variability of the Sexes at Birth," *American Journal of Sociology* 20, no. 11 (1914): 335–70.

ponents of the hypothesis also assumed that mental traits hewed without exception to normal (bell-shaped) distribution. The assumption of a normal distribution of traits was important to the hypothesis because difference in variability would necessarily indicate a difference in range, and it was the greater range of male ability which was actually used to account for superiority of male ability. Hollingworth argued that a normal distribution of mental abilities was itself an assumption, and that if there were deviations from normality in the actual occurrence of traits within the population, differences in variability would give no indication of differences in range.[84] The most serious flaw in the proponents' arguments, however, was the reliance on data showing a preponderance of males among both the eminent and the feebleminded to prove greater male variability in intellectual abilities. Hollingworth found no reason to resort to circuitous biological explanations for these observations when much more straightforward social ones were available. Statistics reporting a larger number of males among the feebleminded could be explained by the fact that institutions were the source of data. She reasoned that feebleminded women could make their way in the world since society allowed and expected dependent behavior in women. A man incapable of financial self-support would be a much more likely candidate for institutionalization than a woman who could perform domestic chores or make a living by prostitution.[85] She reasoned that women were thus more likely to be institutionalized when they were older and no longer useful or self-supporting. Thus the mean age of women admitted to an institution would be greater than that of men admitted at the same time. A survey of sex and age ratios in New York institutions supported her hypothesis: the ratio of females to males increased with the age of the patients.[86]

Hollingworth's strongest stand was taken against those psychologists and armchair philosophers who continued to use the relative infrequency of eminence among women as proof of an innate incompatibility between femaleness and genius or who further reasoned that if women were unlikely to be geniuses, women's minds on the whole must be inferior to men's. She argued that the maternal role had hindered the full participation of women in other than nurturant activities. Because the social role of women was defined in terms of housekeeping and child rearing, "channels where eminence is impossible," and because of the concomitant constraints that the female role placed on women's education and employment, there was simply no valid way to compare the ability of women with that of men, who "have followed the greatest possible range of occupations, and have at the same time procreated

84. Hollingworth, "Variability."
85. Ibid.; Leta S. Hollingworth, "The Frequency of Amentia as Related to Sex," *Medical Record* 84 (October 1913): 753–56.
86. Hollingworth, "The Frequency of Amentia as Related to Sex."

unhindered."[87] She stated emphatically that women should not be forced to choose between career and maternity and that society should at least accept, if not support, women's choice to assume either or both of these roles. Without such options for women, society must expect "restlessness, unhappiness, and strife with the social order on the part of these individuals."[88] She believed that differences in social training made it not only difficult but impossible to compare the inherent capabilities of males and females.

Such emphasis on the importance of nurture over nature in directing development is still associated with the feminist position. In fact, arguments raised by the earlier generation of feminist scholars are identical to those raised today. Early socialization of males is fundamentally different from that of females. Culturally acceptable adult roles available to males and females are fundamentally different from one another. Therefore, description of inherent sex differences is restricted by basic differences in the environment which shape gender-specific behavior. That the arguments of an earlier generation of feminist scholars have not had a lasting impact and have had to be reintroduced with some regularity testifies to the conservatism of social science.

The Qualification of the Variability Hypothesis

With this country's increasing involvement and final entrance into World War I, the efforts of mental testers were redirected into developing and administering tests of ability to army inductees, and concern about sex differences waned. The war also shut off the flow of European literature, which had accounted for much of the theoretical writing on topics pertaining to gender. When Helen Thompson Woolley reviewed the sex-differences literature in 1914, she included references to over eighty articles and books; when Leta Hollingworth completed a similar project in 1918 just after the war's end, she cited only about twenty new pieces.[89]

With the temporary reduction of interest in sex differences went a corresponding decline of interest in the social implications of the variability hypothesis. Most discussions in the literature no longer linked gender differences in variability with gender differences in evolutionary role or innate capacity for achievement, and, in general, the equation of eminence with genius had become suspect. While in an earlier period

87. Hollingworth, "Variability," pp. 526, 524.
88. Ibid., p. 528. See also Stephanie A. Shields, "Ms. Pilgrim's Progress: The Contributions of Leta Stetter Hollingworth to the Psychology of Women," *American Psychologist* 30, no. 8 (1975): 852–57.
89. Helen T. Woolley, "The Psychology of Sex," *Psychological Bulletin* 11, no. 11 (1914): 353–79, and Leta S. Hollingworth, "Comparison of the Sexes in Mental Traits," *Psychological Bulletin* 15, no. 12 (1918): 427–32.

research on variability had attempted to explain differences in adult status and contributions to society, most of the research from the 1920s onward involved testing schoolchildren on a massive scale in order to describe their abilities, with little emphasis on theoretical explanations: variability became just one more dimension along which the sexes were compared.[90] Tests were developed to tap progressively more specific sets of abilities, including not only achievement as separate from intelligence but also components of cognitive processes such as language fluency and spatial ability.[91] Although the overall goal of educational testing was the improvement of education, suggestions like Thorndike's, that education should take into account women's greater tendency to mediocrity, no longer appeared in the literature.

The quantification of variation, which had been an issue since Ellis and Pearson debated the merits of defining variation in terms of anomaly, was becoming even more controversial with the increasing statistical sophistication of social scientists. A variety of expressions purporting to compare variational tendencies appears in the literature from the 1920s onward. Many continued to use Pearson's coefficient of variability (the standard deviation divided by the mean, then typically multiplied by 100 for convenience).[92] Others compared the size of the variance or standard deviation of male and female samples directly or expressed the comparison as a ratio.[93] In one of the most often cited reviews of the variability literature, Quinn McNemar and Lewis Terman devised their own "variability index" (the difference between standard deviations divided by their standard errors).[94] Although a comparison of models underlying the statistical test of variational differences is beyond the scope of this paper, suffice it to say that comparing findings of research studies employing such disparate indexes of variability is like comparing apples and oranges.[95] Large discrepancies were reported in studies employing dif-

90. One facet of education reform was the progressive education movement; see Patricia Graham, *Progressive Education* (New York: Teachers College Press, 1967).

91. The burgeoning literature on sex differences in specific mental abilities is covered in a series of review essays by Chauncey N. Allen: "Studies in Sex Differences," *Psychological Bulletin* 24, no. 5 (1927): 294–304; "Recent Studies in Sex Differences," ibid. 27, no. 5 (1930): 394–407; and "Recent Research in Sex Differences," ibid. 32, no. 5 (1935): 343–54.

92. See, e.g., Helen T. Woolley, *An Experimental Study of Children* (New York: Macmillan Publishing Co., 1926). No sex differences in performance on mental tests were observed.

93. See, e.g., W. D. Commins, "More about Sex Differences," *School and Society* 28, no. 724 (November 1928): 599–600.

94. Quinn McNemar and Lewis M. Terman, "Sex Differences in Variational Tendency," *Genetic Psychology Monographs* 18, no. 1 (1936): 1–66. Factor analysis was even proposed as a solution; see Alice I. Bryan, Harry E. Garrett, and Ruth E. Perl, "A Genetic Study of Several Mental Abilities at Three Age Levels," *Psychological Bulletin* 31, no. 9 (1934): 702–10.

95. For discussion of this issue, see A. Leon Winsor, "The Relative Variability of Boys and Girls," *Journal of Educational Psychology* 18, no. 5 (1927): 327–36, and Florence L. Goodenough, "The Consistency of Sex Differences in Mental Traits at Various Ages," *Psychological Review* 34, no. 6 (1927): 440–62.

ferent statistical expressions of variation, and the resulting confusion made some researchers more cautious in their interpretations of variability studies. Beth Wellman's review in 1933 of the literature on developmental sex difference is typical in its handling of variability: "There seems to be some slight support for the hypothesis of greater variability of boys. The case is by no means clear, however, the findings depending so much on the measuring instrument, the measure of variability used, the selection of the children, and the sex which obtains the higher mean or median score."[96]

There was also a greater general willingness to acknowledge that any gender differences in variability were probably caused by a more complex set of circumstances than a simple biological tendency. It was suggested, for example, that these differences could be explained by sex differences in maturation rate.[97] The appearance of new and more sophisticated biological models of sex differences also eroded support for the hypothesis, which no longer appeared to be an adequate description of or explanation for sex differences in test performance.[98]

A growing emphasis on experimentation and large-scale testing was casting doubt on turn-of-the-century studies of variability which had been made on very small samples, and the results of many of the first studies on large samples tended to contradict the hypothesis. As early as 1911, revised versions of standardized intelligence tests and newly developed standardized tests of special abilities revealed no differences in the variability of boys and girls.[99] In one study designed specifically to compare the sexes on variability, G. W. Fraiser tested over 60,000 thirteen-year-olds in cities across the country. He found no differences in variability, no differences in the proportion of males and females in the extremes of range, and no overall differences in the range of scores.[100] From the years just prior to World War I to the years im-

96. Beth Wellman, "Sex Differences," in *Handbook of Child Psychology*, ed. Carl Murchison, 2d rev. ed. (Worcester, Mass.: Clark University Press, 1933), pp. 626–49, esp. p. 630. See also Goodenough; Winsor.

97. Commins; Lincoln (n. 52 above).

98. A. J. Rosanoff, L. M. Handy, I. A. Rosanoff, and C. V. Inaman-Kane ("Sex Factors in Intelligence," *Journal of Nervous and Mental Disease* 80, no. 1 [1934]: 125–37) suggested that a sex-linked tendency for biological vulnerability could explain the better performance of females on intelligence tests. See also Robert G. Lehrke, "Sex Linkage: A Biological Basis for Greater Male Variability in Intelligence," in *Human Variation*, ed. R. T. Osborne, C. E. Noble, and N. Weyl (New York: Academic Press, 1978), pp. 171–98, and "A Theory of X-Linkage of Major Intellectual Traits," *American Journal of Mental Deficiency* 76, no. 6 (1972): 611–19.

99. S. A. Courtis, "Report on the Courtis Tests in Arithmetic," *New York Committee on School Inquiry* 1 (1911): 391–546, and H. H. Goddard, "Two Thousand Normal Children Measured by the Binet Measuring Scale of Intelligence," *American Journal of Psychology* 18 (1911): 232–59.

100. G. W. Fraiser, "A Comparative Study of the Variability of Boys and Girls," *Journal of Applied Psychology* 3, no. 2 (1919): 151–55.

mediately following it, evidence from the overwhelming majority of studies supported no sex difference in variability whatsoever.[101]

Despite these discouraging results, several influential scientists continued to search for evidence that males are more variable than females in performance on mental tests. In 1926 Thorndike conducted a study of adolescent intelligence that found only very weak support for a sex difference in variability. He no longer treated the hypothesis as fact or expounded on its social applications, and he described his results more cautiously than he had in his earlier writings, noting that "the difference being very small, the results concerning it are not consistent"; nevertheless, he continued to assert that "the general drift is toward a greater male variability."[102]

Around the same time, Lewis Terman conducted an extensive longitudinal study of gifted children, the results of which convinced him that males are more variable in intelligence. Although the highest IQ scores were earned by girls, there were more boys than girls in his sample of children who had been identified as gifted, and he felt compelled to conclude that his results were "in harmony with the hypothesis that exceptionally superior intelligence occurs with greater frequency among boys than among girls."[103] Although children were chosen for testing on the basis of teachers' nominations of children they considered especially bright, Terman denied that bias in the subject-selection process could account for the excess of males.[104]

In 1932 and again in 1947, the Scottish Council for Research in Education gave group intelligence tests to nearly complete samples of eleven-year-old children.[105] In 1932 the mean score of boys was 0.28

101. See, e.g., V. A. Henmon and W. F. Livingston, "Comparative Variability of Different Ages," *Journal of Educational Psychology* 3, no. 1 (1922): 17–29; Lewis M. Terman, *The Measurement of Intelligence* (Boston: Houghton Mifflin Co., 1916), and *The Intelligence of School Children* (n. 3 above).

102. Staff of the Division of Psychology of the Institute of Educational Research, "On the Variability of Boys and Girls," p. 183.

103. Lewis M. Terman et al., *Genetic Studies of Genius*, vol. 1, *Mental and Physical Traits of a Thousand Gifted Children* (Stanford, Calif.: Stanford University Press, 1925), p. 54.

104. Over the years, Terman's initial zeal was tempered as he took an increasingly broad view of the factors which could account for sex differences in performance. Once the staunchest defender of the variability hypothesis, he began to echo the interpretations that Calkins, Woolley, and Hollingworth had put forward years before: "The woman who is a potential poet, novelist, lawyer, physician, or scientist usually gives up any professional ambition she may have had and devotes herself to home, husband, and children. The exclusive devotion to domestic pursuits robs the arts and sciences of a large fraction of the genius that might otherwise be dedicated to them. My data strongly suggest that this loss must be debited to motivational causes and limited opportunity rather than to lack of ability" (Lewis Terman, "Human Intelligence and Achievement" [Sigma Chi address, Stanford University, Spring 1941], reprinted in May V. Seagoe's *Terman and the Gifted* [Los Altos, Calif.: William Kaufmann, Inc., 1975], p. 225).

105. Scottish Council for Research in Education, *The Trend of Scottish Intelligence* (London: University of London Press, 1949).

points higher than that of girls, while in 1947 the mean score of girls was 1.72 points higher than that of boys. In both samples males exhibited a slightly larger variability when the standard deviations of male and female samples were compared. In 1932 the standard deviation of the boys' scores exceeded that of the girls' by 0.90, the difference increasing to 1.24 in 1947. Although the absolute differences between the male and female standard deviations were quite small, the large sample size (over 87,000 in 1932 and nearly 71,000 in 1947) resulted in the differences being statistically significant. In both 1932 and 1947 an additional smaller sample of 1,000 children was drawn for individual testing with a set of intelligence tests. Sex differences in sample variability similar to those obtained with the group test were observed. For some time it was assumed that the slightly greater male variability found in the Scottish survey was attributable to a greater range of male intelligence extending to both higher and lower extremes than the range of female intelligence. Later examination of the data by the prominent differential psychologist Anne Anastasi, however, showed that the variability difference could be attributed to a "slight excess" of males with very low scores.[106]

Research from the 1960s and 1970s provides only equivocal support for the variability hypothesis. While scores on tests of mathematical ability show a consistent trend in the direction of greater male variability, studies of verbal ability show no consistent results. Maccoby and Jacklin note in their review of the literature that in the studies of verbal ability, as in the Scottish studies, results indicating greater male variability are attributable to a high incidence of low-scoring males, with no correspondingly high incidence of high-scoring males. The greater tendency of boys to suffer from learning deficits and developmental problems affects the incidence of very low scores on tests of mental abilities.[107]

Even though it is obvious that results indicating a gender difference in variability must be interpreted with caution and that no generalization about variability differences can apply to all intellectual abilities, a few scientists continue to assert, like the biologist Lionel Penrose, that "the larger range of variation in males than in females for general intelligence is an outstanding phenomenon."[108] One of the most recent advocates of the variability hypothesis is Robert Lehrke, who has suggested that intelligence is a recessive trait on the X chromosome, that is, a sex-linked characteristic, drawing evidence from studies of retardation—both heritability studies and studies comparing the in-

106. Anne Anastasi, "Four Hypotheses with a Dearth of Data: Response to Lehrke's 'A Theory of X-Linkage of Major Intellectual Traits,'" *American Journal of Mental Deficiency* 76, no. 6 (1972): 620–22.

107. Eleanor E. Maccoby and Carol N. Jacklin, *The Psychology of Sex Differences* (Stanford, Calif.: Stanford University Press, 1974), pp. 114–20.

108. Lionel S. Penrose, *The Biology of Mental Defects*, 2d rev. ed. (New York: Grune & Stratton, 1963), p. 186.

cidence of retardation among males and females. After identifying evidence of sex-linked retardation, he reasons that the same gene action is responsible for high IQ, even though he has no biological grounds for assuming that an identical biological operation explains the divergent traits of genius and feeblemindedness.[109] Contemporary arguments favoring the variability hypothesis typically cite the results of the Scottish surveys, Terman's study of genius, and Thorndike's survey of adolescents, even though this evidence is clearly inadequate. Thus the variability hypothesis, though it has undergone some modifications and lost much of its social importance since its origins at the turn of the century, has survived to the present day with relatively little factual support.

Conclusions

The evolution of the variability hypothesis is not unique in the study of sex differences or the history of social science in general. The hypothesis is a classic case of a concept introduced without supporting evidence or rigor in its definition and accepted as if consensus were an adequate test of validity. Darwin originally suggested greater male variability as a tentative hypothesis but later stated it as fact, although the evidence had become no more persuasive. Later evolutionary theorists took up his idea because it complemented the principle of variation as the source of evolutionary change. For the scientist in patriarchal Victorian society, it was perfectly reasonable to equate maleness with the element of species development. With no more evidence than the "common knowledge" that genius was an attribute of males, Ellis went on to assert that males were more variable not only physically but intellectually as well. Even Ellis's main critic, Karl Pearson, did not dispute the allegation of greater male intellectual ability; he only faulted Ellis's inclusion of a secondary sex characteristic in his data. Supporters of Ellis, such as Cattell, simply parroted his conclusions in their own writing without adding evidence of their own.

The success or failure of any scientific proposition affects the prestige and credibility of the individual or group supporting it. But with scientific propositions such as the variability hypothesis, more is at stake than individual reputation. Those who questioned the hypothesis tended to be those who had something to lose if it were accepted. When the status or worth of a class of individuals is tested, as in the study of sex or racial differences, subjective considerations which tend to be disguised or ignored in other research play an obvious and important role. Politics as well as personal and cultural values are inextricably involved in all of science but for the most part are unobtrusive. When the com-

---

109. Lehrke, "Sex Linkage," and "A Theory of X-Linkage of Major Intellectual Traits" (n. 98 above).

petencies of an entire group are at issue, however, the social implications of research may become more salient than the scientific aspect of the question.

The variability hypothesis changed over the years as social needs and scientific methodologies changed. It evolved not by refining the description of the biological mechanisms accounting for sex differences in variability but by changing definitions and measures of variability. The many formulations of the hypothesis can be at least partially explained by the social purposes it satisfied during each period of its history. Its first champion, Ellis, invoked it to support the ideology of male intellectual superiority with a roundabout sort of logic: males are more variable; their greater variability is expressed in the higher incidence of anomalies among males; genius is a kind of psychic anomaly; therefore, genius is a peculiarly male attribute. In the next decade, with the success of women in graduate school and the growth and successes of the feminist movement, belief in overall male mental superiority was no longer tenable. But belief in a form of male mental superiority was maintained by examining the relative proportion of male to female genius. For scientists like Thorndike, "discovery" of the greater proportion of males at the extremes of ability seemed to provide biological justification for placing limits on educational and professional opportunities open to women. Scientists assumed that measures of variation within the distribution of test scores, such as the standard deviation, also expressed information about the relative proportion of cases occurring in the extremes of the distribution. This assumption is justifiable only in the case of the normal (bell-shaped) distribution; if the distribution deviates from normal, a mathematical expression of variation will not be sufficient to determine the distribution's modality or symmetry. Thus, although proponents of the variability hypothesis maintained that their interest lay in describing the extremes of the distribution, the measures of relative variability they employed validly expressed only the dispersion of scores around the mean.

Changes in the formulation of the hypothesis also occurred through the personal and professional alliances and rivalries of the major figures involved in its discussion. Early publicity for the hypothesis was partially stimulated by Pearson's strong attack on Ellis's proposal. Cattell was editor of *Popular Science Monthly* when Ellis printed his rebuttal to Pearson, and the content of Cattell's later writings on the hypothesis suggests that he was strongly influenced by Ellis's ideas. Leta Hollingworth probably responded to Thorndike because she was his student at Teachers College.

Variability differences are still questioned in current reviews of sex differences,[110] but the issue has nothing like the prominence of other

110. Anne Anastasi, *Differential Psychology,* 3d ed. (New York: Macmillan Publishing Co., 1958); Leona E. Tyler, *The Psychology of Human Differences,* 3d ed. (New York: Appleton-Century-Crofts, 1965); and Maccoby and Jacklin, pp. 114–20.

sex-difference questions, such as those involving spatial ability and hemispheric lateralization. Psychometricians have all but abandoned the idea that variation in psychological traits can be inferred from knowledge of physical variation. And overt suggestions that evidence of a sex difference constitutes grounds for limiting women's opportunities are rarely made by scientists today.

The major flaws of the variability hypothesis were the assumptions that there exists always and everywhere a one-to-one relationship between biological impulsion and environmental result, and that whatever biological quirk produces greater male variability in a trait acts equally and with equal directness at all points along the trait's distribution. This disregard for the interaction among influences which direct development is clearly illustrated by Lehrke's assumption that the same genetic action produces both mental retardation and high IQ.

It is difficult and perhaps impossible to avoid an oversimplified resolution of the nature/nurture dilemma; simply acknowledging the complex form of the interactions between heredity and environment does not by itself constitute a scientific explanation for the operation of that interaction. Environmental models, in particular, can serve only at the descriptive level; those using them, moreover, assert that environment shapes behavior without describing precisely how such shaping operates. Behaviorism would have to be considered the theory most successful in its use of the environmental models, but it ignores rather than explains individual differences, and assessment of individual differences is essential to understanding possible gender-related differences. Biological models may have the advantage of appearing to be more testable because their major constructs are derived from tangible physical realities.

A question that remains unanswered is whether a biological model is necessarily a deterministic model. The proponents of the variability hypothesis were, to a man, convinced of the primacy of biological factors in dictating the course of development. In sharp contrast, its critics hewed to the nurture side of the nature/nurture dichotomy. Helen Lambert has described the range of moral values characterizing biological models, noting that the extent and kind of value judgments vary considerably.[111] At one extreme is the blatant assertion that biological dictates possess intrinsic positive value and that their consequences should not be altered. Biology is good; any deviation from it is bad. A milder form of moral judgment suggests that biological imperatives may not be assigned a value, but they will out. That is, social structure will inevitably reflect biological predispositions, no matter what efforts are made to circumvent natural inclinations. Both of these alternatives, as different as they are, look gloomy from a feminist perspective, in that

111. Lambert, "Biology and Equality" (n. 4 above), pp. 113–14.

struggle for change is cast as a doomed enterprise. Whether all biological models ineluctably lead to the reification of "natural" dispositions is open to debate: if one acknowledges the biological bases of behavior, must one also assume their inevitability and immutability? Is a feminist/humanist scientific model nonbiological by definition?

The evolutionary perspective is currently increasing in scientific importance. In contrast, analyses of the impact of the environment on development are much more limited in theoretical scope. Within science there are greater and greater pressures for acknowledgment of the primacy of gradual and cumulative biological forces in affecting development. The prevalence of biological models may some day lead to a state of affairs in which arguments for social change are only as persuasive as their biological rationale, and environmental explanations are looked on as antiscientific excuses. If social influence is not to be trivialized, development of more rigorous and comprehensive models of its operation are essential.

Department of Psychology
University of California, Davis

Animal Sociology and a Natural Economy of the Body Politic, Part I: A Political Physiology of Dominance

Donna Haraway

> "I want to do something very important. Like fly
> into the past and make it come out right." [Dawn, from
> MARGE PIERCY, *Woman on the Edge of Time*]

The concept of the body politic is not new. Elaborate organic images for human society were richly developed by the Greeks. They conceived the citizen, the city, and the cosmos to be built according to the same principles. To perceive the body politic as an organism, as fundamentally alive and as part of a large cosmic organism, was central for them.[1] To see the structure of human groups as a mirror of natural forms has remained imaginatively and intellectually powerful. Throughout the early period of the Industrial Revolution, a particularly important development of the theory of the body politic linked natural and political economy on multiple levels. Adam Smith's theory of the market and of the division of labor as keystones of future capitalist economic thought, with Thomas Malthus's supposed law of the relation of population and resources, together symbolize the junction of natural forces and economic progress in the formative years of capitalist industrialism. The permeation of Darwin's evolutionary theory with this form of political economy has been a subject of considerable analysis from the nineteenth century to the present.[2] Without question, the modern evolutionary concept of a population, as the fundamental natural group, owes much to classical ideas of the body politic, which in turn are inextricably interwoven with the social relationships of production and reproduction.

The union of the political and physiological is the focus of this essay.

---

1. R. G. Collingwood, *The Idea of Nature* (Oxford: Clarendon Press, 1945).
2. R. M. Young, "Malthus and the Evolutionists: The Common Context of Biological and Society Theory," *Past and Present* 43 (1969): 109–41.

This essay originally appeared in *Signs*, vol. 4, no. 1, Autumn 1978.

That union has been a major source of ancient and modern justifications of domination, especially of domination based on differences seen as natural, given, inescapable, and therefore moral. It has also been transformed by the modern biobehavioral sciences in ways we must understand if we are to work effectively for societies free from domination. The degree to which the principle of domination is deeply embedded in our natural sciences, especially in those disciplines that seek to explain social groups and behavior, must not be underestimated. In evading the importance of dominance as a part of the theory and practice of contemporary sciences, we bypass the crucial and difficult examination of the *content* as well as the social function of science. We leave this central, legitimating body of skill and knowledge to undermine our efforts, to render them utopian in the worst sense. Nor must we lightly accept the damaging distinction between pure and applied science, between use and abuse of science, and even between nature and culture. All are versions of the philosophy of science that exploits the rupture between subject and object to justify the double ideology of firm scientific objectivity and mere personal subjectivity. This antiliberation core of knowledge and practice in our sciences is an important buttress of social control.[3]

Recognition of that fact has been a major contribution by feminist theorists. Women know very well that knowledge from the natural sciences has been used in the interests of our domination and not our liberation, birth control propagandists notwithstanding. Moreover, general exclusion from science has only made our exploitation more acute. We have learned that both the exclusion and the exploitation are fruits of our position in the social division of labor and not of natural incapacities.[4] But if we have not often underestimated the principle of domination in the sciences, if we have been less mesmerized than many by the claims to value-free truth by scientists as we most frequently encounter them—in the medical marketplace[5]—we have allowed our

3. R. M. Young, "Science *Is* Social Relations," *Radical Science Journal* 5 (1977): 65–129. This essay has an exceptionally useful bibliography of radical critique of science. See also E. A. Burtt, *The Metaphysical Foundations of Modern Science* (New York: Humanities Press, 1952); Herbert Marcuse, *One Dimensional Man: Studies in the Ideology of Advanced Industrial Society* (Boston: Beacon Press, 1968); Karl Marx and Frederick Engels, *The German Ideology* (New York: International Publishers, 1970).

4. Harry Braverman, *Labor and Monopoly Capital: The Degradation of Work in the Twentieth Century* (New York: Monthly Review Press, 1974). Braverman situates the female work force in the center of his powerful Marxist analysis of modern labor, scientific management, and the deskilling of working people in a period of increasing scientific and technical expertise.

5. Linda Gordon, *Women's Body, Women's Right: A Social History of Birth Control in America* (New York: Viking Press, 1976). Excerpts from this important book may be found in a two-part series by Linda Gordon ("Birth Control: An Historical Study," *Science for the People* 9, no. 1 [1977]: 10–16, "Birth Control and the Eugenicists," ibid., 9, no. 2 [1977]: 8–15). See also James Reed, *From Private Vice to Public Virtue: The Birth Control Movement and American Society since 1830* (New York: Basic Books, 1978).

distance from science and technology to lead us to misunderstand the status and function of natural knowledge. We have accepted at face value the traditional liberal ideology of social scientists in the twentieth century that maintains a deep and necessary split between nature and culture and between the forms of knowledge relating to these two putatively irreconcilable realms. We have allowed the theory of the body politic to be split in such a way that natural knowledge is reincorporated covertly into techniques of social control instead of being transformed into sciences of liberation. We have challenged our traditional assignment to the status of natural objects by becoming antinatural in our ideology in a way which leaves the life sciences untouched by feminist needs.[6] We have granted science the role of a fetish, an object human beings make only to forget their role in creating it, no longer responsive to the dialectical interplay of human beings with the surrounding world in the satisfaction of social and organic needs. We have perversely worshiped science as a reified fetish in two complementary ways: (1) by completely rejecting scientific and technical discipline and developing feminist social theory totally apart from the natural sciences, and (2) by agreeing that "nature" is our enemy and that we must control our "natural" bodies (by techniques given us by biomedical science) at all costs to enter the hallowed kingdom of the cultural body politic as defined by liberal (and radical) theorists of political economy, instead of by ourselves. This cultural body politic was clearly identified by Marx: the marketplace that remakes all things and people into commodities.

A concrete example may help explain what I see as our dangerous misunderstanding, an example which takes us back to the point of union of the political and physiological. In *Civilization and Its Discontents*,[7] Freud developed a theory of the body politic that based human social development on progressive domination of nature, particularly of human sexual energies. Sex as danger and as nature are central to Freud's system, which repeats rather than initiates the traditional reduction of the body politic to physiological starting points. The body politic is in the first

6. Sherry B. Ortner, "Is Female to Male as Nature Is to Culture?" in *Women, Culture, and Society,* ed. Michelle Zimbalist Rosaldo and Louise Lamphere (Stanford, Calif.: Stanford University Press, 1974), pp. 67–87; Simone de Beauvoir, *The Second Sex* (London: Jonathan Cope, 1953). Both Ortner, from the point of view of structuralist anthropology, and de Beauvoir, from the perspective of existentialism, allow the ideology of the nature-culture split to dominate their feminist analyses. Carol P. MacCormack ("Biological Events and Cultural Control," *Signs* 3 [1977]: 93–100) has made use of anthropological theories of Mary Douglas (*Purity and Danger* [London: Routledge & Kegan Paul, 1966] and *Rules and Meanings* [London: Penguin Books, 1973]) in order to challenge the nature-culture distinction. MacCormack analyzes the female Sande sodality of Sierra Leone to stress women's collective construction of their own bodies for assuming their active roles in the body politic. The organicism and functionalism in MacCormack's framework should be critically explored.

7. Sigmund Freud, *Civilization and Its Discontents* (New York: W. W. Norton & Co., 1962).

instance seen to be founded on natural individuals whose instincts must be conquered to make possible the cultural group. Two recent neo-Freudian and neo-Marxist theorists have ironically reworked Freud's position in illuminating ways for the thesis of this essay: one is Norman O. Brown, the other Shulamith Firestone. Freud, Brown, and Firestone are useful tools in a dissection of the theories of the political and physiological organs of the body politic because they all begin their explanations with sexuality, add a dynamic of cultural repression, and then attempt to liberate again the personal and collective body.

Brown, in *Love's Body*,[8] developed an elaborate metaphorical play between the individual and political bodies to show the extraordinary patriarchal and authoritarian structure of our conceptions and experiences of both. The phallus, the head; the state, the body; the brothers, the rebellious overthrow of kingship only to establish the tyranny of the fraternal liberal market—these are Brown's themes. If only the father was head, only the brothers may be citizens. The only escape from the domination that Brown explored was through fantasy and ecstasy, leaving the body politic unchallenged in its fundamental male supremacy and in its reduction to the dynamic of repression of nature. Brown rejected civilization (the body politic) in order to save the body; the solution was necessitated by his root acceptance of the Freudian sexual reductionism and ensuing logic of domination. He turned nature into a fetish worshiped by total return to it (polymorphous perversity). He betrayed the socialist possibilities of a dialectical theory of the body politic that neither worships nor rejects natural science, that refuses to make nature and its knowledge a fetish.

Firestone, in the *Dialectic of Sex*,[9] also faces the implications of Freud's biopolitical theory of patriarchy and repression but tries to transform it to yield a feminist and socialist theory of liberation. She has been immensely important to feminists in this task. I think, however, that she committed the same mistake that Brown did, that of "physiological reduction of the body politic to sex," which fundamentally blocks a liberating socialism that neither fatalistically exploits the techniques given by sciences (while despairing of transforming their content) nor rejects a technical knowledge altogether for fantasy. Firestone located the flaw in women's position in the body politic in our own bodies, in our subservience to the organic demands of reproduction. In that critical sense she accepted a historical materialism based on reproduction and lost the possibility for a feminist-socialist theory of the body politic that would not see our personal bodies as the ultimate enemy. In that step she prepared for the logic of domination of technology—total control of now alienated bodies in a machine-determined future. She made the basic mistake of reducing social relations to natural objects, with the

8. Norman O. Brown, *Love's Body* (New York: Random House, 1966).
9. Shulamith Firestone, *Dialectic of Sex* (New York: William Morrow & Co., 1970).

logical consequence of seeing technical control as a solution. She certainly did not underestimate the principle of domination in the biobehavioral sciences, but she did misunderstand the status of scientific knowledge and practice. That is, she accepted that there are natural objects (bodies) separate from social relations. In that context, liberation remains subject to supposedly natural determinism, which can only be avoided in an escalating logic of counterdomination.

I think it is possible to build a socialist-feminist theory of the body politic that avoids physiological reductionism in both its forms: (1) capitulating to theories of biological determinism of our social position, and (2) adopting the basically capitalist ideology of culture against nature and thereby denying our responsibility to rebuild the life sciences. I understand Marxist humanism to mean that the fundamental position of the human being in the world is the dialectical relation with the surrounding world involved in the satisfaction of needs and thus in the creation of use values. The labor process constitutes the fundamental human condition. Through labor, we make ourselves individually and collectively in a constant interaction with all that has not yet been humanized. Neither our personal bodies nor our social bodies may be seen as natural, in the sense of outside the self-creating process called human labor. What we experience and theorize as nature and as culture are transformed by our work. All we touch and therefore know, including our organic and our social bodies, is made possible for us through labor. Therefore, culture does not dominate nature, nor is nature an enemy. The dialectic must not be made into a dynamic of growing domination.[10] The position, a historical materialism based on production, contrasts fundamentally with the ironically named historical materialism based on reproduction that I have tried to outline above.

One area of the biobehavioral sciences has been unusually important in the construction of oppressive theories of the body political: animal sociology, or the science of animal groups. To reappropriate the biosocial sciences for new practices and theories, a critical history of the physiological politics based on domination that have been central in animal sociology is important. The biosocial sciences have not simply been sexist mirrors of our own social world. They have also been tools in the reproduction of that world, both in supplying legitimating ideologies and in enhancing material power. There are three main reasons for choosing to focus on the science of animal, especially primate, groups.

First, its subject and procedures developed so as to span the nature-

---

10. Nancy Hartsock, "Objectivity and Revolution: Problems of Knowledge in Marxist Theory," and "Social Science, Praxis, and Political Action" (manuscripts available from author, Political Science Department, Johns Hopkins University). Because of the well-developed feminist analysis and avoidance of sexual reductionism, these papers are more useful for a critique of the theory and practice of scientific objectivity than those of Jürgen Habermas or Herbert Marcuse. See J. Habermas, *Toward a Rational Society: Student Protest, Science, and Politics* (Boston: Beacon Press, 1970).

culture split at precisely the same time in American intellectual history, between 1920 and 1940, when the ideology of the autonomy of the social sciences had at last gained acceptance, that is, when the liberal theory of society (based on functionalism and hierarchical systems theories) was being established in the universities. Intrinsic to the new liberal relations of natural and social disciplines was the project of human engineering—that is, the project of design and management of human material for efficient, rational functioning in a scientifically ordered society. Animals played an important role in this project. On one hand, they were plastic raw material of knowledge, subject to exact laboratory discipline. They could be used to construct and test model systems for both human physiology and politics. A model system of, for example, menstrual physiology or socialization processes did not necessarily imply reductionism. It was precisely direct reduction of human to natural sciences that the post-Spencerian, postevolutionary naturalist new ordering of knowledge forbade. The management sciences of the 1930s and after have been strict on that point. It is part of the nature-culture split. On the other hand, animals have continued to have a special status as natural objects that can show people their origin, and therefore their prerational, premanagement, precultural essence. That is, animals have been ominously ambiguous in their place in the doctrine of autonomy of human and natural sciences. So, despite the claims of anthropology to be able to understand human beings solely with the concept of culture and of sociology to need nothing but the idea of the human social group, animal societies have been extensively employed in rationalization and naturalization of the oppressive orders of domination in the human body politic.[11] They have provided the point of union of the physiological and political for modern liberal theorists while they continue to accept the ideology of the split between nature and culture.

Second, animal sociology has been central in the development of the most thorough naturalization of patriarchal division of authority in the body politic and in reduction of the body politic to sexual physiology. Thus this area of the natural sciences is one we need to understand thoroughly and tranform completely to produce a science that might express the social relations of liberation without committing the vulgar Marxist mistake of deriving directly the substance of knowledge from material conditions. We need to understand how and why animal groups have been used in theories of the evolutionary origin of human beings, of "mental illness," of the natural basis of cultural cooperation and competition, of language and other forms of communication, of technology, and especially of the origin and role of human forms of sex

11. See, for example, the University of Chicago 50th anniversary celebration symposium jointly produced by the biological and social sciences departments: Robert Redfield, ed., *Levels of Integration in Biological and Social Systems* (Lancaster, Pa.: Jaques Cattell Press, 1942).

and the family. In short, we need to know the animal science of the body politic as it has been and might be.[12] I believe the result of a liberating science of animal groups would better express who the animals are as well; we might free nature in freeing ourselves.

Third, the levels at which domination has formed an analytical principle in animal sociology allow a critique of the embodiment of social relations in the content and basic procedures of a natural science in such a way as to expose the fallacies of the claim to objectivity, but not in such a way as to permit facile rejection of scientific discipline in our knowledge of animals. We cannot dismiss the layers of domination in the science of animal groups as a film of unfortunate bias or ideology that can be peeled off the healthy objective strata of knowledge below. Neither can we think just anything we please about animals and their meaning for us. We come face to face with the necessity of a dialectical understanding of scientific labor in producing for us our knowledge of nature.

I will restrict my analysis primarily to a few years around World War II and to work on a single group of animals—the primates, in particular, the rhesus monkey, native to Asia but present in droves in scientific laboratories and research stations worldwide. I will focus principally on the work of one person, Clarence Ray Carpenter, who helped found the first major research station for free-ranging monkeys as part of the school of tropical medicine affiliated with Columbia University off Puerto Rico on the tiny island of Cayo Santiago in the late 1930s. These monkeys and their descendants have been central actors in dramatic reconstructions of natural society. Their affiliation with tropical medicine in a neocolonial holding of the United States, which has been so extensively used as an experiment station for capitalist fertility management policies, adds an ironic backdrop appropriate to our subject.

Men like Carpenter moved within a complex scientific world in which it would be incorrect to label most individuals or theories as sexist or whatever. It is not to attach simplistic labels but to unwind the specific social and theoretical structures of an area of life science that we need to examine the interconnections of laboratory heads, students, funding agencies, research stations, experimental designs, and historical setting. Carpenter earned his Ph.D. at Stanford for a study of the effects on sexual behavior of the removal of the gonads of male pigeons in mated pairs. He then received a National Research Council Fellowship in 1931 to study social behavior of primates under the direction of Robert M. Yerkes of the Laboratories of Comparative Psychobiology at Yale University. Yerkes had recently established the first comprehensive research

12. For important comparison and contrast of anarchist and Marxist socialisms on the meaning of nature for the body politic, see Peter Kropotkin, *Mutual Aid* (London: William Heinemann, 1902), and Frederick Engels, *Dialectics of Nature* (New York: International Publishers, 1940).

institution for the psychobiological study of anthropoid apes in the world. For Yerkes, apes were perfect models of human beings. They played a major part in his sense of mission to promote scientific management of every phase of society, an idea typical of his generation. "It has always been a feature for the use of the chimpanzee as an experimental animal to shape it intelligently to specification instead of trying to preserve its natural characteristics. We have believed it important to convert the animal into as nearly ideal a subject for biological research as is practicable. And with this intent has been associated the hope that eventual success might serve as an effective demonstration of the possibility of re-creating man himself in the image of a generally acceptable ideal."[13] He, then, designed primates as scientific objects in relation to his ideal of human progress through human engineering.

Yerkes was interested in the apes in two main regards—their intelligence and their social-sexual life. For him intelligence was the perfect expression of evolutionary position. He saw every living object in terms of the outstanding problem of experimental comparative psychology in America since its inception around 1900: the intelligence test. Species, racial, and individual qualities were fundamentally tied to the central index of intelligence, revealed on the one hand through behavior testing and on the other through the neural sciences. He had designed the army intelligence tests administered to recruits in World War I, tests seen to provide a rational basis for assignment and promotion, to indicate natural merit fitting men for command.[14] His role in the war was entirely compatible with his role as an entrepreneur in primate studies. In both cases he saw himself and his scientific peers working to foster a rational society based on science and preserved from old ignorance, embodied especially in religion and politics.

The social-sexual life of primates was for Yerkes thoroughly inter-

13. R. M. Yerkes, *Chimpanzees: A Laboratory Colony* (New Haven, Conn.: Yale University Press, 1943), p. 10. I am currently working on a book-length treatment of the history of primate studies in the twentieth century. Yerkes is a critical figure in linking foundations, universities, neurophysiology and endocrinology, personnel management, psychopathology, educational testing, personality studies, social and sexual hygiene—in short, the whole complex of science and society in his lifetime. James Reed is undertaking a comprehensive biography of Yerkes.

14. R. M. Yerkes, "What Psychology Contributed to the War," in *The New World of Science* (New York: Century Co., 1920). See also Daniel Kevles, "Testing the Army's Intelligence: Psychologists and the Military in World War I," *Journal of American History* 55 (1968): 565–81. Yerkes and his peers were not using "human engineering" simply as a metaphor. They explicitly saw physiological, biopsychological, and social sciences as key parts of rational management in advanced monopoly capitalism. The sciences inventoried raw materials, and the laboratory functioned as a pilot plant for human engineering (see R. M. Yerkes, "What Is Personnel Research?" *Journal of Personnel Research* 1 [1922]: 56–63). For a history of the project of human engineering, see David F. Noble, *America by Design: Science, Technology and the Rise of Corporate Capitalism* (New York: Alfred A. Knopf, Inc., 1977), esp. chap. 10.

twined with their intelligence. Mind would order and rule lower functions to create society. In a classic study of the origin of the body politic, Yerkes observed that female chimpanzees who were sexually receptive were allowed by the dominant males to have food and "privileges" to which they were ordinarily not entitled.[15] Primate intelligence allowed sexual states to stimulate the beginnings of human concepts of social right and privilege. The sexual reductionism hardly needs emphasis. His study linking sex and power was typical of work in the 1930s, and hardly different from much to this day. In an early feminist critique, Ruth Herschberger[16] marvellously imagined the perspective of Josie, the female chimpanzee whose psychosexual life was of such concern to Yerkes. Josie seems not to have seen her world in terms of trading sex for "privilege," but to Yerkes that economic link of physiology and politics seemed to have been scientifically confirmed to life at the organic base of civilization.

In addition to direct investigation of physiological sex and social behavior in human beings' closest relatives, Yerkes exercised, along with his peers, a tremendous influence on the overall direction of scientific study of sex in this country. He was for twenty-five years chairman of the Rockefeller Foundation–funded National Research Council Committee for Research on Problems of Sex (CRPS). This committee, from 1922 until well after the Second World War when federal funding became massively available for science, provided the financial base for the transformation of human sex into a scientific problem. Fundamental work on hormones and behavior, sex-linked differences in mental and emotional qualities, marital happiness, and finally the Kinsey studies was all funded by the Committee for Research on Problems of Sex. It played a key role in opening up sexual topics for polite discussion and respectable investigation in an era of undoubted prurience and ignorance.[17]

However, the opening was double edged; the committee, in its practice and ideological expressions, was structured on several levels according to the principle of the primacy of sex in organic and social processes. To make sex a scientific problem also made it an object for medical

---

15. R. M. Yerkes, "Social Dominance and Sexual Status in the Chimpanzee," *Quarterly Review of Biology* 14, no. 2 (1939): 115–36.

16. Ruth Herschberger, *Adam's Rib* (New York: Pellegrine & Cudhay, 1948).

17. See Emma Goldman's *Living My Life* (New York: Alfred A. Knopf, Inc., 1931) for her keen analysis of the effects of sexual ignorance on working-class women. See Diana Long Hall ("Biology, Sex Hormones and Sexism in the 1920s," *Philosophical Forum* 5 [1974]: 81–96) for general background on the political context of sex research. She is currently completing an important book on the history of endocrinology, the Committee for Research in Problems of Sex, and related subjects in American society to 1940. For an insiders' discussion, see Sophie Aberle and George W. Corner, *Twenty-five Years of Sex Research: History of the National Research Council Committee for Research in Problems of Sex, 1922–47* (Philadelphia: W. B. Saunders Co., 1953). The complicated network of scientific communities emerges clearly from Diana Long Hall's work.

therapy for all kinds of sexual "illness," most certainly including homosexuality and unhappy marriages. The biochemical and physiological basis of the therapeutic claims immensely strengthened the legitimating power of scientific managers over women's lives. The committee closed the escape holes for those who rejected Freud's kind of sexual reductionism: whether from the psychoanalytic or physical-chemical directions, sex was safely in the care of scientific-medical managers. Monkeys and apes were enlisted in this task in central roles; as natural objects unobscured by culture, they would show most plainly the organic base in relation to which culture emerged. That these "natural objects" were thoroughly designed according to the many-leveled meanings of an ideal of human engineering has hardly been noticed.

Carpenter arrived at Yale's primate laboratories already enmeshed in the web of funding and practice represented by the CRPS. His Ph.D work had been funded by the committee, his postdoctoral fellowship granted by essentially the same men, and his host, Yerkes, was the central figure in a very important network of scientific assumptions and practices. Those scientific networks crucially determined who did science and what science was considered good. From his education, funding, and social environment, there was little reason for Carpenter to reject the basic assumptions that identified reproduction and dominance based on sex with the fundamental organizing principles of a natural body politic. What Carpenter added, however, was significant. Methodologically, he established the demanding skill of naturalistic observation of wild primates in two extraordinarily careful field studies, one on New World howler monkeys and one on Asian gibbons. These studies have deeply influenced the techniques and interpretations of animal sociology. They are worthy of note because they are simultaneously excellent, commanding work and fully reflective of social relations based on dominance in the human world of scientists.[18] Theoretically, Carpenter tied the interpretations of the laboratory disciplines of comparative psychology and sexual physiology to evolutionary and ecological field biology centered on the concepts of population and community. In short, he started to link the elements of natural and political economy in new and important ways. The classic Darwinian conception of natural political economy of populations began to be integrated with the physiological and psychological sciences that greatly

18. C. R. Carpenter, *Naturalistic Behavior of Nonhuman Primates* (University Park: Pennsylvania State University Press, 1964). This book is a collection of Carpenter's major papers covering work from the 1930s on. Carpenter moved from primate studies to concern with educational television in American rural and Third World contéxts. He brought into communications systems work the same functionalist, hierarchical conceptions of organization he used in analyzing primates. See also his "Concepts and Problems of Primate Sociometry," *Sociometry* 8 (1945): 56–61, and "The Applications of Less Complex Instructional Technologies," in *Quality Instructional Television,* ed. W. Schramm (Honolulu: Hawaii University Press, East-West Center, 1972), pp. 191–205.

flourished in the early twentieth century. The integration would be complete only after World War II, when Sherwood Washburn and his students transformed physical anthropology and primate studies by systematically exploiting the evolutionary functionalism of the neo-Darwinian synthesis and the social functionalism of Bronislaw Malinowski's theory of culture.

In addition to linking levels of psychobiological analysis to modern evolutionary theory, Carpenter analyzed primate groups with the tools of early systems theory that were simultaneously providing the technical base for the claim to scientific maturity of the social sciences based on concepts of culture and social group. Carpenter's early social functionalism—with all its remaining ties to an older comparative psychology and to developmental physiology (experimental embryology)—is crucial for examining the connecting chains from physiology to politics, from animal to human. Carpenter himself did not work within the doctrine of autonomy of natural and social sciences. Neither did he permit direct reduction of social to physiological or of human to animal. He elaborated analytical links between levels that were shared by both adherents and opponents of the crucial nature-culture distinction. Indeed, his primate sociology is a useful place to begin to unravel the many varieties of functionalism emerging within biological and social sciences between the two world wars, all based on principles of hierarchical order of the body and body politic. The functionalist disciplines underlay strong ideologies of social control and techniques of medical, educational, and industrial management.

A single experimental manipulation embodies in miniature all the layers of significance of the principle of dominance in Carpenter's seminal work on the animal body politic. In 1938 he collected about 400 rhesus monkeys in Asia and freed them on Cayo Santiago. After a period of social chaos, they organized themselves into six groups containing both sexes and ranging in size from three to 147 animals. The monkeys were allowed to range freely over the thirty-seven acre island and to divide space and other resources with little outside interference. The first major study undertaken of them was of their sexual behavior, including periodicity of estrus, homosexual, autoerotic, and "nonconformist" behavior. Carpenter's conclusions noted that intragroup dominance by males was strongly correlated with sexual activity, and so presumably with evolutionary advantage. All the sexist interpretations with which we have become monotonously familiar were present in the analysis of the study, including such renderings of animal activities as, "Homosexual females who play masculine roles attack females who play the feminine role prior to the formation of a female-female consort relation."[19]

19. Carpenter, *Naturalistic Behavior*, p. 339.

In harmony with the guiding notion of the ties of sex and dominance in the fundamental organization of the rhesus groups, Carpenter performed what on the surface is a very simple experiment, but one which represents the whole complex of layered explanation of the natural body politic from the physiological to the political. After watching the undisturbed group for one week as a control, he removed the "alpha male" (animal judged most dominant on the basis of priority access to food, sex, etc.), named Diablo, from his group. Carpenter then observed the remaining animals for one week, removed the number 2 male, waited another week, removed the number 3 male, waited, restored all three males to the group, and again observed the social behavior. He noted that removal of Diablo resulted in immediate restriction of the territorial range of the group on the island relative to other groups. Social order was seriously disrupted. "The group organization became more fluid and there was an increase in *intra-group conflict* and fights. . . . After a marked disruption lasting three weeks, the group was suddenly restructured when the dominant males were released."[20] Social order was restored, and the group regained its prior favorable position relative to other groups.

Several questions immediately arise. Why did Carpenter not use as a control the removal of other than dominant males from the group to test his organizing hypothesis about the source of social order? Literally, he removed the putative head from the collective animal body. What did this field experiment, this decapitation, mean to Carpenter?

First, it must be examined on a physiological level. Carpenter relied on biological concepts for understanding social bodies. He drew from theories of embryological development that tried to explain the formation of complex whole animals from simpler starting materials of fertilized eggs. One important embryological theory used the concept of fields organized by axes of activity called dominance gradients. A field was a spatial whole formed by the complex interaction of gradients. A gradient was conceived, in this theory, to consist of an ordered series of processes from low to high levels of activity measured, for example, by differential oxygen consumption. Note that at this basic level dominance was conceived as a purely physiological property that could be objectively measured. The slope of a gradient could be shallow or steep. Several gradients making up a field would be organized around a principal axis of greatest slope, the organization center. An organism grew in complexity through integrated multiplication of dominance systems. An appropriate experimental system within developmental physiology designed to test theories of fields, gradients, physiological dominance, and organization centers was the simple hydra. It had only one axis or possible gradient: head to tail. One could cut off the polyp's head, observe

20. Ibid., p. 362.

temporary disorganization of remaining tissue, and see ultimate re-establishment of a new head from among the physiologically "competing" cells. Further, one could remove much or little from the head portion of the activity gradient and test the extent of ensuing organic disorganization.[21]

Carpenter conceived social space to be like the organic space of a developing organism, and so he looked for gradients that organized the social field through time. He found such a physiological gradient of activity in the dominance hierarchy of the males of the social group. He performed the theoretically based experiment of head removal and "observed" ensuing physiological competition among cells or organs (i.e., other points—animals—on the activity-dominance gradient) to re-establish a chief organization center (achieve alpha male status) and restore social harmony. Several consequences flow from these identifications.

First, other groups of animals in the society could be ordered on activity axes as well; females, for example, were found to have a dominance hierarchy of less steepness or lower slope. Young animals had unstable dominance gradients; the observation underlying that interpretation was that ordinary dominance behavior could not be reliably seen and that immature animals did not show constant dominance relations to one another. As unseen "observations" became just as important as evidence as seen ones, a concept of latent dominance followed readily. From this point, it is an easy step to judgments about the amount of dominance that functions to organize social space (call that quantity leadership) and the amount that causes social disruption (call that pathological aggression). Throughout the period around World War II, similar studies of the authoritarian personality in human beings abounded; true social order must rest on a balance of dominance, interpreted as the foundation of cooperation. Competitive aggression became the chief form that organized other forms of social integration. Far from competition and cooperation being mutual opposites, the former is the precondition of the latter—on physiological grounds. If the most active (dominant) regions, the organization centers, of an organism are removed, other gradient systems compete to reestablish organic order: a period of fights and fluidity ensues within the body politic. The chief point is that without an organizing dominance hierarchy, social order supposedly is seen to break down into individualistic, unproductive competition. The control experiment of removing other animals than the dominant males was not done because it did not make sense within the whole complex of theory, analogies to individual organisms, and unexamined assumptions.

21. C. M. Child was the primary gradient field theorist whose ideas entered social theory (see C. M. Child, "Biological Foundations of Social Integration," *Publications of the American Sociological Society* 22 [1928]: 26–42).

The authoritarian personality studies bring us to the second level of explanation of the body politic implicit in Carpenter's experiment: the psychological. The idea of a dominance hierarchy was derived in the first instance from study of "peck orders" in domestic chickens and other birds initiated by the Norwegian Thorlief Schjelderup-Ebbe as early as 1913, but not incorporated into American comparative psychology in any important way until the 1930s. Then animal sociology and psychology, as well as human branches of the disciplines, focused great attention on ideas of competition and cooperation.[22] Society was derived from complex interactions of pairs of individuals, understood and measured by psychological techniques, which constituted the social field space. One looked for axes of dominance as organizing principles on both the physiological and psychological levels.

The third and last level implicit in Carpenter's manipulation is that of natural political economy. The group that loses its alpha male loses in the competitive struggle with other organized organic societies. The result would be reflected in less food, higher infant mortality, fewer offspring, and thus evolutionary disadvantage or even extinction. The market competition implicit in organic evolutionary theory surfaces here. The theory of the function of male dominance nicely joins the political economy aspect of the study of animal behavior and evolution (competitive, division of labor, resource allocation model) with the social integration aspect (cooperative coordination through leadership and social position) with the purely physiological understandings of reproductive and embryological phenomena. All three perspectives link functionalist equilibrium social models—established in social sciences of the period—to explicit ideological, political concerns with competition and cooperation (in labor struggles, for example).[23] Since animal societies are seen to have in simpler form all the characteristics of human societies and cultures, one may legitimately learn from them the base of supposedly natural, integrated community for humanity. Elton Mayo—the influential Harvard,

22. Thorlief Schjelderup-Ebbe, "Social Behavior of Birds," in *Handbook of Social Psychology,* ed. Carl Murchison (Worcester, Mass.: Clark University Press, 1935), 2:947–72.

23. Elton Mayo, *The Human Problems of Industrial Civilization* (New York: Macmillan Co., 1933). Stephen Cross, Johns Hopkins University History of Science Department, is writing a dissertation on industrial sociology-physiology developed in relation to L. J. Henderson's Pareto seminars in the 1930s at Harvard. See Barbara Heyl, "The Harvard Pareto Circle," *Journal of the History of Behavioral Sciences* 4 (1968): 316–34; L. J. Henderson, *Pareto's General Sociology: A Physiologist's Interpretation* (Cambridge, Mass.: Harvard University Press, 1935); and Talcott Parsons, "On Building Social System Theory: A Personal History," *Daedalus* (Fall 1970), pp. 826–81. The theme of cooperation-competition in anthropological focus on personality and culture in the 1930s was pervasive and crucial. See Margaret Mead, ed., *Cooperation and Competition among Primitive Peoples* (New York: McGraw-Hill Book Co., 1937); and Mark A. May and Leonard W. Dobb, *Competition and Cooperation* (New York: Social Science Research Council, 1937). This report was mainly a bibliography on the competition-cooperation theme in the 1930s.

anti-labor union, industrial psychologist-sociologist of the same period—called such a community the "Garden of Industry."[24]

The political principle of domination has been transformed here into the legitimating scientific principle of dominance as a natural property with a physical-chemical base. Manipulations, concepts, organizing principles the entire range of tools of the science—must be seen to be penetrated by the principle of domination. Science cannot be reclaimed for liberating purposes by simply reinterpreting observations or changing terminology, a crass ideological exercise in any case, which denies a dialectical interaction with the animals in the project of self-creation through scientific labor. But the difficult process of remaking the biosocial and biobehavioral sciences for liberation has begun. Not surprisingly, one of the first steps has been to switch the focus from primates as models of human beings to a deeper look at the animals themselves— how they live and relate to their environments in ways that may have little to do with us and that will surely reform our sense of relation to nature in our theories of the body politic. These "revisionist" scientific theories and practices deserve serious attention. Of them, "feminist" perspectives in physical anthropology and primatology have stressed principles of organization for bodies and societies that do not depend on dominance hierarchies. Dominance structures are still seen and examined, but cease to be used as causal explanations of functional organization. Rather, the revisionists have stressed matrifocal groups, long-term social cooperation rather than short-term spectacular aggression, flexible process rather than strict structure, and so on.[25] The scientific and ideological issues are complex; the emerging work is justly controversial.

In our search for an understanding of a feminist body politic, we need the discipline of the natural and social sciences, just as we need every creative form of theory and practice. These sciences will have liberating functions insofar as we build them on social relations not

24. Leon Baritz, *Servants of Power* (Middletown, Conn.: Wesleyan University Press, 1960). This work develops Mayo's industrial mythology in the context of a general criticism of the subservient role to established power played by American social science (see esp. chaps. 5 and 6).

25. For further discussion of revisionist primatologists see the second part of my article, "The Past Is the Contested Zone," in this volume. Other revisionists include Shirley Strum, "Life with the Pumphouse Gang," *National Geographic* 147 (1975): 672–91; Glenn Hausfater, "Dominance and Reproduction in Baboons (*Papio Cynocephalus*): A Quantitative Analysis" (Ph.D. diss., University of Chicago, 1974); Jane Lancaster, *Primate Behavior and the Emergence of Human Culture* (New York: Holt, Rinehart & Winston, 1975); and Cynthia Moss, *Portraits in the Wild: Behavior Studies of East African Mammals* (Boston: Houghton Mifflin Co., 1975). Moss is a science writer; Lancaster and Strum are part of Sherwood Washburn's remarkable lineage. Again, we must remember influential networks in the scientific community.

based on domination. A corollary of that requirement is the rejection of all forms of the ideological claims for pure objectivity rooted in the subject-object split that has legitimated our logics of domination of nature and ourselves. If our experience is of domination, we will theorize our lives according to principles of dominance. As we transform the foundations of our lives, we will know how to build natural sciences to underpin new relations with the world. We, like Dawn in Marge Piercy's *Woman on the Edge of Time*, want to fly into nature, as well as into the past, to make it come out all right. But the sciences are collective expressions and cannot be remade individually. Like Luciente and Hawk, in the same novel, feminists have been clear that "nobody can *make* things come out right," that "it isn't bad to want to help, to want to work, to seize history, . . . But to want to do it alone is less good. To hand history to someone like a cake you baked."[26]

Department of History of Science
Johns Hopkins University

26. Marge Piercy, *Woman on the Edge of Time* (New York: Alfred A. Knopf, Inc., 1976), pp. 188–89.

Feminism and Science

Evelyn Fox Keller

In recent years, a new critique of science has begun to emerge from a number of feminist writings. The lens of feminist politics brings into focus certain masculinist distortions of the scientific enterprise, creating, for those of us who are scientists, a potential dilemma. Is there a conflict between our commitment to feminism and our commitment to science? As both a feminist and a scientist, I am more familiar than I might wish with the nervousness and defensiveness that such a potential conflict evokes. As scientists, we have very real difficulties in thinking about the kinds of issues that, as feminists, we have been raising. These difficulties may, however, ultimately be productive. My purpose in the present essay is to explore the implications of recent feminist criticism of science for the relationship between science and feminism. Do these criticisms imply conflict? If they do, how necessary is that conflict? I will argue that those elements of feminist criticism that seem to conflict most with at least conventional conceptions of science may, in fact, carry a liberating potential for science. It could therefore benefit scientists to attend closely to feminist criticism. I will suggest that we might even use feminist thought to illuminate and clarify part of the substructure of science (which may have been historically conditioned into distortion) in order to preserve the things that science has taught us, in order to be more objective. But

---

EDITORS' NOTE: Evelyn Fox Keller takes up the question of the intermingling of masculine bias with what purports to be objective, scientific statement. While agreeing that a feminist critique of such bias is valid, Keller warns against a relativism possible in feminist theory that, in her view, "dooms women to residing outside of realpolitik modern culture."

This essay originally appeared in *Signs,* vol. 7, no. 3, Spring 1982.

first it is necessary to review the various criticisms that feminists have articulated.

The range of their critique is broad. Though they all claim that science embodies a strong androcentric bias, the meanings attached to this charge vary widely. It is convenient to represent the differences in meaning by a spectrum that parallels the political range characteristic of feminism as a whole. I label this spectrum from right to left, beginning somewhere left of center with what might be called the liberal position. From the liberal critique, charges of androcentricity emerge that are relatively easy to correct. The more radical critique calls for correspondingly more radical changes; it requires a reexamination of the underlying assumptions of scientific theory and method for the presence of male bias. The difference between these positions is, however, often obscured by a knee-jerk reaction that leads many scientists to regard all such criticism as a unit—as a challenge to the neutrality of science. One of the points I wish to emphasize here is that the range of meanings attributed to the claim of androcentric bias reflects very different levels of challenge, some of which even the most conservative scientists ought to be able to accept.

First, in what I have called the liberal critique, is the charge that is essentially one of unfair employment practices. It proceeds from the observation that almost all scientists are men. This criticism is liberal in the sense that it in no way conflicts either with traditional conceptions of science or with current liberal, egalitarian politics. It is, in fact, a purely political criticism, and one which can be supported by all of us who are in favor of equal opportunity. According to this point of view, science itself would in no way be affected by the presence or absence of women.

A slightly more radical criticism continues from this and argues that the predominance of men in the sciences has led to a bias in the choice and definition of problems with which scientists have concerned themselves. This argument is most frequently and most easily made in regard to the health sciences. It is claimed, for example, that contraception has not been given the scientific attention its human importance warrants and that, furthermore, the attention it has been given has been focused primarily on contraceptive techniques to be used by women. In a related complaint, feminists argue that menstrual cramps, a serious problem for many women, have never been taken seriously by the medical profession. Presumably, had the concerns of medical research been articulated by women, these particular imbalances would not have arisen.[1] Similar biases in sciences remote from the subject of women's bodies are more

---

1. Notice that the claim is not that the mere presence of women in medical research is sufficient to right such imbalances, for it is understood how readily women, or any "outsiders" for that matter, come to internalize the concerns and values of a world to which they aspire to belong.

difficult to locate—they may, however, exist. Even so, this kind of criticism does not touch our conception of what science is, nor our confidence in the neutrality of science. It may be true that in some areas we have ignored certain problems, but our definition of science does not include the choice of problem—that, we can readily agree, has always been influenced by social forces. We remain, therefore, in the liberal domain.

Continuing to the left, we next find claims of bias in the actual design and interpretation of experiments. For example, it is pointed out that virtually all of the animal-learning research on rats has been performed with male rats.[2] Though a simple explanation is offered—namely, that female rats have a four-day cycle that complicates experiments—the criticism is hardly vitiated by the explanation. The implicit assumption is, of course, that the male rat represents the species. There exist many other, often similar, examples in psychology. Examples from the biological sciences are somewhat more difficult to find, though one suspects that they exist. An area in which this suspicion is particularly strong is that of sex research. Here the influence of heavily invested preconceptions seems all but inevitable. In fact, although the existence of such preconceptions has been well documented historically,[3] a convincing case for the existence of a corresponding bias in either the design or interpretation of experiments has yet to be made. That this is so can, I think, be taken as testimony to the effectiveness of the standards of objectivity operating.

But evidence for bias in the interpretation of observations and experiments is very easy to find in the more socially oriented sciences. The area of primatology is a familiar target. Over the past fifteen years women working in the field have undertaken an extensive reexamination of theoretical concepts, often using essentially the same methodological tools. These efforts have resulted in some radically different formulations. The range of difference frequently reflects the powerful influence of ordinary language in biasing our theoretical formulations. A great deal of very interesting work analyzing such distortions has been done.[4] Though I cannot begin to do justice to that work here, let me offer, as a single example, the following description of a single-male troop of animals that Jane Lancaster provides as a substitute for the familiar concept of "harem": "For a female, males are a

2. I would like to thank Lila Braine for calling this point to my attention.

3. D. L. Hall and Diana Long, "The Social Implications of the Scientific Study of Sex," *Scholar and the Feminist* 4 (1977): 11–21.

4. See, e.g., Donna Haraway, "Animal Sociology and a Natural Economy of the Body Politic, Part I: A Political Physiology of Dominance"; and "Animal Sociology and a Natural Economy of the Body Politic, Part II: The Past Is the Contested Zone: Human Nature and Theories of Production and Reproduction in Primate Behavior Studies," *Signs: Journal of Women in Culture and Society* 4, no. 1 (Autumn 1978): 21–60.

resource in her environment which she may use to further the survival of herself and her offspring. If environmental conditions are such that the male role can be minimal, a one-male group is likely. Only one male is necessary for a group of females if his only role is to impregnate them."[5]

These critiques, which maintain that a substantive effect on scientific theory results from the predominance of men in the field, are almost exclusively aimed at the "softer," even the "softest," sciences. Thus they can still be accommodated within the traditional framework by the simple argument that the critiques, if justified, merely reflect the fact that these subjects are not sufficiently scientific. Presumably, fair-minded (or scientifically minded) scientists can and should join forces with the feminists in attempting to identify the presence of bias—equally offensive, if for different reasons, to both scientists and feminists—in order to make these "soft" sciences more rigorous.

It is much more difficult to deal with the truly radical critique that attempts to locate androcentric bias even in the "hard" sciences, indeed in scientific ideology itself. This range of criticism takes us out of the liberal domain and requires us to question the very assumptions of objectivity and rationality that underlie the scientific enterprise. To challenge the truth and necessity of the conclusions of natural science on the grounds that they too reflect the judgment of men is to take the Galilean credo and turn it on its head. It is not true that "the conclusions of natural science are true and necessary, and the judgement of man has nothing to do with them";[6] it is the judgment of woman that they have nothing to do with.

The impetus behind this radical move is twofold. First, it is supported by the experience of feminist scholars in other fields of inquiry. Over and over, feminists have found it necessary, in seeking to reinstate women as agents and as subjects, to question the very canons of their fields. They have turned their attention, accordingly, to the operation of patriarchal bias on ever deeper levels of social structure, even of language and thought.

But the possibility of extending the feminist critique into the foundations of scientific thought is created by recent developments in the history and philosophy of science itself.[7] As long as the course of scientific thought was judged to be exclusively determined by its own logi-

5. Jane Lancaster, *Primate Behavior and the Emergence of Human Culture* (New York: Holt, Rinehart & Winston, 1975), p. 34.

6. Galileo Galilei, *Dialogue on the Great World Systems,* trans. T. Salusbury, ed. G. de Santillana (Chicago: University of Chicago Press, 1953), p. 63.

7. The work of Russell Hanson and Thomas S. Kuhn was of pivotal importance in opening up our understanding of scientific thought to a consideration of social, psychological, and political influences.

cal and empirical necessities, there could be no place for any signature, male or otherwise, in that system of knowledge. Furthermore, any suggestion of gender differences in our thinking about the world could argue only too readily for the further exclusion of women from science. But as the philosophical and historical inadequacies of the classical conception of science have become more evident, and as historians and sociologists have begun to identify the ways in which the development of scientific knowledge has been shaped by its particular social and political context, our understanding of science as a social process has grown. This understanding is a necessary prerequisite, both politically and intellectually, for a feminist theoretic in science.

Joining feminist thought to other social studies of science brings the promise of radically new insights, but it also adds to the existing intellectual danger a political threat. The intellectual danger resides in viewing science as pure social product; science then dissolves into ideology and objectivity loses all intrinsic meaning. In the resulting cultural relativism, any emancipatory function of modern science is negated, and the arbitration of truth recedes into the political domain.[8] Against this background, the temptation arises for feminists to abandon their claim for representation in scientific culture and, in its place, to invite a return to a purely "female" subjectivity, leaving rationality and objectivity in the male domain, dismissed as products of a purely male consciousness.[9]

Many authors have addressed the problems raised by total relativism;[10] here I wish merely to mention some of the special problems added by its feminist variant. They are several. In important respects, feminist relativism is just the kind of radical move that transforms the political spectrum into a circle. By rejecting objectivity as a masculine ideal, it simultaneously lends its voice to an enemy chorus and dooms women to residing outside of the realpolitik modern culture; it exacerbates the very problem it wishes to solve. It also nullifies the radical potential of feminist criticism for our understanding of science. As I see it, the task of a feminist theoretic in science is twofold: to distinguish that which is parochial from that which is universal in the scientific impulse,

8. See, e.g., Paul Feyerabend, *Against Method* (London: New Left Books, 1975); and *Science in a Free Society* (London: New Left Books, 1978).

9. This notion is expressed most strongly by some of the new French feminists (see Elaine Marks and Isabelle de Courtivron, eds., *New French Feminisms: An Anthology* [Amherst: University of Massachusetts Press, 1980]), and is currently surfacing in the writings of some American feminists. See, e.g., Susan Griffin, *Woman and Nature: The Roaring Inside Her* (New York: Harper & Row, 1978).

10. See, e.g., Steven Rose and Hilary Rose, "Radical Science and Its Enemies," *Socialist Register 1979*, ed. Ralph Miliband and John Saville (Atlantic Highlands, N.J.: Humanities Press, 1979), pp. 317–35. A number of the points made here have also been made by Elizabeth Fee in "Is Feminism a Threat to Objectivity?" (paper presented at the American Association for the Advancement of Science meeting, Toronto, January 4, 1981).

reclaiming for women what has historically been denied to them; and to legitimate those elements of scientific culture that have been denied precisely because they are defined as female.

It is important to recognize that the framework inviting what might be called the nihilist retreat is in fact provided by the very ideology of objectivity we wish to escape. This is the ideology that asserts an opposition between (male) objectivity and (female) subjectivity and denies the possibility of mediation between the two. A first step, therefore, in extending the feminist critique to the foundations of scientific thought is to reconceptualize objectivity as a dialectical process so as to allow for the possibility of distinguishing the objective effort from the objectivist illusion. As Piaget reminds us:

> Objectivity consists in so fully realizing the countless intrusions of the self in everyday thought and the countless illusions which result—illusions of sense, language, point of view, value, etc.—that the preliminary step to every judgement is the effort to exclude the intrusive self. Realism, on the contrary, consists in ignoring the existence of self and thence regarding one's own perspective as immediately objective and absolute. Realism is thus anthropocentric illusion, finality—in short, all those illusions which teem in the history of science. So long as thought has not become conscious of self, it is a prey to perpetual confusions between objective and subjective, between the real and the ostensible.[11]

In short, rather than abandon the quintessentially human effort to understand the world in rational terms, we need to refine that effort. To do this, we need to add to the familiar methods of rational and empirical inquiry the additional process of critical self-reflection. Following Piaget's injunction, we need to "become conscious of self." In this way, we can become conscious of the features of the scientific project that belie its claim to universality.

The ideological ingredients of particular concern to feminists are found where objectivity is linked with autonomy and masculinity, and in turn, the goals of science with power and domination. The linking of objectivity with social and political autonomy has been examined by many authors and shown to serve a variety of important political functions.[12] The implications of joining objectivity with masculinity are less well understood. This conjunction also serves critical political functions. But an understanding of the sociopolitical meaning of the entire constellation requires an examination of the psychological processes

---

11. Jean Piaget, *The Child's Conception of the World* (Totowa, N.J.: Littlefield, Adams & Co., 1972).

12. Jerome R. Ravetz, *Scientific Knowledge and Its Social Problems* (London: Oxford University Press, 1971); and Hilary Rose and Steven Rose, *Science and Society* (London: Allen Lane, 1969).

through which these connections become internalized and perpetuated. Here psychoanalysis offers us an invaluable perspective, and it is to the exploitation of that perspective that much of my own work has been directed. In an earlier paper, I tried to show how psychoanalytic theories of development illuminate the structure and meaning of an interacting system of associations linking objectivity (a cognitive trait) with autonomy (an affective trait) and masculinity (a gender trait).[13] Here, after a brief summary of my earlier argument, I want to explore the relation of this system to power and domination.

Along with Nancy Chodorow and Dorothy Dinnerstein, I have found that branch of psychoanalytic theory known as object relations theory to be especially useful.[14] In seeking to account for personality development in terms of both innate drives and actual relations with other objects (i.e., subjects), it permits us to understand the ways in which our earliest experiences—experiences in large part determined by the socially structured relationships that form the context of our developmental processes—help to shape our conception of the world and our characteristic orientations to it. In particular, our first steps in the world are guided primarily by the parents of one sex—our mothers; this determines a maturational framework for our emotional, cognitive, and gender development, a framework later filled in by cultural expectations.

In brief, I argued the following: Our early maternal environment, coupled with the cultural definition of masculine (that which can never appear feminine) and of autonomy (that which can never be compromised by dependency) leads to the association of female with the pleasures and dangers of merging, and of male with the comfort and loneliness of separateness. The boy's internal anxiety about both self and gender is echoed by the more widespread cultural anxiety, thereby encouraging postures of autonomy and masculinity, which can, indeed may, be designed to defend against that anxiety and the longing that generates it. Finally, for all of us, our sense of reality is carved out of the same developmental matrix. As Piaget and others have emphasized, the capacity for cognitive distinctions between self and other (objectivity) evolves concurrently and interdependently with the development of psychic autonomy; our cognitive ideals thereby become subject to the same psychological influences as our emotional and gender ideals. Along with autonomy the very act of separating subject from object—objectivity itself—comes to be associated with masculinity. The combined psycho-

13. Evelyn Fox Keller, "Gender and Science," *Psychoanalysis and Contemporary Thought* 1 (1978): 409–33.

14. Nancy Chodorow, *The Reproduction of Mothering: Psychoanalysis and the Sociology of Gender* (Berkeley: University of California Press, 1978); and Dorothy Dinnerstein, *The Mermaid and the Minotaur: Sexual Arrangements and Human Malaise* (New York: Harper & Row, 1976).

logical and cultural pressures lead all three ideals—affective, gender, and cognitive—to a mutually reinforcing process of exaggeration and rigidification.[15] The net result is the entrenchment of an objectivist ideology and a correlative devaluation of (female) subjectivity.

This analysis leaves out many things. Above all it omits discussion of the psychological meanings of power and domination, and it is to those meanings I now wish to turn. Central to object relations theory is the recognition that the condition of psychic autonomy is double edged: it offers a profound source of pleasure, and simultaneously of potential dread. The values of autonomy are consonant with the values of competence, of mastery. Indeed competence is itself a prior condition for autonomy and serves immeasurably to confirm one's sense of self. But need the development of competence and the sense of mastery lead to a state of alienated selfhood, of denied connectedness, of defensive separateness? To forms of autonomy that can be understood as protections against dread? Object relations theory makes us sensitive to autonomy's range of meanings; it simultaneously suggests the need to consider the corresponding meanings of competence. Under what circumstances does competence imply mastery of one's own fate and under what circumstances does it imply mastery over another's? In short, are control and domination essential ingredients of competence, and intrinsic to selfhood, or are they correlates of an alienated selfhood?

One way to answer these questions is to use the logic of the analysis summarized above to examine the shift from competence to power and control in the psychic economy of the young child. From that analysis, the impulse toward domination can be understood as a natural concomitant of defensive separateness—as Jessica Benjamin has written, "A way of repudiating sameness, dependency and closeness with another person, while attempting to avoid the consequent feelings of aloneness."[16] Perhaps no one has written more sensitively than psychoanalyst D. W. Winnicott of the rough waters the child must travel in negotiating the transition from symbiotic union to the recognition of self and other as autonomous entities. He alerts us to a danger that others have missed—a danger arising from the unconscious fantasy that the subject has actually destroyed the object in the process of becoming separate.

Indeed, he writes, "It is the destruction of the object that places the

15. For a fuller development of this argument, see n. 12 above. By focusing on the contributions of individual psychology, I in no way mean to imply a simple division of individual and social factors, or to set them up as alternative influences. Individual psychological traits evolve in a social system and, in turn, social systems reward and select for particular sets of individual traits. Thus if particular options in science reflect certain kinds of psychological impulses or personality traits, it must be understood that it is in a distinct social framework that those options, rather than others, are selected.

16. Jessica Benjamin has discussed this same issue in an excellent analysis of the place of domination in sexuality. See "The Bonds of Love: Rational Violence and Erotic Domination," *Feminist Studies* 6, no. 1 (Spring 1980): 144–74, esp. 150.

object outside the area of control. . . . . After 'subject relates to object' comes 'subject destroys object' (as it becomes external); then may come '*object survives* destruction by the subject.' But there may or may not be survival." When there is, "because of the survival of the object, the subject may now have started to live a life in the world of objects, and so the subject stands to gain immeasurably; but the price has to be paid in acceptance of the ongoing destruction in unconscious fantasy relative to object-relating."[17] Winnicott, of course, is not speaking of actual survival but of subjective confidence in the survival of the other. Survival in that sense requires that the child maintain relatedness; failure induces inevitable guilt and dread. The child is poised on a terrifying precipice. On one side lies the fear of having destroyed the object, on the other side, loss of self. The child may make an attempt to secure this precarious position by seeking to master the other. The cycles of destruction and survival are reenacted while the other is kept safely at bay, and as Benjamin writes, "the original self assertion is . . . converted from innocent mastery to mastery over and against the other."[18] In psychodynamic terms, this particular resolution of preoedipal conflicts is a product of oedipal consolidation. The (male) child achieves his final security by identification with the father—an identification involving simultaneously a denial of the mother and a transformation of guilt and fear into aggression.

Aggression, of course, has many meanings, many sources, and many forms of expression. Here I mean to refer only to the form underlying the impulse toward domination. I invoke psychoanalytic theory to help illuminate the forms of expression that impulse finds in science as a whole, and its relation to objectification in particular. The same questions I asked about the child I can also ask about science. Under what circumstances is scientific knowledge sought for the pleasures of knowing, for the increased competence it grants us, for the increased mastery (real or imagined) over our own fate, and under what circumstances is it fair to say that science seeks actually to dominate nature? Is there a meaningful distinction to be made here?

In his work *The Domination of Nature* William Leiss observes, "The necessary correlate of domination is the consciousness of subordination in those who must obey the will of another; thus properly speaking only other men can be the objects of domination."[19] (Or women, we might add.) Leiss infers from this observation that it is not the domination of physical nature we should worry about but the use of our knowledge of physical nature as an instrument for the domination of human nature. He therefore sees the need for correctives, not in science but in its uses. This is his point of departure from other authors of the Frankfurt

17. D. W. Winnicott, *Playing and Reality* (New York: Basic Books, 1971), pp. 89–90.
18. Benjamin, p. 165.
19. William Leiss, *The Domination of Nature* (Boston: Beacon Press, 1974), p. 122.

school, who assume the very logic of science to be the logic of domination. I agree with Leiss's basic observation but draw a somewhat different inference. I suggest that the impulse toward domination does find expression in the goals (and even in the theories and practice) of modern science, and argue that where it finds such expression the impulse needs to be acknowledged as projection. In short, I argue that not only in the denial of interaction between subject and other but also in the access of domination to the goals of scientific knowledge, one finds the intrusion of a self we begin to recognize as partaking in the cultural construct of masculinity.

The value of consciousness is that it enables us to make choices—both as individuals and as scientists. Control and domination are in fact intrinsic neither to selfhood (i.e., autonomy) nor to scientific knowledge. I want to suggest, rather, that the particular emphasis Western science has placed on these functions of knowledge is twin to the objectivist ideal. Knowledge in general, and scientific knowledge in particular, serves two gods: power and transcendence. It aspires alternately to mastery over and union with nature.[20] Sexuality serves the same two gods, aspiring to domination and ecstatic communion—in short, aggression and eros. And it is hardly a new insight to say that power, control, and domination are fueled largely by aggression, while union satisfies a more purely erotic impulse.

To see the emphasis on power and control so prevalent in the rhetoric of Western science as projection of a specifically male consciousness requires no great leap of the imagination. Indeed, that perception has become a commonplace. Above all, it is invited by the rhetoric that conjoins the domination of nature with the insistent image of nature as female, nowhere more familiar than in the writings of Francis Bacon. For Bacon, knowledge and power are one, and the promise of science is expressed as "leading to you Nature with all her children to bind her to your service and make her your slave,"[21] by means that do not "merely exert a gentle guidance over nature's course; they have the power to conquer and subdue her, to shake her to her foundations."[22] In the context of the Baconian vision, Bruno Bettelheim's conclusion appears inescapable: "Only with phallic psychology did aggressive manipulation of nature become possible."[23]

20. For a discussion of the different roles these two impulses play in Platonic and in Baconian images of knowledge, see Evelyn Fox Keller, "Nature as 'Her' " (paper delivered at the Second Sex Conference, New York Institute for the Humanities, September 1979).

21. B. Farrington, "*Temporis Partus Masculus:* An Untranslated Writing of Francis Bacon," *Centaurus* 1 (1951): 193–205, esp. 197.

22. Francis Bacon, "Description of the Intellectual Globe," in *The Philosophical Works of Francis Bacon,* ed. J. H. Robertson (London: Routledge & Sons, 1905), p. 506.

23. Quoted in Norman O. Brown, *Life against Death* (New York: Random House, 1959), p. 280.

The view of science as an oedipal project is also familiar from the writings of Herbert Marcuse and Norman O. Brown.[24] But Brown's preoccupation, as well as Marcuse's, is with what Brown calls a "morbid" science. Accordingly, for both authors the quest for a nonmorbid science, an "erotic" science, remains a romantic one. This is so because their picture of science is incomplete: it omits from consideration the crucial, albeit less visible, erotic components already present in the scientific tradition. Our own quest, if it is to be realistic rather than romantic, must be based on a richer understanding of the scientific tradition, in all its dimensions, and on an understanding of the ways in which this complex, dialectical tradition becomes transformed into a monolithic rhetoric. Neither the oedipal child nor modern science has in fact managed to rid itself of its preoedipal and fundamentally bisexual yearnings. It is with this recognition that the quest for a different science, a science undistorted by masculinist bias, must begin.

The presence of contrasting themes, of a dialectic between aggressive and erotic impulses, can be seen both within the work of individual scientists and, even more dramatically, in the juxtaposed writings of different scientists. Francis Bacon provides us with one model;[25] there are many others. For an especially striking contrast, consider a contemporary scientist who insists on the importance of "letting the material speak to you," of allowing it to "tell you what to do next"—one who chastises other scientists for attempting to "impose an answer" on what they see. For this scientist, discovery is facilitated by becoming "part of the system," rather than remaining outside; one must have a "feeling for the organism."[26] It is true that the author of these remarks is not only from a different epoch and a different field (Bacon himself was not actually a scientist by most standards), she is also a woman. It is also true that there are many reasons, some of which I have already suggested, for thinking that gender (itself constructed in an ideological context) actually does make a difference in scientific inquiry. Nevertheless, my point here is that neither science nor individuals are totally bound by ideology. In fact, it is not difficult to find similar sentiments expressed by male scientists. Consider, for example, the following remarks: "I have often had cause to feel that my hands are cleverer than my head. That is a crude way of characterizing the dialectics of experimentation. When it is going well, it is like a quiet conversation with Nature."[27] The difference

24. Brown; and Herbert Marcuse, *One Dimensional Man* (Boston: Beacon Press, 1964).

25. For a discussion of the presence of the same dialectic in the writings of Francis Bacon, see Evelyn Fox Keller, "Baconian Science: A Hermaphrodite Birth," *Philosophical Forum* 11, no. 3 (Spring 1980): 299–308.

26. Barbara McClintock, private interviews, December 1, 1978, and January 13, 1979.

27. G. Wald, "The Molecular Basis of Visual Excitation," *Les Prix Nobel en 1967* (Stockholm: Kungliga Boktryckerlet, 1968), p. 260.

between conceptions of science as "dominating" and as "conversing with" nature may not be a difference primarily between epochs, nor between the sexes. Rather, it can be seen as representing a dual theme played out in the work of all scientists, in all ages. But the two poles of this dialectic do not appear with equal weight in the history of science. What we therefore need to attend to is the evolutionary process that selects one theme as dominant.

Elsewhere I have argued for the importance of a different selection process.[28] In part, scientists are themselves selected by the emotional appeal of particular (stereotypic) images of science. Here I am arguing for the importance of selection within scientific thought—first of preferred methodologies and aims, and finally of preferred theories. The two processes are not unrelated. While stereotypes are not binding (i.e., they do not describe all or perhaps any individuals), and this fact creates the possibility for an ongoing contest within science, the first selection process undoubtedly influences the outcome of the second. That is, individuals drawn by a particular ideology will tend to select themes consistent with that ideology.

One example in which this process is played out on a theoretical level is in the fate of interactionist theories in the history of biology. Consider the contest that has raged throughout this century between organismic and particulate views of cellular organization—between what might be described as hierarchical and nonhierarchical theories. Whether the debate is over the primacy of the nucleus or the cell as a whole, the genome or the cytoplasm, the proponents of hierarchy have won out. One geneticist has described the conflict in explicitly political terms:

> Two concepts of genetic mechanisms have persisted side by side throughout the growth of modern genetics, but the emphasis has been very strongly in favor of one of these. . . . The first of these we will designate as the "Master Molecule" concept. . . . This is in essence the Theory of the Gene, interpreted to suggest a totalitarian government. . . . The second concept we will designate as the "Steady State" concept. By this term . . . we envision a dynamic self-perpetuating organization of a variety of molecular species which owes its specific properties not to the characteristic of any one kind of molecule, but to the functional interrelationships of these molecular species.[29]

Soon after these remarks, the debate between "master molecules" and dynamic interactionism was foreclosed by the synthesis provided by

28. Keller, "Gender and Science."

29. D. L. Nanney, "The Role of the Cyctoplasm in Heredity," in *The Chemical Basis of Heredity,* ed. William D. McElroy and Bentley Glass (Baltimore: Johns Hopkins University Press, 1957), p. 136.

DNA and the "central dogma." With the success of the new molecular biology such "steady state" (or egalitarian) theories lost interest for almost all geneticists. But today, the same conflict shows signs of reemerging—in genetics, in theories of the immune system, and in theories of development.

I suggest that method and theory may constitute a natural continuum, despite Popperian claims to the contrary, and that the same processes of selection may bear equally and simultaneously on both the means and aims of science and the actual theoretical descriptions that emerge. I suggest this in part because of the recurrent and striking consonance that can be seen in the way scientists work, the relation they take to their object of study, and the theoretical orientation they favor. To pursue the example cited earlier, the same scientist who allowed herself to become "part of the system," whose investigations were guided by a "feeling for the organism," developed a paradigm that diverged as radically from the dominant paradigm of her field as did her methodological style.

In lieu of the linear hierarchy described by the central dogma of molecular biology, in which the DNA encodes and transmits all instructions for the unfolding of a living cell, her research yielded a view of the DNA in delicate interaction with the cellular environment—an organismic view. For more important than the genome as such (i.e., the DNA) is the "overall organism." As she sees it, the genome functions "only in respect to the environment in which it is found."[30] In this work the program encoded by the DNA is itself subject to change. No longer is a master control to be found in a single component of the cell; rather, control resides in the complex interactions of the entire system. When first presented, the work underlying this vision was not understood, and it was poorly received.[31] Today much of that work is undergoing a renaissance, although it is important to say that her full vision remains too radical for most biologists to accept.[32]

This example suggests that we need not rely on our imagination for a vision of what a different science—a science less restrained by the impulse to dominate—might be like. Rather, we need only look to the thematic pluralism in the history of our own science as it has evolved. Many other examples can be found, but we lack an adequate understanding of the full range of influences that lead to the acceptance or rejection not only of particular theories but of different theoretical orientations. What I am suggesting is that if certain theoretical inter-

30. McClintock, December 1, 1978.

31. McClintock, "Chromosome Organization and Genic Expression," *Cold Spring Harbor Symposium of Quantitative Biology* 16 (1951): 13–44.

32. McClintock's most recent publication on this subject is "Modified Gene Expressions Induced by Transposable Elements," in *Mobilization and Reassembly of Genetic Information,* ed. W. A. Scott, R. Werner, and J. Schultz (New York: Academic Press, 1980).

pretations have been selected against, it is precisely in this process of selection that ideology in general, and a masculinist ideology in particular, can be found to effect its influence. The task this implies for a radical feminist critique of science is, then, first a historical one, but finally a transformative one. In the historical effort, feminists can bring a whole new range of sensitivities, leading to an equally new consciousness of the potentialities lying latent in the scientific project.

Visiting Professor of Mathematics and Humanities
Northeastern University

The Cartesian Masculinization
of Thought

Susan Bordo

On November 10, 1619, Descartes had a series of dreams—bizarre, richly imaginal sequences manifestly full of anxiety and dread. He interpreted these dreams—which most readers would surely regard as nightmares[1]— as revealing to him that mathematics is the key to understanding the universe. Descartes' resolute and disconcertingly positive interpretation has become a standard textbook anecdote, a symbol of the seventeenth-century rationalist project. That project, in the official story told in most philosophy and history texts, describes seventeenth-century culture as Descartes described his dream: in terms of intellectual beginnings, fresh confidence, and a new belief in the ability of science—armed with the discourses of mathematics and the "new philosophy"—to decipher the language of nature.

Recent scholarship, however, has detected a certain instability, a dark underside, to the bold rationalist vision. Different writers describe it in different ways. Richard Bernstein speaks of the great "Cartesian anxiety" over the possibility of intellectual and moral chaos;[2] Karsten Harries

The ideas in this paper have been presented in various forms and various contexts, from informal discussion to colloquium presentations to class lectures. To all my friends, colleagues, and students who have helped me explore and clarify those ideas, I express my appreciation. In particular, I would like to thank Mario Moussa and Carolyn Merchant for their insightful editorial suggestions and helpful criticisms of earlier drafts of this paper.

1. For an account of these dreams, see Karl Stern, *The Flight from Woman* (New York: Noonday Press, 1965), 80–84. Descartes' original account is, unfortunately, lost.

2. Richard Bernstein, "Philosophy in the Conversation of Mankind," *Review of Metaphysics* 32, no. 4 (1980): 762.

This essay originally appeared in *Signs*, vol. 11, no. 3, Spring 1986.

speaks of the (Cartesian) "dread of the distorting power of perspective";[3] Richard Rorty reminds us that the seventeenth-century ideal of a perfectly mirrored nature is also an "attempt to escape" from history, culture, and human finitude.[4] Looking freshly at Descartes' *Meditations*, one cannot help but be struck by the manifest epistemological anxiety of the earlier *Meditations* and by how unresolute a mode of inquiry they embody: the dizzying vacillations, the constant requestioning of the self, the determination, if only temporary, to stay within confusion and contradiction, to favor interior movement rather than clarity and resolve.

All that, of course, is ultimately left behind by Descartes, as firmly as his bad dreams (as he tells his correspondent, Elizabeth of Bohemia) were conquered by the vigilance of his reason. The model of knowledge that Descartes bequeathed to modern science, and of which he is often explicitly described as the father, is based on clarity, dispassion, and detachment. Yet the transformation from the imagery of nightmare (the *Meditations'* demons, dreamers, and madmen) to the imagery of objectivity remains unconvincing. The sense of experience conveyed by the first two *Meditations*—what Karl Stern calls the sense of "reality founded on uncertainty"[5]—is not quite overcome for the reader by the positivity of the later *Meditations*. Descartes' critics felt this in his own time. Over and over, the objection is raised: given the power of the first two *Meditations*, how can you really claim to have extricated yourself from the doubt and from the dream?

This instability was shared by the intellectual culture that Descartes was to transform so decisively. Indeed, I will propose in this essay that the *Meditations* be read as a mirror of that culture, a reflection both of its anxieties and its responses to those anxieties. I will further suggest that the categories of modern developmental psychology—specifically those that have developed around the issues of separation and individuation—can provide an illuminative psychocultural framework for a fresh reading of the *Meditations* and of the major philosophical and cultural transformations of the seventeenth century.

Drawing on the work of Margaret Mahler, Jean Piaget, and Norman O. Brown, I will undertake a reexamination of Cartesian doubt and of Descartes' seeming triumph over the epistemological insecurity of the first two *Meditations*. I will suggest that we view the "great Cartesian anxiety," although manifestly expressed in epistemological terms, as anxiety over separation—from the organic female universe of the Middle

---

3. Karsten Harries, "Descartes, Perspective and the Angelic Eye," *Yale French Studies*, no. 49 (1973), 29.

4. Richard Rorty, *Philosophy and the Mirror of Nature* (Princeton, N.J.: Princeton University Press), 9.

5. Stern, 99.

Ages and the Renaissance. Cartesian objectivism, correspondingly, will be explored as a defensive response to that separation anxiety, an aggressive intellectual "flight from the feminine" rather than (simply) the confident articulation of a positive new epistemological ideal. The more concrete, political, and institutional expressions of such a seventeenth-century flight from the feminine have been chronicled by a number of authors. This essay will explore its philosophical expression, in what I will describe as the Cartesian *re*-birthing and re-imaging of knowledge and the world as masculine.

The notion that the project of modern science crystallizes "masculin-ist" modes of thinking has been a prominent theme in some recent writing: "[What] we encounter in Cartesian rationalism," says Karl Stern, "is the pure masculinization of thought."[6] The scientific model of know-ing, says Sandra Harding, represents a "super-masculinization of rational knowledge."[7] "The specific consciousness we call scientific, Western and modern," claims James Hillman, "is the long sharpened tool of the mascu-line mind that has discarded parts of its own substance, calling it 'Eve,' 'female' and 'inferior.'"[8] Understanding the development of Cartesian objectivism, and modern science in general, in terms of the cultural "drama of parturition" described in this essay will give some textual and historical support to these insights and clarify their importance.

Separation and Individuation Themes in the Meditations

The need for God's guarantee, in the *Meditations*, is a need for a principle of continuity and coherence for what is experienced by Des-cartes as a disastrously fragmented and discontinuous mental life. For Descartes, indeed, discontinuity is the central fact of human experience. Nothing—neither certainty, nor temporal existence itself—endures past the present moment without God. Time—both external and internal—is so fragmented that "in order to secure the continued existence of a thing, no less a cause is required than that needed to produce it at the first."[9] This means not only that our continued existence is causally dependent on God (HR 1:158–59), but that God is required to provide continuity and unity to our inner life as well. That inner life, without God, "is always of

6. Ibid., 104.
7. Sandra Harding "Is Gender a Variable in Conceptions of Rationality?" (paper delivered at the Fifth International Colloquium on Rationality, Vienna, 1981).
8. James Hillman, *The Myth of Analysis* (New York: Harper & Row, 1972), 250.
9. René Descartes, *Philosophical Works*, vols. 1 and 2, ed. Elizabeth Haldane and G. R. T. Ross (Cambridge: Cambridge University Press, 1961), 2:56 (hereafter referred to as HR 1 and HR 2).

the present moment";[10] two and two may equal four right now, while we are attending to it, but we need God to assure us that two and two will always form four, whether we are attending to it or not. Even the most forcefully experienced insights—save the *cogito*—become open to doubt once the immediacy of the intuition passes:

> For although I am of such a nature that as long as I understand anything very clearly and distinctly, I am naturally impelled to believe it to be true, yet because I am also of such a nature that I cannot have my mind constantly fixed on the same object in order to perceive it clearly, and as I often recollect having formed a past judgement without at the same time properly recollecting the reasons that led me to make it, it may happen meanwhile that other reasons present themselves to me, which would easily cause me to change my opinion, if I were ignorant of the facts of the existence of God, and thus I should have no true and certain knowledge, but only vague and vacillating opinion. [HR 1:183–84]

This strong sense of the fragility of human cognitive relations with the object world is closely connected to the new Cartesian sense (which Descartes shared with the culture around him) of what Stephen Toulmin has called "the inwardness of mental life": the sense of experience as occurring deeply within and bounded by a self.[11] According to many scholars of the era, such a sense was not prominent in the medieval experience of the world:

> When we think casually, we think of consciousness as situated at some point in space . . . even those who achieve the intellectual contortionism of denying that there is such a thing as consciousness, feel that this denial comes from inside their own skins. . . . This was not the background picture before the scientific revolution. The background picture then was of man as a microcosm within the macrocosm. It is clear that he did not feel himself isolated by his skin from the world outside to quite the same extent that we do. He was integrated or mortised into it, each different part of him being united to a different part of it by some invisible thread. In his relation to his environment, the man of the middle ages was rather less like an island, rather more like an embryo.[12]

---

10. Poulet, quoted in Marshall McLuhan, *The Gutenberg Galaxy* (Toronto: University of Toronto Press, 1962), 241.

11. Stephen Toulmin, "The Inwardness of Mental Life," *Critical Inquiry* 6 (Autumn 1979): 1–16.

12. Owen Barfield, *Saving the Appearances: A Study in Idolatry* (New York: Harcourt Brace Jovanovich, 1965), 78.

During the Renaissance, as Claudio Guillen argues, European culture became "interiorized."[13] The portrayal of the "inner life"—both as a dramatic problem and as a subject of literary exploration—becomes an issue. How can the playwright depict the experience of a character? In Shakespeare, a new theme emerges: the hidden substance of the self—the notion that the experience of individuals is fundamentally opaque, even inaccessible to others, who can only take an outer view of it. It is in the Renaissance, too, that philosophers begin to image mental life as an inner arena or space inside, deeply interior and, at the same time, capable of objectification and examination. Montaigne is a striking example: "I turn my gaze inward, I fix it there and keep it busy. . . . I look inside myself; I continually observe myself, I take stock of myself, I taste myself. . . . I roll about in myself."[14]

The *Meditations*, however, in both form and content, remain the most thoroughgoing and compelling examples we have of confrontation with the "inwardness of mental life." Augustine's *Confessions* embody a stream of consciousness, to be sure, but they very rarely confront that stream as an object of exploration. Descartes provides the first real phenomenology of the mind, and one of the central results of that phenomenology is the disclosure of the deep epistemological alienation that attends the sense of mental interiority: the enormous gulf that must separate what is conceived as occurring "in here" from that which, correspondingly, must lie "out there." The central inquiry of *Meditation 2*, and the first formulation of what was, unfortunately, to become the primary epistemological question for philosophers until Kant, is "whether any of the objects of which I have ideas *within me* exist *outside of me*" (HR 1:161; emphasis mine). Under such circumstances, *cogito ergo sum* is, indeed, the only emphatic reality, for to be assured of its truth, we require nothing but confrontation with the inner stream itself. Beyond the direct and indubitable "I am," the meditation on the self can lead to no other truths without God to bridge the gulf between the "inner" and the "outer."

Consider, in this connection, the difference between the ancient Greek and medieval view of the nature of error and the Cartesian view. For Descartes, error consists in the judgment that "inner" reality—modes of thought—"are similar or conformable to the things which are outside." (HR 1:160). For the Greeks and medievals, not only is such uncertainty about the capacity of the "inner" to lead reliably to the "outer" foreign, but the very notion of location in these matters is inappropriate. Reason

13. Claudio Guillen, *Literature as System* (Princeton, N.J.: Princeton University Press, 1971), 306–10.

14. Michel de Montaigne, *Essays*, trans. and ed. Donald Frame (New York: St. Martin's Press, 1963), 273.

was a human faculty, resisting metaphors of locatedness, neither inside nor outside the human being.[15] Completely absent is the image, so striking in Montaigne and Galileo as well as Descartes, of an unreliable, distorting "inner space"; rather, there are two worlds (or, for Aristotle, it might be more correct to say two aspects of the same world) and two human faculties—intellect and sense—appropriate to each. Error is the result of confusion between these worlds—the sensible and the unchanging—not the result of inner misrepresentation of the "external" world.

For Descartes, contrastingly, an epistemological chasm separates a highly self-conscious self from a universe that now lies decisively outside the self. This profound Cartesian experience of self as inwardness ("I think, therefore I am") and its corollary—the heightened sense of distance from the "not-I"—inspires a more psychological consideration of the popular imagery that describes the transition from Middle Ages to Renaissance in terms of birth.

Ortega y Gasset describes the "human drama which began in 1400 and ends in 1650" as a "drama of parturition."[16] Arthur Koestler compares the finite universe to a nursery and later, to a womb.[17] Owen Barfield, as we have seen, speaks of the medieval as an embryo. Such imagery may be more appropriate than any of these authors intended. As individuals, according to Margaret Mahler, our true psychological birth comes when we begin to experience our separateness from the mother, when we begin to individuate from her. That process, whose stages are described in detail by Mahler,[18] involves a slowly unfolding reciprocal delineation of self and world. For Mahler (as for Piaget, in describing cognitive development), as subjectivity becomes ever more internally aware, so the object world (via its principal representative, the mother) becomes ever more external and autonomous. Thus, the normal adult experience or "being both fully 'in' and at the same time basically separate from the world out there" is developed from an original state of unity with the mother.[19]

This is not easy for the child, for every major step in the direction of individuation also revives an "eternal longing" for the "ideal state of self" in which mother and child were one and recognition of our ever-increasing distance from it. "Side by side with the growth of [the child's] emotional life, there is a noticeable waning . . . of his [previous] relative

---

15. In Greek, as Richard Rorty points out, there is no way to divide "conscious states" from events in an external world (Rorty [n. 4 above], 47).

16. José Ortega y Gasset, *Man and Crisis*, trans. Mildred Adams (New York: W. W. Norton & Co., 1958), 184.

17. Arthur Koestler, *The Sleepwalkers* (New York: Grosset & Dunlap, 1959), 218.

18. Margaret Mahler, "On the First Three Phases of the Separation-Individuation Process," *International Journal of Psychoanalysis* 53 (1972): 333–38.

19. Ibid., 333.

obliviousness to the mother's presence. Increased separation anxiety can be observed . . . [in] a seemingly constant concern with the mother's whereabouts."[20] Although we become more or less reconciled to our separateness, the process of individuation and its anxieties "reverberates throughout the life cycle. It is never finished, it can always be reactivated."[21]

May not such a process reverberate, too, on the cultural level? Perhaps some cultural eras compensate for the pain of individuation better than others through a mother imagery of the cosmos (such as was dominant, e.g., throughout the Chaucerian and Elizabethan eras) that assuages the anxiety of our actual separateness as individuals. On the other hand, during periods in which long-established images of symbiosis and cosmic unity break down (as they did during the period of the scientific revolution), may we not expect an increase in self-consciousness and anxiety over the distance between self and world—a constant concern, to paraphrase Mahler, over the whereabouts of the world? All these, as I have suggested, are central motifs in the *Meditations*.[22]

Recall, in this connection, Descartes' concern over the inability of the mind to be "constantly fixed on the same object in order to perceive it clearly" (and thus, without God's guarantee, to be assured only of "vague and vacillating opinion"; HR 1:183–84). The original model of epistemological security (which Descartes knows cannot be fulfilled—thus, the

20. Ibid., 337.
21. Ibid., 33.
22. It might be argued that theories of separation and individuation belong to a peculiarly modern discourse, one in which the very concept of self that was nascent and still tenuous in the Cartesian era is a fully developed given and central focus. Moreover, the dynamics of separation and individuation describe individual development on the level of infant object relations. These categories cannot be transported so glibly as to describe cultural developments in an advanced historical era. I do not wish to be understood as making any general theoretical claims about the relationship of phylogeny and ontogeny or any empirical claims about the actual evolutionary progress of the race. I take the psychology of infancy presented here not so much as a scientific theory—a genetic schema to be mapped onto the progress of events in a particular era—but more as a hermeneutic aid, which provides clues to interpreting cultural developments. My use of psychological categories forces us to recognize the thoroughly historical character of precisely those categories of self and innerness that describe the modern sense of relatedness to the world. They do so because they do not presuppose these categories as givens, but view them as developmental accomplishments. To be sure, the development that someone like Mahler has in mind is individual development. But the originally undifferentiated experience of the infant—the psychological/cognitive "state of nature" out of which we develop into fully separate, self-conscious beings—nonetheless argues for the possibility of human modes of relatedness to the world in which separateness of self and world is less sharply delineated than it is within the accepted norms of modern experience. Such modes, it has been suggested, were characteristic of the prescientific experience of the world. (See, in particular, Owen Barfield and Morris Berman, *The Reenchantment of the World* [New York: Cornell University Press, 1981].)

need for God) is a constant state of mental vigilance over the object; in the absence of that, nothing can be certain. To put this in more concrete terms: no previously reached conclusions, no past insights, no remembered information can be trusted. Unless the object is present and immediately in sight, it ceases to be available to the knower.

Consider this epistemological instability in connection with Piaget's famous experiments on the development of "permanent object concept" in children. From them, we learn that the developing child does not at first perceive objects as having enduring stability, "firm in existence though they do not directly affect perception." Instead, the object world is characterized by "continuous annihilations and resurrections," depending upon whether or not the object is within the child's perceptual field. When an object leaves the child's sight, it (effectively) leaves the universe.[23] In a sense, that is Descartes' dilemma, too, and the reason he needs God. This is not to say that Descartes saw the world as a child does. He, of course, had "permanent object concept." In speaking of perceiving objects, he is talking not about rudimentary perception but of the intellectual apprehension of the essences of things. The structural similarity between his "doubt by inattention" (as Robert Alexander calls it)[24] and the developing child's perceptual deficiencies is suggestive, however, of the newness, the tenuousness (albeit on a far more sophisticated plane than that of the child) of the Cartesian experience of self-confronting world. Neither the self nor objects are stable, and the lack of stability in the object world is, indeed, experienced as concern over the whereabouts of the world. Re-union with the (mother) world is, however, impossible; only God the father can now provide the (external) reassurance Descartes needs.

I have spoken of re-union here because (without making any unsupportable claims about how medievals "saw" the world) it seems clear that for the medieval aesthetic and philosophical imagination, the categories of self and world, inner and outer, human and natural were not as rigorously opposed as they came to be during the Cartesian era. The most striking evidence for this comes, not only from the organic, holistic imagery of the cosmos and the animistic science that prevailed until the seventeenth century,[25] but from medieval art as well. That art, which seems so distorted and spatially incoherent to a modern viewer, does so precisely because it does not represent the point of view of a detached,

23. Jean Piaget, *The Construction of Reality in the Child* (New York: Random House, 1954), 103.

24. Robert Alexander, "Metaphysical Doubt and Its Removal," in *Cartesian Studies*, ed. R. J. Butler (New York: Barnes & Noble, 1972), 121.

25. See Carolyn Merchant, *The Death of Nature* (San Francisco: Harper & Row, 1980); Morris Berman; and Brian Easlea, *Witch-Hunting, Magic and the New Philosophy* (Atlantic Highlands, N.J.: Humanities Press, 1980).

discretely located observer confronting a visual field of separate objects. The latter mode of representation—that of the perspective painting—had become the dominant artistic convention by the seventeenth century. In the medieval painting, by contrast, the fiction of the fixed beholder is entirely absent; instead, the spectator, as art historian Samuel Edgerton describes the process, is invited to become "absorbed within the visual world . . . to walk about, experiencing structures, almost tactilely, from many different sides, rather than from a single, overall vantage."[26] Often, sides of objects that could not possibly be seen at once (from one perceptual point of view) are represented as though the (imagined) movement of the subject in relating to the object—touching it, considering it from all angles—constitutes the object itself. The re-created experience is of the world and self as an unbroken continuum.[27]

Owen Barfield suggests that the reason perspective was not discovered before the Renaissance was because they did not need it: "Before the scientific revolution the world was more like a garment men wore about them than a stage on which they moved. In such a world the convention of perspective was unnecessary. . . . It was as if the observers were themselves in the picture. Compared with us, they felt themselves and the objects around them and the words that expressed those objects, immersed together in something like a clear lake of—what shall we say?—of 'meaning,' if you choose."[28] By extreme contrast, consider Pascal's despair at what seems to him an arbitrary and impersonal "allotment" in the "infinite immensity of spaces of which I know nothing and which know nothing of me. . . . There is no reason for me to be here rather than there, now rather than then. Who put me here?"[29] Pascal's sense of homelessness and abandonment, his apprehension of an almost personal indifference on the part of the universe, is closely connected here to an acute anxiety at the experience of personal boundedness and locatedness, of "me-here-now" (and *only* "here-now"). A similar anxiety, as I have suggested, is at the heart of Descartes' need for a God to sustain both his existence and his inner life from moment to moment, to provide a reassurance of permanence and connection between self and world. Once, such connection had not been in question.

26. Samuel Edgerton, Jr., *The Renaissance Rediscovery of Linear Perspective* (New York: Harper & Row, 1975), 9. See this work for an excellent discussion of the contrasting aesthetic spaces of medieval and Renaissance art and the different perceptual world suggested by them.

27. The correspondence between central features of the pictorial space of medieval art and the real perceptual space of children, as studied by Piaget, are striking and suggestive.

28. Barfield (n. 12 above), 94–95.

29. Blaise Pascal, *Pensées*, trans. A. J. Krailsheimer (Harmondsworth: Penguin Books, 1966), 68.

Cultural Re-birth: The "Father of Oneself" Fantasy

> If a kind of Cartesian ideal were ever completely fulfilled, i.e., if the whole of nature were only what can be explained in terms of mathematical relationships—then we would look at the world with that fearful sense of alienation, with that utter loss of reality with which a future schizophrenic child looks at his mother. A machine cannot give birth. [Karl Stern, *The Flight from Woman*]

> Descartes envisages for himself a kind of rebirth. Intellectual salvation comes only to the twice-born. [Harry Frankfurt, *Demons, Dreamers and Madmen*][30]

The dialectics of separation and individuation offer a way of seeing the Cartesian era empathically and impressionistically, through association and image, allowing the psychological categories normally reserved for our understanding of individuals to come alive on the level of history and culture. As such, they become a means of imaginative rehabilitation of the psychological values that mediate and "deepen" events (as James Hillman puts it),[31] which makes them resonate in a particular way for the human beings living them. The story that emerges through such an imaginative rehabilitation of the Cartesian era is a drama of parturition—from a universe that had ceased to beat with the same heart as the human being long before Descartes declared it to be pure *res extensa*.

If then, the transition from Middle Ages to Renaissance can be looked on as a kind of protracted birth—from which the human being emerges as a decisively separate entity, no longer continuous with the universe with which it had once shared a soul—so the possibility of objectivity, strikingly, is conceived by Descartes as a kind of *re*-birth, on one's own terms, this time.

Most of us are familiar with the dominant Cartesian themes of starting anew, alone, without influence from the past or other people, with the guidance of reason alone. The product of our original and actual birth, childhood, being ruled by the body, is the source of all obscurity and confusion in our thinking. For, as body, we are completely reactive and nondiscriminative, unable to make the most basic distinctions between an inner occurrence and an external event. One might say, in fact, that the distinction has no meaning at all for the body. This is why infancy—when the mind, "newly united" to the body, was "swamped" or "immersed" within it (HR 1:237)—is for Descartes primarily a period of egocentrism

---

30. Harry Frankfurt, *Demons, Dreamers and Madmen* (New York: Bobbs-Merrill, 1970), 16.

31. James Hillman, *Re-Visioning Psychology* (New York: Harper & Row, 1972), x.

(in the Piagetian sense): of complete inability to distinguish between subject and object. It is this feature of infancy that is responsible for all the "childhood prejudices" that later persist in the form of adult philosophical confusion between primary and secondary qualities, the "preconceptions of the senses," and the dictates of reason. As children, we judged subjectively, determining "that there was more or less reality in each body, according as the impressions made [on our own bodies] were more or less strong. So, we attributed much greater reality to rocks than air, believed the stars were actually as small as "tiny lighted candles," and believed that heat and cold were properties of the objects themselves (HR 1:250). These "prejudices" stay with us, "we quite forget we had accepted them without sufficient examination, admitting them as though they were of perfect truth and certainty"; thus, "it is almost impossible that our judgments should be as pure and solid as they would have been if we had had complete use of our reason since birth and had never been guided except by it" (HR 1:88).

It is crucial to note that it is the lack of differentiation between subject and object, between self and world, that is construed here as the epistemological threat. The medieval sense of relatedness to the world had not depended on such "objectivity" but on continuity between the human and physical realms, on the interpenetrations, through meanings and associations, of self and world.[32] Now, a clear and distinct sense of the boundaries of the self has become the ideal; the lingering of infantile subjectivism has become the impediment to solid judgment. The state of childhood, moreover, can be revoked through a deliberate and methodical reversal of all the prejudices of childhood—and one can begin anew with reason as one's only parent. This is precisely what the *Meditations* attempt to do.

The precise form of our infantile prejudices, as we have seen, is the inability to distinguish properly what is happening solely "inside" the subject from what has an external existence. "Swamped" inside the body, one simply did not have a perspective from which to discriminate, to examine, to judge. In *Meditation* 1, Descartes re-creates that state of utter entrapment by luring the reader through the continuities between madness, then dreaming—that state each night when each of us loses our adult clarity and detachment—and finally to the possibility that the whole of our existence may be like a dream, a grand illusion so encompassing that there is no conceivable perspective from which to judge its correspondence with reality. This, in essence, is the Evil Demon hypothesis—a specter of complete enclosedness and entrapment within the self. The difference, of course, is that in childhood, we assumed that what we felt

32. This is not the place to detail or defend this characterization of the medieval sense of relatedness to the world, about which much has been written. See Barfield; and C. S. Lewis, *The Discarded Image* (Cambridge: Cambridge University Press, 1964).

was a measure of external reality; now, as mature Cartesian doubters, we reverse that prejudice. We assume nothing. We refuse to let our bodies mystify us: "I shall close my eyes, I shall stop my ears, I shall call away all my senses" (HR 1:157). We begin afresh. The result, in the *Meditations*, is a securing of all the boundaries that, in childhood, are so fragile: between the "inner" and the "outer," between the subjective and the objective, between self and world.

The separation of knower from known, which modern philosophy and science came to regard as a given—a condition for knowledge—was for Descartes a project, not a foundation, to be discovered. Crucial to that project was the Cartesian refashioning of the ontological orders of the human and the natural into two distinct substances—the spiritual and the corporeal—that share no qualities (other than being created), permit of interaction but no merging, and, indeed, are each defined precisely in opposition to the other (HR 1:190). The mutual exclusion of *res extensa* and *res cogitans* made possible the conceptualization of complete intellectual transcendence of the body, organ of the deceptive senses and distracting "commotion" in the heart, blood, and animal spirits. (The body, as we have seen, is the chief impediment to human objectivity, for Descartes.) It also established the utter diremption—detachment, dislocation—of the natural world from the realm of the human. It now became inappropriate to speak, as the medievals had done, in anthropocentric terms about nature, which for Descartes, is pure *res extensa*, totally devoid of mind and thought. More important, it means that the values and significances of things in relation to the human realm must now be understood as purely a reflection of how we feel about them, having nothing to do with their "objective" qualities. "Thus," says Whitehead in his famous sardonic criticism of seventeenth-century philosophy, "the poets are entirely mistaken. They should address their lyrics to themselves and should turn them into odes of self-congratulation. . . . Nature is a dull affair, soundless, scentless, colourless; merely the hurrying of material, endlessly, meaninglessly."[33] For the model of knowledge that results, neither bodily response (the sensual or the emotional) nor associational thinking, exploring the various personal or spiritual meanings the object has for us, can tell us anything about the object *"itself."* That can only be grasped, as Charles Gillispie puts it, "by measurement rather than sympathy."[34]

It is in this sense that the dominant philosophic and scientific culture of the seventeenth century indeed inaugurated "a truly masculine birth of

33. Alfred North Whitehead, *Science and the Modern World* (1925; reprint, Toronto: Collier Macmillan, 1967), 54.
34. Charles Gillispie, *The Edge of Objectivity* (Princeton, N.J.: Princeton University Press, 1960), 42.

time," as Francis Bacon proclaimed it.[35] The notion has been fleshed out by a number of feminist writers.[36] Here "masculine" describes not a biological category but a cognitive style, an epistemological stance. Its key term is *detachment*: from the emotional life, from the particularities of time and place, from personal quirks, prejudices, and interests, and most centrally, from the object itself. This masculine orientation toward knowledge, which Evelyn Fox Keller sees epitomized in the modern scientific ideal of objectivity, depends on a clear and distinct determination of the boundaries between self and world: "The scientific mind is set apart from what is to be known, that is, from nature, and its autonomy is guaranteed . . . by setting apart its modes of knowing from those in which that dichotomy is threatened. In this process, the characterization of both the scientific mind and its modes of access to knowledge as masculine is indeed significant. Masculine here connotes, as it so often does, autonomy, separation, and distance . . . a radical rejection of any comingling of subject and object."[37] Situating this masculine birth—or more precisely, *re*-birth—within the context of the cultural separation anxieties described earlier, it appears not only as an intellectual orientation but as a mode of denial as well, a reaction formation to the loss of "being-one-with-the-world" brought about by the disintegration of the organic, centered cosmos of the Middle Ages and Renaissance. The Cartesian reconstruction of the world is a defiant gesture of independence from the female cosmos—a gesture that is at the same time compensation for a profound loss.

The project of growing up, as developmental theorists emphasize, is primarily a project of separation, of learning to deal with the fact that mother and child are no longer one.[38] One mode of dealing with that separation is through the denial of any longing for that lost union through an assertion of self against the mother and all that she represents and a rejection of all dependency on her. In this way, the pain of separateness is assuaged, paradoxically, by an even more definitive separation—but one that is chosen this time and aggressively pursued. It is therefore experienced as autonomy rather than helplessness in the face of the discontinuity between self and mother.

35. Francis Bacon, *The Philosophy of Francis Bacon*, ed. B. Farrington (Liverpool: Liverpool University Press, 1970), 130. For discussions of Bacon's use of sexual metaphors, see Carolyn Merchant (n. 25 above), 164–90; Evelyn Fox Keller, *Reflections on Gender and Science* (New Haven, Conn.: Yale University Press, 1985), 33–42; Genevieve Lloyd, *The Man of Reason* (Minneapolis: University of Minnesota Press, 1984), 10–17.
36. On science and the masculine, see esp. Keller. On masculinity as a cognitive style, see esp. Nancy Chodorow, *The Reproduction of Mothering* (Berkeley: University of California Press, 1978); and Carol Gilligan, *In a Different Voice* (Cambridge, Mass.: Harvard University Press, 1982).
37. Keller, 79.
38. Norman O. Brown, *Life against Death* (New York: Random House, 1959).

Within the context of such ideas, Norman O. Brown reinterprets the oedipal desire to possess the mother sexually as a fantasy of "becoming the father of oneself" (rather than the helpless child of the mother).[39] Sexual activity here (or, rather, the fantasy of it) becomes a means of denying the actual passivity of having been born from that original state of union into "a body of limited powers, and a time and place [one] never chose,"[40] and at the mercy of the now-alien will of the mother. The mother is still "other," but she is an other whose power has been harnessed by the will of the child. The pain of separateness is thus compensated for by the peculiar advantages of separateness: the possibility of mastery and control over those on whom one is dependent.

The Cartesian project of starting anew through the revocation of one's actual childhood—during which one was "immersed" in body and nature—and the re-creation of a world in which absolute separateness (both epistemological and ontological) from body and nature are keys to control rather than sources of anxiety, can be seen as a "father of oneself" fantasy on a highly abstract plane. The sundering of the organic ties between person and nature—originally experienced as epistemological estrangement, as the opening up of a chasm between self and world—is reenacted, this time with the human being as the engineer and architect of the separation.

A new theory of knowledge, thus, is born, one which regards all sense experience as illusory and insists that the object can only truly be known by the perceiver who is willing to purge the mind of all obscurity, all irrelevancy, all free imaginative associations, and all passionate attachments. (This Descartes believed eminently possible, given the right method: "Even those who have the feeblest souls can acquire a very absolute dominion over all their passions if sufficient industry is applied in training and guiding them" [HR 1:356].) A new world is constructed, one in which all generativity and creativity fall to God the spiritual father rather than to the female flesh of the world. For Plato and Aristotle, and throughout the Middle Ages, the natural world has been "mother"— passive, receptive, *natura naturata* to be sure, but living and breathing nonetheless.[41] Now, in the same brilliant stroke that insured the objectivity of science—the mutual opposition of the spiritual and the corporeal—the formerly female earth becomes inert *res extensa*: dead, mechanically interacting matter.

"She" becomes "it"—and "it" can be understood. Not through sympathy, of course, but by virtue of the very *object*-ivity of the "it." At the

39. Ibid., 127.
40. Simone de Beauvoir, *The Second Sex* (New York: Alfred A. Knopf, 1957), 146.
41. See esp. Plato, *Timaeus*; and Aristotle, *On The Generation of Animals*, 72a222. For an excellent discussion of the imagery of earth as mother, see Carolyn Merchant.

same time, the wound of separateness is healed through the denial that there ever was any union: for the mechanists, unlike Donne, the female world-soul did not die; rather, the world *is* dead. There is nothing to mourn, nothing to lament. Indeed, the new epistemological anxiety is not over loss but is evoked by the memory or suggestion of union: empathic, associational, or emotional response obscures objectivity, feeling for nature muddies the clear lake of the mind. The otherness of nature is now what allows it to be known.

The Seventeenth-Century Flight from the Feminine

The historical research of such writers as Carolyn Merchant, Brian Easlea, Barbara Ehrenreich and Dierdre English, and Adrienne Rich has forced us to recognize the years between 1550 and 1650 as a particularly gynophobic century. The prevailing ideas of the era now appear as obsessed with the untamed natural power of female generativity and a dedication to bringing it under forceful cultural control. Nightmare fantasies of female power over reproduction and birth run throughout the era, from Kramer and Sprenger's *Malleus Maleficarum*, which accuses witches of every imaginable natural and supernatural crime involving conception and birth, to Boyle's characterization of nature as "God's great pregnant Automaton,"[42] whose secrets are deliberately and slyly concealed from the scientist. There were the witch hunts themselves, which, aided more politely by the gradual male takeover of birthing, virtually purged the healing arts of female midwives.[43] The resulting changes in obstetrics came to identify birth—as Bacon identified nature itself—with the potentiality of disorder and the need for forceful male control.[44]

It was not only in practice that women were being denied an active role in the processes of conception and birth. Mechanist reproductive theory as well had "happily" (as Easlea sarcastically puts it) made it "no longer necessary to refer to any women" at all in its descriptions of conception and gestation.[45] Denied even her limited, traditional Aristotelian role of supplying (living) menstrual material, the woman becomes instead the mere container for the temporary housing and incubation of already formed human beings, originally placed in Adam's semen by God and parcelled out, over the ages, to all his male descendants. The specifics

42. Quoted in Easlea (n. 25 above), 214.
43. See in particular Barbara Ehrenreich and Dierdre English, *For Her Own Good* (New York: Doubleday Anchor, 1979), 33–68; and Adrienne Rich, *Of Woman Born* (New York: Bantam Books, 1976), 124–48.
44. Rich, 133.
45. Easlea, 49.

of mechanistic reproductive theory are a microcosmic recapitulation of the mechanistic vision itself, within which God the father is the sole creative, formative principle in the cosmos. We know, from what now must be seen as almost paradigmatic examples of the power of belief over perception, that tiny horses and men were actually "seen" by mechanist scientists examining sperm under their microscopes.[46]

What can account for this upsurge of fear of female generativity? No doubt many factors—economic, political, and institutional—are crucial; but the "drama of parturition" described in connection with Descartes can provide an illuminative psychocultural framework within which to situate seventeenth-century gynophobia.

The culture in question, in the wake of the dissolution of the medieval intellectual and imaginative system, had lost a world in which the human being could feel nourished by the sense of oneness, of continuity between all things. The new, infinite universe was an indifferent home, an alien will, and the sense of separateness from her was acute. Not only was she "other" but she seemed a perverse and uncontrollable other: during the years 1550–1650, a century that had brought the worst food crisis in history, violent wars, plague, and devastating poverty, the Baconian imagery of nature as an unruly and malevolent virago is no paranoid fantasy. More important, the cruelty of the world could no longer be made palatable by the old medieval sense of organic justice—that is, justice on the level of the workings of a whole with which one's identity merged and that, while perhaps not fully comprehensible, was nonetheless to be trusted. Now there seemed no organic unity, but only "I" and "she"—an unpredictable and seemingly arbitrary "she" whose actions could not be understood in any of the old, sympathetic ways.

"She" is "other"; and "otherness" itself becomes dreadful—particularly the otherness of the female, whose powers have always been mysterious to men and evocative of the mystery of existence itself. Like the infinite universe, which threatens to swallow the individual "like a speck,"[47] the female, with her strange rhythms, long acknowledged to have their chief affinities with the rhythms of the natural (now alien) world, becomes a reminder of how much lies outside the grasp of man. "The quintessential incarnation" of that which appears to man as "mysterious, powerful and not himself," as Dinnerstein says, is "the woman's fertile body."[48] Now, with the universe appearing to man more decisively "not-himself" than ever before, both its mystery and the mystery of the female require a more decisive "solution" than had been demanded by the organic worldview.

46. For an extremely interesting discussion of this, see Hillman, "On Psychological Femininity," in *The Myth of Analysis* (n. 8 above), 215–58.

47. Pascal (n. 30 above), 59.

48. Dorothy Dinnerstein, *The Mermaid and the Minotaur* (New York: Harper & Row, 1977), 125.

The project that fell to empirical science and "rationalism" was to tame the female universe. Empirical science did this through aggressive assault and violation of her "secrets"; rationalism, through the philosophical neutralization of her vitality. The barrenness of matter correlatively insured the revitalization of human hope of conquering nature (through knowledge in this case rather than through force). The mystery of the female, however, could not be bent to man's control simply through philosophical means, more direct and concrete means of "neutralization" were required for that project. It is within this context that witch hunting and the male takeover of the processes of reproduction and birth, whatever their social and political causes, can be seen to have a profound psychocultural dimension as well.

The Contemporary Revaluation of the Feminine

The recent scholarly emergence and revaluation of epistemological and ethical perspectives that have been identified as feminine in classical as well as contemporary writing (as, e.g., in the work of Carol Gilligan, Sarah Ruddick, and Nancy Chodorow) claim a natural foundation for knowledge, not in detachment and distance, but in closeness, connectedness, and empathy. They find the failure of connection (rather than the blurring of boundaries) as the principle cause of breakdown in understanding.

An appreciation of the historical nature of the masculine model of knowledge to which this "different voice" is often contrasted helps to underscore that the embodiment of these gender-related perspectives in actual men and women is a cultural, not a biological, phenomenon. There have been cultures in which (using our terms now, not theirs) men thought more like women, and there may be a time in the future when they do so again. For the prescientific understanding of the world, detachment is not an epistemological value. Rather, it is precisely because the scientific and intellectual revolutions of the seventeenth century changed that, that we can today describe those revolutions as effecting a "masculinization" of thought. The conclusion is not, however, that the categories of "masculinity" and "femininity" are mythologies, useless and reactionary hypostatizations. The sexual division of labor within the family in the modern era has indeed fairly consistently reproduced these gender-related perspectives along sexual lines. The central importance of Nancy Chodorow's work, for example, has been to show that boys tend to grow up learning to experience the world like Cartesians, while girls do not, because of developmental asymmetries resulting from female-dominated infant care, rather than from biology, anatomy, or "nature."

This sociological emphasis and understanding of gender as a social construction is one crucial difference between the contemporary feminist revaluation of the "feminine" and the nineteenth-century doctrine of female moral superiority. A still more central difference is the contemporary feminist emphasis on the insufficiency of any ethics or rationality—feminine or masculine—that operates solely in one mode without drawing on the resources and perspective of the other.[49] The nineteenth-century celebration of a distinctively feminine sensibility and morality, by contrast, functioned in the service of pure masculinized thought by defining itself as a separate entity. This was, of course, precisely what the seventeenth-century masculinization of thought had accomplished—the exclusion of feminine modes of knowing, not from culture in general, but from the scientific and philosophical arenas, whose objectivity and purity needed to be guaranteed. Romanticizing "the feminine" within its own sphere is no alternative to Cartesianism because it suggests that the feminine has a "proper" (domestic) place. If Dorothy Dinnerstein and others are right, it is precisely the suppression of the feminine that is the deepest root of our modern cultural woes.[50] The historical identification of rationality and intelligence with the masculine modes of detachment, distance, and clarity has disclosed its limitations, and it is necessary (and inevitable) that feminine modes should now appear as revealing more innovative, more humane, and more hopeful perspectives. Clearly, the (unmythologizing) articulation of "the feminine"—and its potential contribution to ethics, epistemology, science, education, and politics—is one of the most important movements of the twentieth century.

Department of Philosophy
Le Moyne College

49. See esp. the final chapter in Gilligan's *In a Different Voice*, 151–74. In this chapter, it becomes clear that Gilligan is calling, not for a feminization of knowledge, from which more masculinist modes are excluded, but for the recognition that each mode, cut off from the other, founders on its own particular reefs, just as it offers its own partial truths about human experience. See also Keller's discussion of "dynamic objectivity." 115–26 (n. 35 above).

50. Dinnerstein. See also Hillman, "On Psychological Femininity."

Hand, Brain, and Heart: A Feminist Epistemology for the Natural Sciences

Hilary Rose

Science it would seem is not sexless; she is a man, a father
and infected too. [Virginia Woolf, *Three Guineas*]

This paper starts from the position that the attitudes dominant within
science and technology must be transformed, for their telos is nuclear
annihilation. It first examines the achievements during the 1970s of
those who sought to analyze and critique capitalist science's existing
forms and systems of knowledge and goes on to argue that their critiques
(which may well have been developed with a conscious opposition to
sexism) are theoretically sex blind. Their analysis of the division of labor
stops short at the distinction between the manual and mental labor associated with production. Indifferent to the second system of production—reproduction—the analysis excludes the relationship of science to patriarchy, to the sexual division of labor in which caring labor is
primarily allocated to women in both unpaid and paid work. Transcendence of this division of labor set up among hand, brain, and heart
makes possible a new scientific knowledge and technology that will en-
able humanity to live in harmony rather than in antagonism with nature,

I would like to thank the editorial collective of *donnawomanfemme* for publishing an
earlier version of this essay as "Dominio ed Esclusione" (see 17 [1981]: 9–28). Too many
members of the invisible college of feminist scientists have provided encouragement and
helpful comments for me to list them here.

This essay originally appeared in *Signs*, vol. 9, no. 1, Autumn 1983.

including human nature. The necessity and magnitude of the task must be recognized.

The Need for a New Science

Over the past dozen years the critique of science and technology has focused attention on the ways in which existing science and technology are locked into capitalism and imperialism as a system of domination. This denunciation has served two functions. Negatively, it has facilitated the growth of an antipathy to science that rejects all scientific investigation carried out under any conditions and at any historical time. More positively, it has set itself the difficult task of constructing, in a prefigurative way, both the forms and the content of a different, alternative science—one that anticipates the science and technology possible in a new society and, at the same time, contributes through innovatory practice to the realization of that society. This paper aligns itself with such a venture, while recognizing that—with its false starts as well as real achievements, its perilous balancing between atheoretical activism and abstract theoreticism—the project is not without its contradictions and difficulties. Feminism is just beginning to recapture the full force of Virginia Woolf's insight: science it would seem—to rephrase her—is neither sexless nor classless; she is a man, bourgeois, and infected too.

The trouble with science and technology from a feminist perspective is that they are integral not only to a system of capitalist domination but also to one of patriarchal domination; yet to try to discuss science under both these systems of domination is peculiarly difficult.[1] Historically, it has been women outside science, such as the novelist and essayist Virginia Woolf, or the ex-scientist, now writer, Ruth Wallsgrove, or the sociologist Liliane Stéhélin who have dared to speak of science as male, as part of a phallocentric culture.[2] For women inside science, protest has

1. Since this paper was written, the debate concerning the possibility of moving beyond dualism has sharpened with the publication of Lydia Sargent's edited collection *Women and Revolution: A Discussion of the Unhappy Marriage of Marxism and Feminism* (Boston: South End Press, 1981). While I am attracted to the theoretical project of a unified explanation, as elaborated within this collection, e.g., by Iris Young, the political and theoretical conditions for its achievement seem to me to be premature. The fruitful analysis of gender, class, and ethnic divisions of labor is likely to continue in both dualistic and unitary forms. See, e.g., the special issue entitled "Development and the Sexual Division of Labor": *Signs: Journal of Women in Culture and Society*, vol. 7, no. 2 (Winter 1981).

2. Ruth Wallsgrove, "The Masculine Face of Science," in *Alice through the Looking Glass*, ed. Brighton Women and Science Group (London: Virago, 1980); Liliane Stéhélin, "Sciences, Women and Ideology," in *The Radicalisation of Science*, ed. Hilary Rose and Steven Rose (London: Macmillan Publishers, 1976); also Hilary Rose and Steven Rose, eds., *Ideology of/in the Sciences* (Boston: Schenkman Publishing Co., 1979).

been much more difficult. Numbers are few and developing the network among isolated women is intractable work. Yet as we enter the eighties an invisible college of feminist scientists is beginning to assemble.[3] One of the new voices breaking through from within the laboratory is that of Rita Arditti, who, having worked in the competitive and macho world of genetics, was radicalized by the antiwar movement.[4] She became a feminist through this experience and now argues that nothing less than a new science will serve.

Apart from a handful who wrote pioneering papers, feminists in the early days of the movement avoided the discussion of science, often retreating into a total rejection of science as the monolithic enemy. Doing science became an activity in which no serious feminist would engage. But there arose a growing political threat as, during the 1970s, a new wave of biological determinism sought to renaturalize women. It has required women biologists—or women who will enter the terrain of biological knowledge—to contest its claims. Doing biology is thus no longer seen as hostile but as helpful to women's interests, and increasingly it is possible to go forward from this essentially defensive purpose to the much more positive goal of seeking to show how a feminist knowledge of the natural world offers an emancipatory rather than an exterminatory science. The task of developing a feminist critique of existing science and of moving toward an as yet unrealized feminist natural science is at once more difficult and more exciting than the academically respectable activity of making descriptive reports on women's position within science. There is a watershed between work carried out within a transformative view of women's destiny and projects shaped by the main variants of structural functionalism still extant. For

3. See, e.g., Ann Arbor Science for the People Collective, eds., *Sociobiology as a Social Weapon* (Minneapolis: Burgess Publishing Co., 1977); Susan Griffin, *Woman and Nature: The Roaring inside Her* (New York: Harper & Row Publishers, 1978); Lila Leibowitz, *Females, Males and Families: A Biosocial Approach* (North Scituate, Mass.: Duxbury Press, 1978); Ethel Tobach and Betty Rosoff, eds., *Genes and Gender I: On Hereditarianism and Women* (New York: Gordian Press, 1978); Ruth Hubbard and Marian Lowe, eds., *Genes and Gender II: Pitfalls in Research on Sex and Gender* (New York: Gordian Press, 1979); Ethel Tobach and Betty Rosoff, eds., *Genes and Gender III: Genetic Determinism and Children* (New York; Gordian Press, 1980); *Signs: Journal of Women in Culture and Society* (special issue entitled "Women, Science, and Society," esp. the papers by Donna Haraway, Adrienne Zihlman, Helen Lambert, Marian Lowe, and Ruth Bleier), vol. 4, no. 1 (Autumn 1978); Carolyn Merchant, *The Death of Nature: Women, Ecology and the Scientific Revolution* (San Francisco: Harper & Row Publishers, 1980); Ruth Hubbard, Mary Sue Henifin, and Barbara Fried, eds., *Women Look at Biology Looking at Women* (Cambridge, Mass.: Schenkman Publishing Co., 1979); Janet Sayers, *Biological Politics: Feminist and Anti-Feminist Perspectives* (London: Tavistock Publications, 1982).

4. Rita Arditti, "Feminism and Science," in *Science and Liberation*, ed. Rita Arditti, Pat Brennan, and Steve Cavrak (Boston: South End Press, 1979); James Watson's *The Double Helix* (New York: Atheneum Publishers, 1968) was to provide an even more demystified account of how science really gets done—one more honest (not least in its acceptance of machismo) than perhaps he himself had intended.

feminist analysis, unlike developed Marxist theory—though like early Marxism—calls for interpretation constantly tested not simply against the demands of theory, but always and incessantly against the experience of the specific oppression of women.

Here I shall first, briefly, set out the theoretical achievements of the radical critique of science, making plain the weaknesses that stem from its one-sided materialism. Second, drawing on the fast-developing feminist analysis of the links between women's paid and unpaid labor, I shall suggest not only why women are by and large excluded from science, but also what kind of science the exclusion produces. Last, and most tentatively, I pick out some of the examples of a new science that have been developed through the feminist movement.

The Radical Critique of Science

The critique of science was to explode into practice and to struggle into theory during the radical movement of the late 1960s and early 1970s. The numerous issues contained in the class and social struggles of the movement were frequently narrowed and constrained as the theoreticians filtered the wealth of lived experience through the abstract categories of theory. From an early rhetoric that attacked with a certain evenhandedness the class society, imperialism, racism, and sexism (those who were black, colonized, or women might well have had doubts about their equal prioritization in practice as well as in rhetoric), the theoreticians were to develop two main lines of analysis. The first considered the political economy of science, and the second took up the relationship between science and ideology. While the two are linked at many points, work in political economy was more coherently developed; work on the debate over science and ideology was and remains more problematic. The need to reply immediately to the ideological attacks of a racist or sexist science accelerated as the crisis deepened; it was difficult simultaneously to resist the attack and to analyze the issues. Indeed, the hostility to science within the movement so conflated science and ideology that it aided the growth of attitudes that rejected science altogether.[5]

The socialist tradition, at least up to the sixties, believed that the advances of science would automatically create problems that capitalist society could not solve; hence in some way science was at least "neutral," at best allied to those working for a new and socially just society. Such a neutral science was seen as uninfluenced by class, race, gender, national-

5. Hilary Rose, "Hyper-reflexivity: A New Danger for the Countermovements," in *Countermovements in the Sciences: Sociology of Science Yearbook*, ed. Helga Nowotny and Hilary Rose (Dordrecht: Reidel, 1979); and Hilary Rose and Steven Rose, "Radical Science and Its Enemies," in *Socialist Register*, ed. Ralph Miliband and John Saville (Atlantic Highlands, N.J.: Humanities Press, 1979).

ity, or politics; it was the abstract accumulation of knowledge—of facts, theories, and techniques—which could be "used" or "abused" by society. The experiences of the sixties and seventies overthrew such notions of science. What the sixties' radicals discovered in their campaigns against an abused, militarized, and polluting science was that those in charge of neutral science were overwhelmingly white, male, and privileged occupants of positions in advanced industrialized society. The antihuman technologies that science generated were being used for the profit of some and the distress of many. Thus, the politics of experience brought the radical movement's attitudes toward science into a confrontation with the orthodox Marxist analysis of science.

The latter had claimed that there was an inevitable contradiction between the productive forces unleashed by science and the capitalist order. Hence, science could not be used to its full creative potential within capitalism, and attempts to control the forces of technological innovation would ultimately lead to the destruction of capital. Embodied in Bernalism (though perhaps exaggerated there from what is strictly to be found in the work of John Desmond Bernal), this old belief of the Left that science, technology, and socialism have a necessary relationship was, in time, abandoned, but no critical alternative to bourgeois science was set in its place. Hope for that lay buried in the cupboard of the Lysenko affair, and fashioning a critique of science has been carried out in relative isolation from the mainstream of Marxist scholarship.[6]

The Lysenko affair epitomizes the period from the 1930s to the 1940s in the Soviet Union during which there was an attempt to develop a specifically proletarian interpretation of all culture, including the natural sciences. Against the more cautious views of geneticists, Trofim Lysenko advanced the thesis that acquired characteristics are inherited. He set his social origins as a peasant (and thus his experiential knowledge) against the aristocratic origins (and therefore abstract knowledge) of his leading opponent, Nikolai I. Vavilov. The debate was resolved by Lysenko's falsified statistics on the amounts of grain produced according to his theory and this resolution was sustained by the imposition of Stalinist terror. In 1940 Vavilov was arrested and Lysenko set in his place as director of the Institute of Genetics.[7]

6. The writings of John Desmond Bernal were central among socialist scientists of the thirties, esp. *The Social Functions of Science* (London: Routledge & Kegan Paul, 1939). Bernalist ideas remain only in old Left theories—whether those of official communism or of the Trotskyist faction—extending as it were from the Soviet academician Mikhail Millionshchikov to the fourth international theorist Ernest Mandel. See Mikhail Millionshchikov in *The Scientific and Technological Revolution,* ed. Robert Daglish (Moscow: Progress Publishers, 1972), pp. 13–18; and Ernest Mandel, *Late Capitalism* (London: New Left Books, 1975).

7. The Lysenko affair was neglected for almost thirty years by Marxists. In the mid-seventies the American biologists Richard Lewontin and Richard Levins and the French philosopher Dominique Lecourt broke the long silence. Lewontin and Levins, "The

Marxist scholarship shrank from analyzing the circumstances and implications of this failed cultural revolution and retreated to the position that there is only one science—by implication, bourgeois science. Thus, when the radical movement turned to Marxist analysis of the natural sciences, it found embarrassed silences. Nor was the movement helped by the special status of science within Marxism as a body of thought—from Marx's and Engels's claims for a scientific socialism, to Lenin's enthusiasm for Frederick Taylor's application of scientific methods to the production process, to Louis Althusser's influential project to depersonalize Marxist analysis so as to make it truly scientific.

The Myth of the Neutrality of Science

While here I have deliberately focused on the writing of those who have been influential within the radical science movement, over the last decade or so there has also been a dramatic shift in the history, philosophy, and sociology of science. A sophisticated form of externalism holding the thesis that scientific knowledge is structured through its social genesis has become common to all three, so that research is aimed at demonstrating how interests fashion knowledge.[8] Thus, while Thomas Kuhn's work marked the beginnings of the thaw for an age that seemed forever frozen in the timeless certainties of positivism and the Vienna circle, it was a mathematician, historian of science, and political radical—Jerome Ravetz—who posed the question of why and how science is a social problem.[9] To answer this, Ravetz examined the circumstances in which scientists actually produce scientific knowledge. Abandoning the heady and very abstract Popperian theory of "bold conjectures and refutation,"[10] he asked: What do scientists actually do? Through an examination of the production of science from the seventeenth to the nineteenth century he argued that, whereas in the

---

Problem of Lysenkoism," in *Ideology of/in the Sciences* (n. 2 above); Lecourt, *Proletarian Science? The Case of Lysenko* (London: New Left Books, 1977).

8. This project has been particularly associated with the Edinburgh school and the "strong" program in the sociology of knowledge, and would include the work of Barry Barnes, David Bloor, and Donald McKenzie. Ultimately epistemologically problematic, the approach has facilitated a critical view of science. For a feminist use of a sophisticated externalism that avoids the excesses of the strong program, see Elizabeth Fee, "Nineteenth-Century Craniology: The Study of the Female Skull," *Bulletin of the History of Medicine* 53 (1980): 415–33; Donna Haraway, "Animal Sociology and the Natural Economy of the Body Politic Part I," and "Part II," *Signs: Journal of Women in Culture and Society* 4, no. 1 (1978): 21–60.

9. Jerome R. Ravetz, *Scientific Knowledge and Its Social Problems* (New York: Oxford University Press, 1971).

10. Karl R. Popper, *Conjectures and Refutations: The Growth of Scientific Knowledge* (London: Routledge & Kegan Paul, 1963).

early period science was considered a craft, it increasingly adopted industrialized methods of production as it entered the twentieth century. Where the craft worker had labored alone, or with a couple of apprentices, the new system required substantial capital, a large group of scientists, a clear division of labor between them, and common goals to be set and managed by a scientific director.

Ravetz held that this industrialization of science has produced its uncritical character. In an essentially romantic and libertarian political response, he called for the deinstitutionalization of science. When tied to the state and industry, science must inevitably lose its critical force and become an agent of oppression. If a true science is to reachieve the liberating role it had in the time, say, of Galileo, it must, like Galileo, once more stand in opposition to institutionalized science. It must become critical. The problem with Ravetz's position is that it is idealist in both senses. Although we have seen the welcome development of a handful of deinstitutionalized scientific ventures, it would be unrealistic to consider their contribution a sign of the restructuring of science, since their access to the means of scientific production is minimal.[11]

Less attracted than Ravetz by alternatives outside science, and more oriented toward contesting existing science, others within the radical science movement were nonetheless pursuing the same theoretical concerns. They were revolted by the genocidal science that the United States employed in the war in Southeast Asia and by the expanding new technologies of urban repression at home. They asked, How can science claim to be ideologically pure, value-free, and above all neutral, when even a well-regarded text entitled *The Scientific Method* offers as an example of scientific development the making and testing of napalm on a university playing field, without any references to ethical or political problems? From the "use and abuse" model, in which science remained fundamental, basic, and pure, though possibly abused by political others, the New Left—to the equal concern of both the scientific establishment and the Old Left—had laid siege to the myth of the neutrality of science itself.[12]

A new political economy of science, associated with the physicist Marxists Marcello Cini, Michelangelo de Maria, and Giovanni Cicotti,[13] was to argue that bringing science into the capitalist mode of production meant that knowledge itself, as the product of scientific labor, had been

11. These ventures are documented primarily in alternative magazines, particularly those associated with anarchism, but see also Nowotny and Rose, eds. (n. 5 above).

12. Hilary Rose and Steven Rose, "The Myth of the Neutrality of Science," in Arditti, Brennan, and Cavrak, eds. (n. 4 above).

13. Giovanni Cicotti, Marcello Cini, and Michelangelo de Maria, "The Production of Science in Advanced Capitalist Countries," in *The Political Economy of Science*, ed. Hilary Rose and Steven Rose (London: Macmillan Publishers, 1976). (This essay unfortunately is not included in the American edition, published by G. K. Hall in 1979.)

made a commodity. In this analysis, scientific knowledge is no longer timeless but has value only at a particular time and a particular place. For industry the patent laws already encompassing physics and chemistry and presently reaching into the burgeoning area of biotechnology are designed to police ownership patterns.[14] For the basic sciences the rewards go to those who publish the knowledge first. The very process of diffusion reduces the value of the knowledge (typically produced in the elite institutions of the metropolitan countries) by the time it is transferred to the nonelite institutions in the periphery. The value of the knowledge as it passes from the center of production to the periphery declines as surely as that of a car as it moves from second to third hand.

The change in the mode of scientific production, its loss of criticality, and its subjugation to the laws of commodity production are features of the sciences most closely integrated with the reproduction of social and economic power. The physical sciences, above all physics itself, are at once the most arcane and the most deeply implicated in the capitalist system of domination. At the same time, the physical sciences more or less successfully exclude any more than small numbers of women. These industrialized sciences would appear to be highly resistant to feminist reconceptualization, not least because the success of feminist theory has lain in areas such as history, philosophy, and sociology—all characterized by little capital equipment per worker and by craft methods of production.

The Social Origins of Science as Alienated Knowledge

While many within the radical science movement were influenced by the writings of the Frankfurt school, it was Alfred Sohn Rethel, as part of that tradition, who was to seek to explain the social origins of the highly abstract and alienated character of scientific knowledge.[15] Drawing on historical material, he suggested that abstraction arises with the circulation of money; but he went on to argue that the alienated and abstract character of scientific knowledge has its roots in the profound division of intellectual and manual labor integral to the capitalist social formation. Scientific knowledge and its production system are of a piece with the abstract and alienated labor of the capitalist mode of production itself. The Cultural Revolution, with its project of transcending the division of mental and manual labor, was seen by Sohn Rethel and indeed by many or most of the New Left as offering a model of immense historical significance. They saw within this movement not only the possibility of

14. David Noble, *America by Design* (Cambridge, Mass.: MIT University Press, 1979).
15. Alfred Sohn Rethel, *Intellectual and Manual Labour* (London: Macmillan Publishers, 1978).

transcending hierarchical and antagonistic social relations, but also the means for creating a new science and technology not directed toward the domination of nature or of humanity as part of nature. Especially at present, when assessment of the experience of the Cultural Revolution is problematic, it is important to affirm our need of the project it undertook.

In a world where the alienation of science and technology confronts us in the pollution of the seas, the cities, the countryside—and in the fear of nuclear holocaust—such a longing cannot be dismissed as merely romantic. Its realization may rather be a guarantor of our survival. Certainly aerospace workers in Britain—not easily equated with romantic intellectuals—have in their practice come to conclusions very similar to those of Sohn Rethel. Beginning with their opposition to the threat of redundancy and with a moral distaste for being so deeply involved in the manufacture of war technology, the workers went on to design, and in some cases to make, socially useful technologies such as the road-rail bus. In this they have simultaneously both contested the division of mental and manual labor in the production of technology and, through the unity of hand and brain, begun the long struggle to transform the commodity itself.[16]

The Second System of Domination

Despite the advances made through the critique of science pursued during the 1970s, the critique is, in a theoretical sense, sex blind. It is not that the critics have been insensitive to the problems of sexism and racism; many have honorable records in trying to contest them. It is rather that the theoretical categories make it impossible for them to explain why science is not only bourgeois but male. For it is unequivocally clear that the elite of science—its managers and the constructors of its ideology—are men. Within science, as within all other aspects of production, women occupy subordinate positions, and the exceptional women who make it in this man's world only prove the rule. Yet this exclusion of half of humanity means that the 1970s' critics of science, while they grappled with the structuring of science and technology under capitalism, failed to grasp the significance of—or even to recognize—the structuring under patriarchy, that second and pervasive system of domination. Indeed the radical science movement itself was to reflect in its practice much of the sexism of the social order it opposed. Nonetheless, the critique laid successful siege to claims that science and

16. Mike Cooley, *Architect or Bee?* (Slough, England: Hand and Brain Press, 1980); Dave Elliot and Hilary Wainwright, *The Lucas Plan: A New Trade Unionism in the Making* (London: Allison & Bushy, 1982).

technology transcend history, and made plain as well the class character of science within a capitalist social formation. Science as an abstraction came to be analyzed as an ideology having a specific historical development within the making of capitalism. Demystifying science exposed the myths that had served to integrate science and mask its internal contradictions and external functions.

Yet within the radical critique of science there remained a disjuncture between actual struggle, on one hand, and theorizing, on the other. Looking back over the writing of the sixties and early seventies, it is difficult not to feel that, as the critical work became more theoretical, more fully elaborated, so women and women's interests receded. Thus, this writing gives no systematic explanation of the gender division of labor within science, nor, despite its denunciation of scientific sexism, does it explain why science so often works to benefit men. By attributing the exclusion of women to ideology, it ignores the possibility that there is a materialist explanation, nowhere hazarding the suggestion that it is in the interests of men to subordinate women within, as well as without, the production system of science. It is taken for granted that the domination/subordination, oppressor/oppressed relationship between men and women is either irrelevant or is explained by the production process.

Feminism as Materialism

Yet this prioritization of the production process ignores that other materialist necessity of history—reproduction. The preoccupation with production as a social process with a corresponding social division of labor and the neglect of reproduction as an analogous process with its division of labor perpetuates a one-sided materialism. It cannot help us understand our circumstances, let alone transcend them.

Reproduction has, meanwhile, been a central focus of the feminist movement in both pragmatic social struggle and theoretical explanation. It is not by chance that the movement has been concerned with abortion, birth control, sexuality, housework, child care—indeed, with all those matters that had been trivialized into silence in the long period since an earlier wave of feminism. Even though now most Left journals will find space occasionally or even regularly for articles by feminists, the exercise remains relatively tokenistic, as the "important" articles dealing with the present crisis show few signs of integrating feminist theory. More than ever, feminists must insist on the significance of the division of labor and prevent the "renaturalization" of women's labor. Economic, political, and ideological pressures can make it seem only right to restore woman to her "natural" place. Science as the great legitimator is, as usual, offering its services.

The Labor of Love

If we are to understand the character of a science denied the input of women's experiences, feminists must return to the sexual division of labor within the household, which, in science as elsewhere, finds its ironic echo in paid labor. Women's work is of a particular kind—whether menial or requiring the sophisticated skills involved in child care, it always involves personal service. Perhaps to make the nature of this caring, intimate, emotionally demanding labor clear, we should use the ideologically loaded term "love." For without love, without close interpersonal relationships, human beings, and it would seem especially small human beings, cannot survive. This emotionally demanding labor requires that women give something of themselves to the child, to the man. The production of people is thus qualitatively different from the production of things. It requires caring labor—the labor of love.

If we return to Sohn Rethel's emancipatory project of overcoming the division between mental and manual labor, the significance of woman's caring labor for the production of science becomes clear. He saw the division of labor which lies between men, yet took for granted the allocation of caring labor to women. For while intellectual labor is allocated to a minority of men, the majority are assigned manual, highly routine labor. For women, the division of labor is structured along different lines: even those few who become intellectuals are, in the first instance, assigned domestic labor in which caring informs every act. Both the emancipatory theoretical project of Sohn Rethel and the emancipatory practice of the aerospace workers seek to overcome the division of labor and urge a new science and a new technology; nonetheless their projects still lie within the production of commodities. They seek the unity of hand and brain but exclude the heart. A theoretical recognition of caring labor as critical for the production of *people* is necessary for any adequate materialist analysis of science and is a crucial precondition for an alternative epistemology and method that will help us construct a new science and a new technology. Thus, while Sohn Rethel's proposal seeks to overcome capitalist social relations, it leaves untouched the patriarchal relations between the sexes. In the production of knowledge, this limitation carries within it the implication that, even if knowledge thus produced were less abstract and its reification overcome, it would still reflect only the historically masculine concern with production. The historically feminine concern with reproduction would remain excluded.

Sohn Rethel's neglect of the caring labor of women means that the theorist of the transcendence of the division of labor implicitly has joined forces with the far-from-emancipatory program of sociobiology, which argues that woman's destiny is in her genes. The sociobiological thesis that women are genetically programmed for monogamous heterosexual relationships and motherhood receives tacit endorsement through the androcentric preoccupation of Marxist thought. If we return to Marx's vision of the postrevolutionary society—where we fish and hunt before dinner and make social criticism after the dinner—it is clear that Sohn Rethel, like Marx, has made the tacit assumption that the usual invisible laborer cooks the meal.[17]

Nor is it enough merely to add a female dimension to a basically productionist argument by bringing in the caring contribution of women. Such an additive process runs the danger of denying the social genesis of women's caring skills, which are extracted from them by men primarily within the home but also in the work place. It moves toward the essentialist thought that women are "naturally" more caring. The problem for materialists is to admit biology—that is, a constrained essentialism—while giving priority to social construction, without concluding at the same time that human beings are infinitely malleable. The dialectical relationship between both systems of production—the production of things and the production of people—holds the explanation not only of why there are so few women in science, but also, and equally or even more importantly, of why the knowledge produced by science is so abstract and depersonalized.

But at this point it is necessary to explore briefly the nature of a gender-segregated labor market. This is not to give long lists of the sexual divisions of labor within the general economy or specifically within science, but rather to look for explanations of how the segregated labor market has historically come about and to examine the connections between women's paid and unpaid labor.

The Social Origins of the Segregated Labor Market

Examining this problem within capitalist society, feminists point to the central practice of paying a "family wage" to the male breadwinner, who thus receives from capital an income sufficient to reproduce not only his own labor power, but also that of his wife and dependents.[18] Indisputably, the family wage as it emerged during the nineteenth century in the most unionized and better-paid sectors of the economy served to improve the conditions of an entire class fraction—but at the price of enforcing women's and children's dependence on men. Protective legislation excluding women and children from certain kinds and conditions of work (ostensibly for their sake),[19] led to their systematic exclusion from the leading sectors of the economy where the organized

17. Mary O'Brien's feminist rendering of the vision proposes child care as one of the daytime activities. See *The Politics of Reproduction* (London: Routledge & Kegan Paul, 1981).
18. Hilary Land, "The Family Wage," *Feminist Review* 6 (1980): 55–78.
19. Jane Humphries, "Protective Legislation, the Capitalist State and Working Class Men: The Case of the 1842 Mine Regulation Act," *Feminist Review* 7 (1981): 1–33.

(male) breadwinners and capital could battle out the higher wage levels together. During the nineteenth century, although many women were, in fact, breadwinners (not least because of the high rates of widowhood, to say nothing of the large numbers of single women), ideologically they were marginalized in their claims for equal participation and equal pay within the labor market. Today, despite the obvious increase in single-parent families, the family wage retains its ideological grip. Indeed, in a world crisis where unemployment grows apace, the ideology of the family wage and the need to defend the male breadwinner look to be making a comeback. Science and technology as a labor market follow precisely the patterns of this general segregated form. Nor, despite some gains here and there, have educational reform, legislation for equal opportunity and equal pay, the ritual invocation of Marie Curie, or the photographs of a half-dozen smiling women engineering students overcome this structuring of the scientific labor market. The structure is almost always beneficial to men and, according to the trade cycle, is one of manipulable convenience to capital.

Now while it is true that capital in conditions of boom, such as the conditions that characterized Western capitalism in the 1960s, looks to women (and other marginal workers) as a source of labor power and talks of opening up the entire labor market to women,[20] in practice the labor market remains intensely segregated. Even in years of expansion women have remained in an exceedingly narrow range of clerical and service occupations. In Britain, for example, segregation within the labor market was more marked in 1971 than it was in 1901.[21] Nor is a segregated labor market necessarily to capital's advantage; in the post-Sputnik years, the United States, anxious to boost its numbers of scientists and engineers, looked to women as a possible supply source. Focusing on resocialization strategies and publicizing successful role models, the state was nonetheless by and large unsuccessful in opening the field to women. It seems to take little less than a nation state at war for the dual labor market to be significantly eroded. A relatively buoyant labor market plus pressure from below, which existed in the sixties and early seventies, produces some, but not radical, concessions. In present conditions of recession, women are losing jobs faster than men, and within many fields, including science, there are proportionately more unemployed women than men. Yet it is hard to see a male-dominated trade-union movement defending women's paid labor with the same energy it gives to defending that of men. Even the analysis of unemployment by the social sciences constructs it as almost exclusively a problem of men. Few studies of unemployed women have been made

20. Sheila Kammerman and Alfred Kahn, eds., *Family Policy: Government and Families in Fourteen Countries* (New York: Columbia University Press, 1978).
21. Catherine Hakim, "Sexual Divisions within the Labour Force: Occupational Segregation," *Department of Employment Gazette* (November 1978), pp. 1264–68.

(though to be fair, under pressure from feminist critiques, this is beginning to change), and media accounts focus almost entirely on male laborers' loss of work and self-respect and their economic difficulties.

Science as a production system in miniature faithfully reflects the segregated labor market in general. It excludes women—except those in exceptionally favorable circumstances—from occupying elite positions within the production of knowledge. Most women in natural science and engineering are relegated to those tasks that most markedly parallel their primary task as wife-mother.[22] If we examine the full labor force—not just the scientists but also the technicians, secretarial staff, and cleaning personnel—we see that the majority of women are still carrying out menial and personal service work. Neither chance nor biology explains men's occupancy of the leadership positions within science. Only the most exceptional women from highly privileged class backgrounds—who can, therefore, transfer their domestic labor to other women—are able to get into science. And even then, as the increasing number of biographical studies of successful women in science make clear, they often do so only through the personal influence of men—typically husbands, fathers, or lovers. While patronage is an important mechanism of advancement within science, a woman has a much more narrowly defined set of potential patrons, linked to her through her sexuality.

Women Scientists in the Men's Laboratories

Women who manage to get jobs in science have to handle a peculiar contradiction between the demands on them as caring laborers and as abstract mental laborers. Many resolve this by withdrawing or letting themselves be excluded from science; others become essentially honorary men, denying that being a woman creates any problems at all.[23] This sex blindness is particularly evident in the autobiographical accounts of successful women in the sciences, such as those in the 1965 symposium entitled *Women in the Scientific Professions*. It has taken Anne Sayre's passionate defense of Rosalind Franklin to demystify this

22. I regard the evidence that there is no or little institutionalized sexism in Jonathan Coles, *Fair Science* (New York: Free Press, 1979) as too weak to modify this argument. Gaye Tuchman's review summarizes the weaknesses of Coles's work (see *Social Policy* 11, no. 1 [May/June 1980]: 59–64).

23. This view is held by most of the women natural scientists contributing to either the New York symposium of 1965 or the more recent one held by Unesco. However, it is my impression that, in private discussion, distinguished women scientists often have another interpretation; they thus perhaps hold public and private accounts according to which domain they are in. See *Women in the Scientific Professions* (New York: New York Academy of Science, 1965). Alice Rossi's paper in this otherwise ideologically "correct" collection contains a remarkable pioneering discussion.

sexlessness and to insist that the woman scientist is always working in the men's laboratory.[24]

Evelyn Fox Keller, writing of her experiences as a student physicist and later as a research worker, echoes this theme—the continuous, subtle, and not-so-subtle exclusion mechanisms deployed against women scientists. She writes that as a student, she had to be careful to enter a lecture room with or after other students; if she entered first and sat down, men students found it threatening to sit near this low-status person—a woman student—and she was often surrounded by a "sea of seats." On one occasion when she solved a mathematical problem the male university teacher was so incredulous that Keller, like Naomi Weisstein in a similar situation, was quite gently asked who (i.e., which man) did it for her, or where she got (i.e., stole) the solution. Keller's experiences are not, however, unique; what is new is that they are discussed.[25]

Thus a woman scientist is cut in two. Her involvement with the abstraction of scientific practice as it has developed under capitalism and patriarchy, on one hand, is in painful contradiction with her caring labor, on the other. As Ruth Wallsgrove writes, "A woman, especially if she has any ambition or education, receives two kinds of messages: the kind that tells her what it is to be a successful person; and the kind that tells her what it is to be a 'real' woman."[26] Small wonder that women, let alone feminists, working in natural science and engineering are rarities. It is difficult enough to suppress half of oneself to pursue knowledge of the natural world as a woman; it is even more difficult to develop a feminist epistemology.[27] Part of that feminist epistemology involves creation of a practice of feeling, thinking, and writing that opposes the abstraction of male and bourgeois scientific thought.

Reconceptualizing Science

Feminist theorizing about science is of a piece with feminist theoretical production. Unlike the alienated abstract knowledge of science, feminist methodology seeks to bring together subjective and objective ways of knowing the world. It begins with and constantly returns

24. Anne Sayre, *Rosalind Franklin and D.N.A.: A Vivid View of What It Is Like to Be a Gifted Woman in an Especially Male Profession* (New York: W. W. Norton & Co., 1975).

25. Evelyn Fox Keller, "The Anomaly of a Woman in Physics," in *Working It Out: 23 Women, Writers, Scientists and Scholars Talk about Their Lives,* ed. Sara Ruddick and Pamela Daniels (New York: Pantheon Books, 1977); Naomi Weisstein, "Adventures of a Woman in Science," in *Working It Out.*

26. Wallsgrove (n. 2 above), p. 237.

27. Feminism has been quick to spell out its methodology, but slower when it comes to epistemology; see, e.g., Helen Roberts, ed., *Feminist Methodology* (London: Routledge & Kegan Paul, 1981).

to the subjective shared experience of oppression. It is important to stress *shared* experience, since the purely personal account of oppression, while casting some brilliant insights, may tell us more about the essentially idiosyncratic character of individual experience than about the general experience of all or even most women. Nonetheless, within feminist theoretical production, experience, the living participating "I," is seen as a dimension that must be included in an adequate analysis.[28] The very fact that women are, by and large, shut out of the production system of scientific knowledge, with its ideological power to define what is and what is not objective knowledge, paradoxically has offered feminists a fresh page on which to write.[29] Largely ignored by the oppressors and their systems of knowledge, feminists have necessarily theorized from practice and returned theory to practice.

While it would be false to suggest that all work claiming to be feminist achieves this dialectical synthesis, there is a sense in which theoretical writing looks and must look to the women's movement rather than to the male academy. Working from the experience of the specific oppression of women fuses the personal, the social, and the biological. It is not surprising that, within the natural sciences, it has been in biology and medicine that feminists have sought to defend women's interests and advance feminist interpretations. To take an example: menstruation, which so many women experience as distressing or at best uncomfortable, has generated a tremendous amount of collective discussion, study, and writing. A preeminent characteristic of these investigations lies in their fusing of subjective and objective knowledge in such a way as to make new knowledge. Cartesian dualism, biological determinism, and social constructionism fade when faced with the necessity of integrating and interpreting the personal experience of bleeding, pain, and tension.

Many of the slogans as well as titles of books and pamphlets arising from the movement speak to this necessary fusion. "A woman's right to choose" makes immediate sense to women. It is the demand for women to recover the control over their own bodies, a control that male-dominated medical professions and the profit motive have appropriated. Self-examination and self-health-care groups not only offer prefigurative social forms of health care, but also prefigurative forms of knowledge about natural science. The rightly best-selling book *Our Bodies, Ourselves* seeks to reclaim our sense of wholeness—the experiential unity of personal identity. In a similar vein, *For Her Own Good* not only affirms woman's capacities to understand her interests, but also exposes the male

28. See Ruth Hubbard's discussion of evolutionary theory as an example; Hubbard, Henifin, and Fried, eds. (n. 3 above), pp. 7–36.

29. Elizabeth Fee's "Is Feminism a Threat to Scientific Objectivity" (paper presented at the American Association for the Advancement of Science meeting, Toronto, January 4, 1981) pursues parallel themes.

professionalization of medicine in which alienated forms of both knowledge and care have driven out nonalienated female forms.[30] In this situation a feminist biology does not attempt to be objective and external to the female biological entity; it attempts to make over biological knowledge in order to overcome women's alienation from our own bodies, our own selves.

Here I can only pick out particular texts, but any reading of the abundant literature of the movement, particularly at the grass-roots level, reveals a feminism seeking to understand and contest the alienated forms of caring labor and to transform them into nonalienated forms. These moments, in which skill-sharing and skill-enhancing work collectively replaces the private drudgery and sexual servicing of the wife-mother, can only be fragmentary within a society that is systematically capitalist and patriarchal. Even the concept of prefigurative forms is too definite for the fragile but infinitely precious anticipations of the future such moments offer. Nonetheless the future is dialectically contained within the present—that is, insofar as humanity has a future, a prospect that can by no means be taken for granted in the eighties.

The creative energy of the women's movement in simultaneously fashioning new organizational forms and new knowledge is almost taken for granted by the movement itself. It makes fresh syntheses of theoretical significance for reconceptualizing the natural, as well as the social, sciences with the disarming charm of Molière's hero who discovered that he had been speaking prose all his life. While feminist theoreticians are increasingly exploring the epistemological transformations of feminist work in the social sciences, the implications for the natural sciences are only beginning to be articulated.[31] Yet because of the significance of science and technology as major instruments of both ideological and material oppression, the need for a feminist science is increasingly acute.[32] Socialist critical thought of the seventies explored the division of mental and manual labor and its implications for alienated knowledge in the production of things. Feminism points to the third and hidden division of caring labor in the alienated reproduction

---

30. Boston Women's Health Care Collective, *Our Bodies, Ourselves* (New York: Random House, 1971); Barbara Ehrenreich and Deirdre English, *For Her Own Good: 150 Years of the Experts' Advice to Women* (Garden City, N.Y.: Doubleday, Anchor Press, 1978).

31. In this emerging debate some, such as Griffin (n. 3 above), seek to replace men's objectivity with women's subjectivity; others, such as Rose ("Hyper-reflexivity" [n. 5 above]) and Evelyn Fox Keller ("Feminism and Science," *Signs: Journal of Women in Culture and Society* 7, no. 3 [Spring 1982]: 589–602), seek a synthesis combining subjective and objective ways of knowing the natural world.

32. Though it is beyond the scope of this article, the need for a specifically feminist technology is beginning to be expressed particularly in connection with the "new technology." Feminist collectives working on computing and word processing are increasingly evident. See, e.g., a recent issue of *Scarlet Woman*, vol. 14 (1982), available c/o 177 St. Georges Rd., N. Fitzroy, Victoria 3068, Australia.

of human beings themselves. Bringing caring labor and the knowledge that stems from participation in it to the analysis becomes critical for a transformative program equally within science and within society. The baby socks, webs of wool, photos, and flowers threaded into wire fences by the thousands of women peace activists ringing Greenham Common speak for this knowledge of the integration of hand, brain, *and* heart.

Applied Social Studies
University of Bradford

The Instability of the Analytical Categories of Feminist Theory

Sandra Harding

Feminist theory began by trying to extend and reinterpret the categories of various theoretical discourses so that women's activities and social relations could become analytically visible within the traditions of intellectual discourse.[1] If women's natures and activities are as fully social as are men's, then our theoretical discourses should reveal women's lives with just as much clarity and detail as we presume the traditional approaches reveal men's lives. We had thought that we could make the categories and concepts of the traditional approaches objective or Archimedean where they were not already.

As we all have come to understand, these attempts revealed that neither women's activities nor gender relations (both inter- and intra-gender relations) can be added to these theoretical discourses without distorting the discourses and our subject matters. The problem here is not a simple one, because liberal political theory and its empiricist epistemology, Marxism, critical theory, psychoanalysis, functionalism, structural-

1. My thinking about these issues has been greatly improved by the comments of Margaret Andersen and the anonymous reviewers for *Signs: Journal of Women in Culture and Society*, as well as by discussions over the last several years with many of the feminist science critics cited in this paper. I am grateful for support for this research and the larger project of which it is a part provided by the National Science Foundation, a Mina Shaughnessy Fellowship from the Fund for the Improvement of Post-Secondary Education, University of Delaware Faculty Research Grants, and a Mellon Fellowship at the Wellesley Center for Research on Women. For the larger project, see *The Science Question in Feminism* (Ithaca, N.Y.: Cornell University Press, 1986).

This essay originally appeared in *Signs*, vol. 11, no. 4, Summer 1986.

ism, deconstructionism, hermeneutics, and the other theoretical frameworks we have explored both do and do not apply to women and to
gender relations. On the one hand, we have been able to use aspects or
components of each of these discourses to illuminate our subject matters.
We have stretched the intended domains of these theories, reinterpreted
their central claims, or borrowed their concepts and categories to make
visible women's lives and feminist views of gender relations. After our
labors, these theories often do not much resemble what their nonfeminist
creators and users had in mind, to put the point mildly. (Think of the
many creative uses to which feminists have put Marxist or psychoanalytic
concepts and categories; of how subversive these revised theories are of
fundamental tendencies in Marxism and Freudianism.) On the other
hand, it has never been women's experiences that have provided the
grounding for any of the theories from which we borrow. It is not
women's experiences that have generated the problems these theories
attempt to resolve, nor have women's experiences served as the test of the
adequacy of these theories. When we begin inquiries with women's experiences instead of men's, we quickly encounter phenomena (such as
emotional labor or the positive aspects of "relational" personality structures) that were made invisible by the concepts and categories of these
theories. The recognition of such phenomena undermines the legitimacy
of the central analytical structures of these theories, leading us to wonder
if we are not continuing to distort women's and men's lives by our
extensions and reinterpretations. Moreover, the very fact that we borrow
from these theories often has the unfortunate consequence of diverting
our energies into endless disputes with the nonfeminist defenders of
these theories: we end up speaking not to other women but to patriarchs.

Furthermore, once we understand the destructively mythical character of the essential and universal "man" which was the subject and
paradigmatic object of nonfeminist theories, so too do we begin to doubt
the usefulness of analysis that has essential, universal woman as its subject
or object—as its thinker or the object of its thought. We have come to
understand that whatever we have found useful from the perspective of
the social experience of Western, bourgeois, heterosexual, white women
is especially suspect when we begin our analyses with the social experiences of any other women. The patriarchal theories we try to extend and
reinterpret were created to explain not men's experience but only the
experience of those men who are Western, bourgeois, white, and heterosexual. Feminist theorists also come primarily from these categories—not
through conspiracy but through the historically common pattern that it is
people in these categories who have had the time and resources to theorize, and who—among women—can be heard at all. In trying to develop
theories that provide the one, true (feminist) story of human experience,
feminism risks replicating in theory and public policy the tendency in the

patriarchal theories to police thought by assuming that only the problems of *some* women are human problems and that solutions for them are the only reasonable ones. Feminism has played an important role in showing that there are not now and never have been any generic "men" at all—only gendered men and women. Once essential and universal man dissolves, so does his hidden companion, woman. We have, instead, myriads of women living in elaborate historical complexes of class, race, and culture.

I want to talk here about some challenges for theorizing itself at this moment in history, and, in particular, for feminist theorizings. Each has to do with how to use our theories actively to transform ourselves and our social relations, while we and our theories—the agents and visions of reconstruction—are themselves under transformation. Consider, for instance, the way in which we focus on some particular inadequate sexist or earlier feminist analysis and show its shortcomings—often with brilliance and eloquence. In doing so, we speak from the assumptions of some other discourse feminism has adopted or invented. These assumptions always include the belief that we can, in principle, construct or arrive at the perspective from which nature and social life can be seen as they really are. After all, we argue that sexist (or earlier feminist) analyses are wrong, inadequate, or distorting—not that they are equal in scientific or rational grounding to our criticisms.

However, we sometimes claim that theorizing itself is suspiciously patriarchal, for it assumes separations between the knower and the known, subject and object, and the possibility of some powerful transcendental, Archimedean standpoint from which nature and social life fall into what we think is their proper perspective. We fear replicating—to the detriment of women whose experiences have not yet been fully voiced within feminist theory—what we perceive as a patriarchal association between knowledge and power.[2] Our ability to detect androcentrism in traditional analyses has escalated from finding it in the content of knowledge claims to locating it in the forms and goals of traditional knowledge seeking. The voice making *this* proposal is itself super-Archimedean, speaking from some "higher" plane, such that Archimedes' followers in contemporary intellectual life are heard as simply part of the inevitable flux and imperfectly understood flow of human history. (And this is true

2. See, e.g., Maria C. Lugones and Elizabeth V. Spelman, "Have We Got a Theory for You! Feminist Theory, Cultural Imperialism and the Demand for 'the Women's Voice,'" *Hypatia: A Journal of Feminist Philosophy* (special issue of *Women's Studies International Forum*) 6, no. 6 (1983): 573–82; many of the selections in *New French Feminisms*, ed. Elaine Marks and Isabelle de Courtivron (New York: Schocken Books, 1981); Jane Flax, "Gender as a Social Problem: In and For Feminist Theory" *American Studies/Amerika Studien* (June 1986); Donna Haraway, "A Manifesto for Cyborgs: Science, Technology, and Socialist Feminism in the 1980's," *Socialist Review* 80 (1983): 65–107.

even when the voice marks its own historical particularity, its femininity.) When it is unreflective, this kind of postmodernism—a kind of absolute relativism—itself takes a definitive stand from yet further outside the political and intellectual needs that guide our day-to-day thinking and social practices. In reaction we wonder how we can not want to say *the way things really are* to "our rulers" as well as to ourselves, in order to voice opposition to the silences and lies emanating from the patriarchal discourses and our own partially brainwashed consciousnesses. On the other hand, there is good reason to agree with a feminist postmodernist suspicion of the relationship between accepted definitions of "reality" and socially legitimated power.

How then are we to construct adequate feminist theory, or even *theories*—whether postmodern or not? Where are we to find the analytical concepts and categories that are free of the patriarchal flaws? What are the analytical categories for the absent, the invisible, the silenced that do not simply replicate in mirror-image fashion the distorting and mystifying categories and projects of the dominant discourses? Again, there are two ways to look at this situation. On the one hand, we can use the liberal powers of reason and the will, shaped by the insights gained through engaging in continuing political struggles, to piece what we see before our eyes in contemporary social life and history into a clear and coherent conceptual form, borrowing from one androcentric discourse here, another one there, patching in between in innovative and often illuminating ways, and revising our theoretical frameworks week by week as we continue to detect yet further androcentrisms in the concepts and categories we are using. We can then worry about the instability of the analytical categories and the lack of a persisting framework from which we continue to build our accounts. (After all, there should be some progress toward a "normal" discourse in our explanations if we are to create a coherent guide to understanding and action.) On the other hand, we can learn how to embrace the instability of the analytical categories; to find in the instability itself the desired theoretical reflection of certain aspects of the political reality in which we live and think; to use these instabilities as a resource for our thinking and practices. No "normal science" for us![3] I recommend we take the second course, an uncomfortable goal, for the following reason.

The social life that is our object of study and within which our analytical categories are formed and tested is in exuberant transformation.[4] Reason, will power, reconsidering the material—even politi-

---

3. See Thomas S. Kuhn, *The Structure of Scientific Revolutions* (Chicago: University of Chicago Press, 1970). "Normal science" was Kuhn's term for a "mature science," one where conceptual and methodological assumptions are shared by the inquirers in a field.

4. Perhaps it has always been. But the emergence of "state patriarchy" from the "husband patriarchy" of the first half of the century, the rising of people of color from

cal struggle—will not slow these changes in ways over which our feminisms should rejoice. It would be a delusion for feminism to arrive at a master theory, at a "normal science" paradigm with conceptual and methodological assumptions that we presume all feminists accept. Feminist analytical categories *should* be unstable—consistent and coherent theories in an unstable and incoherent world are obstacles to both our understanding and our social practices.

We need to learn how to see our theorizing projects as illuminating "riffing" between and over the beats of patriarchal theories, rather than as rewriting the tunes of any particular one (Marxism, psychoanalysis, empiricism, hermeneutics, deconstructionism, to name a few) so that it perfectly expresses *what we think at the moment we want to say*. The problem is that we do not know and should not know just what we want to say about a number of conceptual choices with which we are presented—except that the choices themselves create no-win dilemmas for our feminisms.

In the field in which I have been working—feminist challenges to science and epistemology—this situation makes the present moment an exciting one in which to live and think, but a difficult one in which to conceptualize a definitive overview. That is, the arguments between those of us who are criticizing science and epistemology are unresolvable within the frameworks in which we have been posing them. We need to begin seeing these disputes not as a process of naming issues to be resolved but instead as opportunities to come up with better problems than those with which we started. The destabilization of thought often has advanced understanding more effectively than restabilizations, and the feminist criticisms of science point to a particularly fruitful arena in which the categories of Western thought need destabilization. Though these criticisms began by raising what appeared to be politically contentious but theoretically innocuous questions about discrimination against women in the social structure of science, misuses of technology, and androcentric bias in the social sciences and biology, they have quickly escalated into ones that question the most fundamental assumptions of modern, Western thought. They therefore implicitly challenge the theoretical constructs within which the original questions were formulated and might be answered.

Feminisms are totalizing theories. Because women and gender relations are everywhere, the subject matters of feminist theories are not containable within any single disciplinary framework or any set of them. "The scientific worldview" has also taken itself to be a totalizing theory—anything and everything worth understanding can be explained or inter-

---

colonized subjugations, and the ongoing shifts in international capitalism all insure that this moment, at any rate, is one of exuberant transformation. See Ann Ferguson, "Patriarchy, Sexual Identity, and the Sexual Revolution," *Signs: Journal of Women in Culture and Society* 7, no. 1 (1981): 158–99, for discussion of the shifts in forms of patriarchy.

preted within the assumptions of modern science. Of course there is another world—the world of emotions, feelings, political values, of the individual and collective unconscious, of social and historical particularity explored in novels, drama, poetry, music, and art, and the world within which we all live most of our waking and dreaming hours under constant threat of its increasing reorganization by scientific rationality.[5] One of the projects of feminist theorists is to reveal the relationships between these two worlds—how each shapes and informs the other. In examining feminist criticisms of science, then, we must consider all that science does not, the reasons for these exclusions, how these shape science precisely through their absences—both acknowledged and unacknowledged.

Instead of fidelity to the assumption that coherent theory is a desirable end in itself and the only reliable guide to action, we can take as our standard fidelity to *parameters* of dissonance within and between assumptions of patriarchal discourses. This approach to theorizing captures what some take to be a distinctively women's emphasis on contextual thinking and decision making and on the processes necessary for gaining understanding in a world not of our own making—that is, a world that does not encourage us to fantasize about how we could order reality into the forms we desire.[6] It locates the ways in which a valuably "alienated consciousness," "bifurcated consciousness," "oppositional consciousness" might function at the level of active theory making—as well as at the level of skepticism and rebellion. We need to be able to cherish certain kinds of intellectual, political, and psychic discomforts, to see as inappropriate and even self-defeating certain kinds of clear solutions to the problems we have been posing.

"Bad Science" or "Science as Usual"?

Are sexist assumptions in substantive scientific research the result of "bad science" or simply "science as usual"? The first alternative offers hopes of reforming the kind of science we have; the second appears to deny this possibility.

5. Milan Kundera, in the article "The Novel and Europe" (*New York Review of Books*, vol. 31, no. 12 [July 19, 1984]), asks if it is an accident that the novel and the hegemony of scientific rationality arose simultaneously.

6. This emphasis is expressed in different ways by Sara Ruddick, "Maternal Thinking," *Feminist Studies* 6, no. 2 (Summer 1980): 342–67; Carol Gilligan, *In a Different Voice: Psychological Theory and Women's Development* (Cambridge, Mass.: Harvard University Press, 1982); Dorothy Smith, "Women's Perspective as a Radical Critique of Sociology," *Sociological Inquiry* 44, no. 1 (1974): 7–13; and "A Sociology for Women," in *The Prism of Sex: Essays in the Sociology of Knowledge*, ed. J. Sherman and E. T. Beck (Madison: University of Wisconsin Press, 1979).

It is clear that feminist criticisms of the natural and social sciences have identified and described science badly practiced—that is, science distorted by masculine bias in problematics, theories, concepts, methods of inquiry, observations, and interpretations of results of research.[7] There are facts of the matter, these critics claim, but androcentric science cannot locate them. By identifying and eliminating masculine bias through more rigorous adherence to scientific methods, we can get an objective, de-gendered (and in that sense, value-free) picture of nature and social life. Feminist inquiry represents not a substitution of one gender loyalty for the other—one subjectivism for another—but the transcendence of gender which thereby increases objectivity.

In this argument, we use empiricist epistemology because its ends are the same as ours: objective, value-neutral results of research. This feminist empiricism argues that sexism and androcentrism are social biases. Movements for social liberation "make it possible for people to see the world in an enlarged perspective because they remove the covers and blinders that obscure knowledge and observation."[8] Thus the women's movement creates the opportunity for such an enlarged perspective— just as did the bourgeois revolution of the fifteenth to seventeenth centuries, the proletarian revolution of the nineteenth century, and the revolutions overthrowing European and U.S. colonialism in recent decades. Furthermore, the women's movement creates more women scientists and more feminist scientists (men as well as women), who are more likely than nonfeminist men to notice androcentric bias.

Feminist empiricism offers a powerful explanation—though a misleading one—for the greater empirical adequacy of so much of feminist research. It has the virtue of answering the question of how a political movement such as feminism could be contributing to the growth of objective scientific knowledge. In making this argument, however, we avert our eyes from the fact that this appeal to empiricism in fact subverts empiricism in three ways. (1) For empiricism, the social identity of the observer is supposed to be irrelevant to the quality of research results. Feminist empiricism argues that women (or feminists, men and women)

7. See, e.g., the *Signs* review essays in the social sciences, and the papers in Brighton Women and Science Group, *Alice through the Microscope* (London: Virago Press, 1980); Ruth Hubbard, M. S. Henifin, and Barbara Fried, eds., *Biological Woman: The Convenient Myth* (Cambridge, Mass.: Schenkman Publishing Co., 1982); Marian Lowe and Ruth Hubbard, eds., *Woman's Nature: Rationalizations of Inequality* (New York: Pergamon Press, 1983); Ethel Tobach and Betty Rosoff, eds., *Genes and Gender I, II, III, IV* (New York: Gordian Press, 1978, 1979, 1981, 1984) (Hubbard and Lowe are the guest editors for vol. 2 in the series, subtitled *Pitfalls in Research on Sex and Gender*); Ruth Bleier, *Science and Gender: A Critique of Biology and Its Theories on Women* (New York: Pergamon Press, 1984).

8. Marcia Millman and Rosabeth Moss Kanter, "Editorial Introduction," in *Another Voice: Feminist Perspectives on Social Life and Social Science* (New York: Anchor Books, 1975), vii.

as a group are more likely to produce unbiased, objective results of inquiry than are men (or nonfeminists) as a group. (2) We claim that a key origin of androcentric bias lies in the selection of problems for inquiry and in the definition of what is problematic about them. Empiricism insists that its methodological norms are meant to apply only to the context of justification and not to the context of discovery where problematics are identified and defined. Hence we have shown the inadequacy, the impotence, of scientific methods to achieve their goals. (3) We often point out that it is exactly following the logical and sociological norms of inquiry which results in androcentric results of research—appealing to the already existing (Western, bourgeois, homophobic, white, sexist) scientific community for confirmation of the results of research; generalizing to all humans from observations only of males. Our empiricist criticisms of "bad science" in fact subvert the very understandings of science they are meant to reinforce.

These problems suggest that the most fundamental categories of scientific thought are male biased. Many of the critics of "bad science" also make this second criticism though it undercuts the assumptions of the first.[9] Here they point to historians' descriptions of how sexual politics have shaped science, and science, in turn, has played a significant role in advancing sexual politics. Each has provided a moral and political resource for the other.[10] Furthermore, they show that "pure science"—inquiry immune from the technological and social needs of the larger culture—exists only in the unreflective mental life of some individual

9. This tension between the two kinds of criticisms is pointed out by Helen Longino and Ruth Doell, "Body, Bias and Behavior: A Comparative Analysis of Reasoning in Two Areas of Biological Science," *Signs* 9, no. 2 (1983): 206–27; and by Donna Haraway, "In the Beginning Was the Word: The Genesis of Biological Theory," *Signs* 6, no. 3 (1981): 469–81. Longino and Doell think "feminists do not have to choose between correcting bad science or rejecting the entire scientific enterprise" (208) and that "only by developing a more comprehensive understanding of the operation of male bias in science, as distinct from its existence, can we move beyond these two perspectives in our search for remedies" (207). Longino and Doell's analysis is helpful indeed in creating this understanding, but since they do not come to grips with the criticisms of "science as usual," my remedy parts from theirs. Haraway does not propose a solution to the dilemma.

10. See, e.g., Elizabeth Fee, "Nineteenth Century Craniology: The Study of the Female Skull," *Bulletin of the History of Medicine* 53, no. 3 (1979): 415–33; Susan Griffin, *Woman and Nature: The Roaring inside Her* (New York: Harper & Row, 1978); Diana Long Hall, "Biology, Sex Hormones and Sexism in the 1920's," *Philosophical Forum* 5 (1973–74): 81–96; Donna Haraway, "Animal Sociology and a Natural Economy of the Body Politic, Parts 1, 2," *Signs* 4, no. 1 (1978): 21–60; Ruth Hubbard, "Have Only Men Evolved?" in Hubbard, Henifin, and Fried, eds. (n. 7 above); L. J. Jordanova, "Natural Facts: A Historical Perspective on Science and Sexuality," in *Nature, Culture and Gender*, ed. Carol MacCormack and Marilyn Strathern (New York: Cambridge University Press, 1980); Carolyn Merchant, *The Death of Nature: Women, Ecology and the Scientific Revolution* (New York: Harper & Row, 1980); Evelyn Fox Keller, *Reflections on Gender and Science* (New Haven, Conn.: Yale University Press, 1985).

scientists and in the rhetoric of science apologists. That is, one does not have to impugn the motives of individual physicists, chemists, or sociologists in order to make a convincing case that the scientific enterprise is structurally and symbolically part and parcel of the value systems of those cultures that maintain it. This argument poses difficulties for us, nonetheless, since if the very concepts of nature, of dispassionate, value-free, objective inquiry, and of transcendental knowledge are androcentric, white, bourgeois, and Western, then no amount of more rigorous adherence to scientific method will eliminate such bias, for the methods themselves reproduce the perspectives generated by these hierarchies and thus distort our understandings.

While these new understandings of the history of science and sexuality expand our understanding immensely, they do not tell us whether a science apparently so inextricably intertwined with the history of sexual politics can be pried loose to serve more inclusive human ends—or whether it is strategically worthwhile to try to do so. Is history destiny? Would the complete elimination of androcentrisms from science leave no science at all? But isn't it important to try to degender science as much as we can in a world where scientific claims are *the* model of knowledge? How can we afford to choose between redeeming science or dismissing it altogether when neither choice is in our best interest?

Successor Science or Postmodernism

The dilemma that arises in criticisms of "bad science" and of "science as usual" reappears at a metalevel in feminist theory's conflicting tendencies toward postmodernism and what I shall call the feminist successor science projects. Feminist empiricism explains (albeit subversively) the achievements of feminist inquiry—of that purported contradiction in terms: a politicized scientific inquiry—by appeal to the familiar empiricist assumptions. In contrast, the feminist standpoint epistemologies articulate an understanding of scientific knowledge seeking that replaces, as successor to, the Enlightenment vision captured by empiricism.[11] Both the

11. Important formulations of the epistemology for a feminist "successor science" have been provided by Jane Flax, "Political Philosophy and the Patriarchal Unconscious: A Psychoanalytic Perspective on Epistemology and Metaphysics," in *Discovering Reality: Feminist Perspectives on Epistemology, Metaphysics, Methodology and Philosophy of Science*, ed. Sandra Harding and Merrill B. Hintikka (Dordrecht: D. Reidel Publishing Co., 1983); Nancy Hartsock, "The Feminist Standpoint: Developing the Ground for a Specifically Feminist Historical Materialism," in Harding and Hintikka, eds., and chap. 10 of *Money, Sex and Power* (Boston: Northeastern University Press, 1983); Hilary Rose, "Hand, Brain, and Heart: A Feminist Epistemology for the Natural Sciences," *Signs* 9, no. 1 (1983): 73–90, and "Is a Feminist Science Possible?" (paper presented at MIT, Cambridge, Massachusetts, 1984); D. Smith, "Women's Perspective as a Radical Critique of Sociology," and "A Sociology for Women" (both in n. 6 above).

standpoint and postmodern tendencies within feminist theory place feminism in an uneasy and ambivalent relationship to patriarchal discourses and projects (just as did feminist empiricism). There are good reasons to think of both as imperfect and converging tendencies toward a postmodernist reality, but there are also good reasons to nourish the tendencies in each which conflict.

The feminist standpoint epistemologies use for feminist ends the Marxist vision in which science can reflect "the way the world is" and contribute to human emancipation. Feminist research claims in the natural and social sciences do appear to be truer to the world, and thus more objective than the sexist claims they replace. They provide an understanding of nature and social life that transcends gender loyalties and does not substitute one gender-loyal understanding for another. Furthermore, these feminist appeals to truth and objectivity trust that reason will play a role in the eventual triumph of feminism, that feminism correctly will be perceived as more than a power politic—though it is that, too. The successor science tendencies aim to provide more complete, less false, less distorting, less defensive, less perverse, less rationalizing understandings of the natural and social worlds.

This is already a radical project, for the Enlightenment vision explicitly denied that women possessed the reason and powers of dispassionate, objective observation required by scientific thinking. Women could be objects of (masculine) reason and observation but never the subjects, never the reflecting and universalizing human minds. Only men were in fact envisioned as ideal knowers, for only men (of the appropriate class, race, and culture) possessed the innate capacities for socially transcendant observation and reason. The ends and purposes of such a science turned out to be far from emancipatory for anyone.

Marxism reformulated this Enlightenment vision so that the proletariat, guided by Marxist theory and by class struggle, became the ideal knowers, the group capable of using observation and reason to grasp the true form of social relations, including our relations with nature.[12] This Marxist successor to bourgeois science was, like its predecessor, to provide one social group—here, the proletariat—with the knowledge and power to lead the rest of the species toward emancipation. Marxism's epistemology is grounded in a theory of labor rather than a theory of innate (masculine) faculties; so just as not all human faculties are equal in the bourgeois version, here not all labor is equal. It was through struggle in the workplace that the proletariat would generate knowledge. In neither

---

12. Friedrich Engels, "Socialism: Utopian and Scientific," in *The Marx and Engels Reader*, ed. R. Tucker (New York: W. W. Norton & Co., 1972); George Lukács, "Reification and the Consciousness of the Proletariat," *History and Class Consciousness* (Cambridge, Mass.: MIT Press, 1968).

socialist practice nor Marxist theory were any women ever conceptualized as fundamentally defined by their relation to the means of production, regardless of their work force participation. They were never thought of as full-fledged members of the proletariat who could reason and thus know how the world is constructed. Women's distinctive reproductive labor, emotional labor, "mediating" labor thus disappeared within the conceptual framework of Marxist theory, leaving women invisible as a class or social group of agents of knowledge. (Other forms of nonwage or nonindustrial labor similarly disappeared from the center of this conceptual scheme, mystifying the knowing available to slaves and colonized peoples.)

This standpoint tendency in feminist epistemology is grounded in a successor theory of labor or, rather, of distinctively human activity, and seeks to substitute women or feminists (the accounts differ) for the proletariat as the potentially ideal agents of knowledge. Men's (sexists') perceptions of themselves, others, nature, and the relations between all three are characteristically not only partial but also perverse.[13] Men's characteristic social experience, like that of the bourgeoisie, hides from them the politically imposed nature of the social relations they see as natural. Dominant patterns in Western thought justify women's subjugation as necessary for the progress of culture, and men's partial and perverse views as uniquely and admirably human. Women are able to use political struggle and analysis to provide a less partial, less defensive, less perverse understanding of human social relations—including our relations with nature. The standpoint theorists argue that this analysis, not feminist empiricism, accounts for the achievements of feminist theory and research because it is politically engaged theory and research from the perspective of the social experience of the subjugated sex/gender.

The second line of thought, one that can be found within many of these very same writings, expresses a profound skepticism toward the Enlightenment vision of the power of "the" human mind to reflect perfectly a readymade world that is out there for the reflecting. Many feminists share a rejection of the value of the forms of rationality, of dispassionate objectivity, of the Archimedean perspective, which were to be the means to knowledge. Here they are ambivalently related to such other skeptics of modernism as Nietzsche, Wittgenstein, Derrida, Foucault, Lacan, Feyerabend, Rorty, Gadamer, and the discourses of semi-

---

13. Hartsock, especially, discusses the perversity of the androcentric vision (n. 11 above). I shall subsequently refer to the men vs. women dichotomy since that is the way most of these standpoint theorists put the issue. However, I think these categories are inadequate even for the standpoint projects: it is feminists vs. nonfeminists (sexists) we should be discussing here.

otics, psychoanalysis, structuralism, and deconstructionism.[14] What is striking is how the successor science idea and the postmodern skepticism of science are both embraced by these theorists, though the concepts are diametrically opposed in the nonfeminist discourses.[15]

From the perspective of this postmodern tendency in feminist thinking, the feminist successor science project can appear still too firmly rooted in distinctively masculine modes of being in the world. As one theorist puts the issue, "Perhaps 'reality' can have 'a' structure only from the falsely universalizing perspective of the master. That is, only to the extent that one person or group can dominate the whole, can 'reality' appear to be governed by one set of rules or be constituted by one privileged set of social relations."[16] How can feminism radically redefine the relationship between knowledge and power if it creates yet another epistemology, yet another set of rules for the policing of thought?

However, this postmodern project can appear viciously utopian from the perspective of the successor science tendency.[17] It seems to challenge the legitimacy of trying to describe the way the world is from a distinctively feminist perspective. It can appear of a piece with masculine and bourgeois desire to justify one's activities by denying one's social, embodied location in history; to attempt to transcend one's objective location in politics by appeal to a *mea culpa*, all-understanding, bird's-eye view (the transcendental ego in naturalistic garb) of the frailty of mere humans. That is, in its uneasy affiliation with nonfeminist postmodernism, the

14. Jane Flax discusses this postmodern strain in feminist theory in "Gender as a Social Problem: In and For Feminist Theory" (n. 2 above) and cites these as among the key skeptics of modernism: Friedrich Nietzsche, *On the Genealogy of Morals* (New York: Vintage, 1969), and *Beyond Good and Evil* (New York: Vintage, 1966); Jacques Derrida, *L'écriture et la Différence* (Paris: Editions du Seuil, 1967); Michel Foucault, *The Order of Things* (New York: Vintage, 1973), and *The Archaeology of Knowledge* (New York: Harper & Row, 1972); Jacques Lacan, *Speech and Language in Psychoanalysis* (Baltimore: Johns Hopkins University Press, 1968), and *The Four Fundamental Concepts of Psychoanalysis* (New York: W. W. Norton & Co., 1973); Paul Feyerabend, *Against Method* (New York: Schocken Books, 1975); Richard Rorty, *Philosophy and the Mirror of Nature* (Princeton, N.J.: Princeton University Press, 1979); Hans-Georg Gadamer, *Philosophical Hermeneutics* (Berkeley: University of California Press, 1976); Ludwig Wittgenstein, *On Certainty* (New York: Harper & Row, 1972), and *Philosophical Investigations* (New York: Macmillan Publishing Co., 1970). See also Jean-François Lyotard, *The Postmodern Condition: A Report on Knowledge*, trans. G. Bennington and B. Massumi (Minneapolis: University of Minnesota Press, 1984).

15. However, different weight is given to one or the other tendency by each theorist. Nevertheless, all are explicitly aware of the tension in their own work between the two kinds of criticisms of modern, Western epistemology. It is another project to explain how each attempts to resolve this tension. See Harding (n. 1 above) for further discussion of these theorists' work.

16. Flax, "Gender as a Social Problem" (n. 2 above), 17.

17. Flax appears to be unaware of this problem. Engels distinguishes utopian and scientific socialisms (n. 12 above).

feminist postmodernist tendency appears to support an inappropriate relativist stance by the subjugated groups, one that conflicts with feminism's perception that the realities of sexual politics in our world demand engaged political struggle. It appears to support an equally regressive relativism for those mildly estranged members of the subjugating groups with doubts about the legitimacy of their own objective privilege and power (see list above of nonfeminist skeptics of modernism). It is worth keeping in mind that the articulation of relativism as an intellectual position emerges historically only as an attempt to dissolve challenges to the legitimacy of purportedly universal beliefs and ways of life. It is an objective problem, or a solution to a problem, only *from the perspective of the dominating groups*. Reality may indeed appear to have many different structures from the perspectives of our different locations in social relations, but some of those appearances are ideologies in the strong sense of the term: they are not only false and "interested" beliefs but also ones that are used to structure social relations for the rest of us. For subjugated groups, a relativist stance expresses a false consciousness. It accepts the dominant group's insistence that their right to hold distorted views (and, of course, to make policy for all of us on the basis of those views) is intellectually legitimate.

Are not the policing of thought in the service of political power and the retreat to purportedly politically innocent, relativistic, mere interpretations of the world the two sides of the Enlightenment and bourgeois coin to which feminism is opposed? Is it not true—as these theorists all argue in different ways—that men's and women's different kinds of interactions with nature and social life (different "labor") provide women with distinctive and privileged scientific and epistemological standpoints? How can feminism afford to give up a successor science project if it is to empower all women in a world where socially legitimated knowledge and the political power associated with it are firmly lodged in white, Western, bourgeois, compulsorily heterosexual, men's hands? Yet how can we give up our distrust of the historic links between this legitimated knowledge and political power?

One way to see these two tendencies in feminist theory is as converging approaches to a postmodernist world—a world that will not exist until both (conflicting) tendencies achieve their goals. From this perspective, at its best postmodernism envisions epistemology in a world where thought does not need policing. It recognizes the existence today of far less than the ideal speech situation, but disregards (or fails to acknowledge) the political struggles necessary to bring about change. The standpoint tendency attempts to move us toward that ideal world by legitimating and empowering the "subjugated knowledges" of women, without which that postmodern epistemological situation cannot come into existence. It fails nonetheless to challenge the modernist intimacies between knowledge and power, or the legitimacy of assuming there can be a single, feminist story of reality. Whether or not this is a useful way to see the relationship between the two tendencies, I am arguing that we must resist the temptation to explain away the problems each addresses and to choose one to the exclusion of the other.

The Feminist Standpoint and Other "Others"

Feminist successor science projects stand in an uneasy relation to other emancipatory epistemologies insofar as the former seek to ground a uniquely legitimate and distinctive science and epistemology on the shared characteristics of women's activity. Hilary Rose locates these grounds in the way women's labor unifies mental, manual, and caring labor. Nancy Hartsock focuses on the deeper opposition to the dualities of mental versus manual labor to be found in women's daily, concrete activities both in domestic life and wage labor. Jane Flax identifies the relatively more reciprocal sense of self women bring to all their activities. She suggests that the small gap between men's and women's concepts of self, others, and nature prefigures the possible larger gap between the defensively dualistic knowledge characteristic of male-dominant social orders and the relational and contextual knowledge possible in a future society of "reciprocal selves." Dorothy Smith argues that women's social labor is concrete rather than abstract, that it cannot be articulated to either administrative forms of ruling or the categories of social science, and that it has been socially invisible—combining to create a valuably alienated and bifurcated consciousness in women.[18] However, other emancipatory perspectives claim as resources for their politics and epistemologies similar aspects of their own activity.

On the one hand, of course, feminism is right to identify women and men as classes in opposition at this moment in history. Everywhere in the world we find these two classes, and virtually everywhere the men subjugate the women in one way or another.[19] Furthermore, even male feminists receive benefits from an institutionalized sexism they actively struggle to eliminate. Objectively, no individual men can succeed in renouncing sexist privilege any more than individual whites can succeed in renouncing racist privilege—the benefits of gender and race accrue regardless of the wishes of the individuals who bear them. Gender, like race and class, is not a voluntarily disposable individual characteristic.

18. Flax (n. 11 above); Hartsock (both items cited in n. 11 above); Rose (n. 11 above); Smith (n. 6 above).

19. "Virtually everywhere" to give the benefit of the doubt to anthropologists' claims about "egalitarian cultures." See, e.g., Eleanor Leacock, *Myths of Male Dominance* (New York: Monthly Review Press, 1981).

After all, fundamentally our feminisms address the extraction and transfer of social benefits from women to men *as groups* of humans, on a worldwide scale. Thus the standpoint theorists, in identifying the common aspects of women's social experience cross-culturally, contribute something important to our work.

On the other hand, the distinctive characteristics of women's activities that Rose, Hartsock, Flax, and Smith identify for our culture are probably to be found also in the labor and social experience of other subjugated groups. There are suggestions in the literature on Native Americans, Africans, and Asians that what feminists call feminine versus masculine personalities, ontologies, ethics, epistemologies, and worldviews may be what these other liberation movements call non-Western versus Western personalities and worldviews.[20] Thus, should there not also be Native American, African, and Asian sciences and epistemologies, based on the distinctive historical and social experience of these peoples? Would not such successor sciences and epistemologies provide similar analyses to those of the standpoint theorists? (I set aside the crucial and fatal complication for this way of thinking—the facts that one-half of these peoples are women and that most women are not Western.) On what grounds would the feminist sciences and epistemologies be superior to these others? What is and should be the relationship of the feminist projects to these other emancipatory knowledge-seeking projects?

It is a vast overgeneralization to presume that all Africans, let alone all colonized peoples, share distinctive personalities, ontologies, ethics, epistemologies, or worldviews. But is it any worse than the presumption that there are commonalities to be detected in *all women's* social experiences or worldviews? Let us note that we are thinking here about perspectives as inclusive as those referred to in such phrases as the "feudal worldview," the "modern worldview," or the "scientific worldview." Moreover, we women also claim an identity we were taught to despise;[21] around the globe we insist on the importance of our social experience as *women*, not just as gender-invisible members of class, race, or cultural

20. Russell Means, "Fighting Words on the Future of the Earth," *Mother Jones* (December 1980): 167; Vernon Dixon, "World Views and Research Methodology," in *African Philosophy: Assumptions and Paradigms for Research on Black Persons*, ed. L. M. King, V. Dixon, and W. W. Nobles (Los Angeles: Fanon Center Publication, Charles R. Drew Postgraduate Medical School, 1976) (but see also Paulin Hountondji, *African Philosophy: Myth and Reality* [Bloomington: Indiana University Press, 1983]); Joseph Needham, "History and Human Values: A Chinese Perspective for World Science and Technology," in *Ideology of/in the Natural Sciences*, ed. Hilary Rose and Steven Rose (Boston: Schenkman Publishing Co., 1979). I have discussed this situation more fully in "The Curious Coincidence of African and Feminine Moralities," in *Women and Moral Theory*, ed. Diana Meyers and Eva Kittay (Totowa, N.J.: Rowman & Allenheld, 1986), and in chap. 7 of Harding (n. 1 above).

21. Michele Cliff, *Claiming an Identity They Taught Me to Despise* (Watertown, Mass.: Persephone Press, 1980).

groups. Similarly, Third World peoples claim their colonized social experience as the grounding for a shared identity and as a common source of alternative understandings. Why is it not reasonable to explore how the experience of colonization itself shapes personalities and worldviews? How can white Western women insist on the legitimacy of what we think we share with all women and not acknowledge the equal legitimacy of what colonized peoples think they share with each other? In short, we cannot resolve this problem for the feminist standpoint by insisting on the cultural particularity of individuals in other cultures while at the same time arguing for the gender similarities of women cross-culturally.

One resolution of this dilemma for the standpoint tendency would be to say that feminist science and epistemology will be valuable in their own right alongside and as a part of these other possible sciences and epistemologies—not superior to them. With this strategy we have relinquished the totalizing, "master theory" character of our theory making which is at least an implicit goal of much feminist theorizing, and we have broken away from the Marxist assumptions that informed the feminist successor science projects. This response to the issue has managed to retain the categories of feminist theory (unstable though they be) and simply set them alongside the categories of the theory making of other subjugated groups. Instead of the "dual systems" theory with which socialist feminists wrestle,[22] this response gives us multisystems theory. Of course, it leaves bifurcated (and perhaps even more finely divided) the identities of all except ruling-class white Western women. There is a fundamental incoherence in this way of thinking about the grounds for feminist approaches to knowledge.

Another solution would be to renounce the goal of unity around shared social experiences in favor of solidarity around those goals that can be shared.[23] From this perspective, each standpoint epistemology—feminist, Third World, gay, working class—names the historical conditions producing the political and conceptual oppositions to be overcome but does not thereby generate universal concepts and political goals. Because gender is also a class and racial category in cultures stratified by class and race as well as by gender, no particular women's experience can uniquely generate the groundings for the visions and politics that will emancipate us from gender hierarchy. A variety of social groups are currently struggling against the hegemony of the Western, white, bourgeois, homophobic androcentric worldview and the politics it both generates and justifies. Our internal racial, sexual, and class struggles,

---

22. Iris Young, "Beyond the Unhappy Marriage: A Critique of the Dual Systems Theory," in *Women and Revolution*, ed. L. Sargent (Boston: South End Press, 1981).

23. See Bell Hooks, *Feminist Theory from Margin to Center* (Boston: South End Press, 1983), esp. chap. 4; and Haraway, "A Manifesto for Cyborgs" (n. 2 above).

and the differences in our cultural histories which define for us who we are as social beings, prevent our federating around our shared goals. It is history that will resolve or dissolve this problem, not our analytic efforts. Nevertheless, white, Western, bourgeois feminists should attend to the need for a more active theoretical and political struggle against our own racism, classism, and cultural centrism as forces that insure the continued subjugation of women around the world.

Culture Versus Nature and Gender Versus Sex

Historians and anthropologists show that the way contemporary Western society draws the borders between culture and nature is clearly both modern and culture bound.[24] The culture versus nature dichotomy reappears in complex and ambiguous ways in a number of other oppositions central to modern, Western thinking: reason versus the passions and emotions; objectivity versus subjectivity; mind versus the body and physical matter; abstract versus concrete; public versus private—to name a few. In our culture, and in science, masculinity is identified with culture and femininity with nature in all of these dichotomies. In each case, the latter is perceived as an immensely powerful threat that will rise up and overwhelm the former unless the former exerts severe controls over the latter.

This series of associated dualisms has been one of the primary targets of feminist criticisms of the conceptual scheme of modern science. It is less often recognized, however, how the dualism reappears in feminist thinking about gender, sex, or the sex/gender system. In preceding sections, I have talked about eliminating gender as if the social could be cleanly separated from the biological aspects of our sexual identities, practices, and desires. In feminist discourses, this mode of conceptualizing sexuality is clearly an advance over the biological determinist assumption that gender differences simply follow from sex differences. Since biological determinism is alive and flourishing in sociobiology, endocrinology, ethology, anthropology and, indeed, most nonfeminist discourses, I do not want to devalue the powerful analytical strategy of insisting on a clean separation between the known (and knowable) effects of biology and of culture. Nevertheless, a very different picture of sexual identities, practices, and desires emerges from recent research in biology, history, anthropology, and psychology.[25] Surprisingly, it could also be

24. See esp. the responses to Sherry Ortner's "Is Female to Male as Nature Is to Culture?" (in *Woman, Culture and Society*, ed. M. Z. Rosaldo and L. Lamphere [Stanford, Calif.: Stanford University Press, 1974]) in MacCormack and Strathern, eds. (n. 10 above).

25. See references cited in nn. 7, 10 above.

called biological determinism, though what is determined on this account is the plasticity rather than the rigidity of sexual identity, practice, and desire. Our species is doomed to freedom from biological constraints in these respects, as existentialists would put the issue.

The problem for feminist theory and practice here is twofold. In the first place, we stress that humans are *embodied* creatures—not Cartesian minds that happen to be located in biological matter in motion. Female embodiment is different from male embodiment. Therefore we want to know the implications for social relations and intellectual life of that different embodiment. Menstruation, vaginal penetration, lesbian sexual practices, birthing, nursing, and menopause are bodily experiences men cannot have. Contemporary feminism does not embrace the goal of treating women "just like men" in public policy. So we need to articulate what these differences are. However, we fear that doing so feeds into sexual biological determinism (consider the problems we have had articulating a feminist perspective on premenstrual syndrome and work-related reproductive hazards in ways that do not victimize women). The problem is compounded when it is racial differences between women we want to articulate.[26] How can we choose between maintaining that our biological differences ought to be recognized by public policy and insisting that biology is not destiny for either women or men?

In the second place, we have trouble conceptualizing the fact that the culture versus nature dichotomy and its siblings are not simply figments of thought to be packed up in the attic of outmoded ideas. The tendency toward this kind of dualism is an ideology in the strongest sense of the term, and such tendencies cannot be shucked off by mental hygiene and will power alone. The culture/nature dichotomy structures public policy, institutional and individual social practices, the organization of the disciplines (the social vs. the natural sciences), indeed the very way we see the world around us. Consequently, until our dualistic practices are changed (divisions of social experience into mental vs. manual, into abstract vs. concrete, into emotional vs. emotion denying), we are forced to think and exist within the very dichotomizing we criticize. Perhaps we can shift the assumption that the natural is hard to change and that the cultural is more easily changed, as we see ecological disasters and medical technologies on the one hand, and the history of sexism, classism, and racism on the other.[27] Nonetheless, we should continue insisting on the distinction between culture and nature, between gender and sex (especially in the face of biological determinist backlash), even as we analytically and experientially notice how inextricably they are intertwined in individuals and in cultures. These dichotomies are empirically false, but we cannot afford to

---

26. Inez Smith Reid, "Science, Politics, and Race," *Signs* 1, no. 2 (1975): 397–422.
27. Janice G. Raymond makes this point in "Transsexualism: An Issue of Sex-Role Stereotyping," in Tobach and Rosoff, eds., vol. 2 (n. 7 above).

dismiss them as irrelevant as long as they structure our lives and our consciousnesses.

Science as Craft: Anachronism or Resource?

Traditional philosophies of science assume an anachronistic image of the inquirer as a socially isolated genius, selecting problems to pursue, formulating hypotheses, devising methods to test the hypotheses, gathering observations, and interpreting the results of inquiry. The reality of most scientific research today is quite different, for these craft modes of producing scientific knowledge were replaced by industrialized modes in the nineteenth century for the natural sciences, and by the mid-twentieth century for the vast majority of social science research. Consequently, philosophy of science's rules and norms for individual knowledge seekers are irrelevant to the conduct of, and understanding of, most of contemporary science, as a number of science critics have pointed out.[28]

However, it is precisely in areas of inquiry that remain organized in craft ways where the most interesting feminist research has appeared.[29] Perhaps all of the most revolutionary claims have emerged from research situations where individual feminists (or small groups of them) identify a problematic phenomenon, hypothesize a tentative explanation, design and carry out evidence gathering, and then interpret the results of this research. In contrast, when the conception and execution of research are performed by different social groups of persons, as is the case in the vast majority of mainstream natural science and much social science research, the activity of conceptualizing the research is frequently performed by a privileged group and the activity of executing the research by a subjugated group. This situation insures that the conceptualizers will be able to avoid challenges to the adequacy of their concepts, categories, methods, and interpretations of the results of research.

This kind of analysis reinforces the standpoint theorists' argument that a prescriptive theory of knowledge—an epistemology—should be based on a theory of labor or human activity, not on a theory of innate faculties as empiricist epistemology assumes. In fact, the feminist epistemologies mentioned above are all grounded in a distinctive theory of human activity, and in one that gains support from an examination of the preconditions for the emergence of modern science in the fifteenth to seventeenth centuries. Feminists point to the unification of mental,

28. Jerome Ravetz, *Scientific Knowledge and Its Social Problems* (New York: Oxford University Press, 1971); Rose and Rose, eds. (n. 20 above); Rita Arditti, Pat Brennan, Steve Cafrak, eds., *Science and Liberation* (Boston: South End Press, 1980).

29. Hilary Rose in particular has pointed this out in "Hand, Brain, and Heart," and in "Is a Feminist Science Possible?" (n. 11 above). Perhaps all new research paradigms must be established through craft activity, as Kuhn argued.

manual, and emotional labor in women's work which provides women with a potentially more comprehensive understanding of nature and social life. As women increasingly are drawn into and seek men's work— from law and policy-making to medicine and scientific inquiry—our labor and social experience violate the traditional distinctions between men's and women's work, thus permitting women's ways of understanding reality to begin to shape public understandings. Similarly, it was a violation of the feudal division of labor that made possible the unity of mental and manual labor necessary to create science's new experimental method.[30]

Traditional philosophy of science's prescriptive image of the scientific inquirer, as craftsman, then, is irrelevant as a model for the activity that occupies the vast majority of scientific workers today. This image instead reflects the practices of the very few scientifically trained workers who are engaged in the construction of new research models. However, since the scientific worldview that feminism criticizes was constructed to explain the activity, results, and goals of the *craft labor* that constituted science in an earlier period, and since contemporary feminist craft inquiry has produced some of the most valuable new conceptualizations, it looks like we need to think more carefully about which aspects of the scientific worldview to reject and retain. Perhaps the mainstream enterprise of today is not scientific at all in the original sense of the term! Can it be that feminism and similarly estranged inquiries are the true offspring of Copernicus, Galileo, and Newton? Can this be true while at the same time these offspring undermine the epistemology that Hume, Locke, Descartes, and Kant developed to explain the birth of modern science? Once again, we are led to what I propose should be regarded as fruitful ambivalence toward the science we have. We should cultivate both "separatist" craft-structured inquiry *and* infuse the industrially structured sciences with feminist values and goals.

These are some of the central conceptual instabilities that emerge in considering the feminist criticism of science. Several of them arise in feminist theorizing more generally. I have been arguing that we cannot resolve these dilemmas in the terms in which we have been posing them and that instead we should learn how to regard the instabilities themselves as valuable resources. If we can learn how to use them, we can match Archimedes' greatest achievement—his inventiveness in creating a new kind of theorizing.

Department of Philosophy
University of Delaware

---

30. Edgar Zilsel, "The Sociological Roots of Science," *American Journal of Sociology* 47, no. 4 (1942): 545–60.

About the Contributors

SALLY G. ALLEN (LIVINGSTON) is a partner in Endowment Planners, a consulting company in Springfield, Massachusetts. In 1987, she will be a visiting professor in the School of Humanities and Arts at Hampshire College where she formerly taught.

SUSAN BORDO is assistant professor of philosophy at Le Moyne College. Her article in this collection is based on her book *The Flight to Objectivity: Essays on Cartesianism and Culture* (Albany: State University of New York Press, 1987). She is coeditor with Alison Jaggar of *Feminist Reconstructions of Being and Knowing* (New Brunswick, N.J.: Rutgers University Press, in press). Her research on the role of the body in the symbolization and reproduction of gender is described in her work in progress, tentatively titled *Food, Fashion and Power*.

SANDRA L. CHAFF is director of and archivist at the Archives and Special Collections on Women in Medicine at the Medical College of Pennsylvania. She is also a research instructor in the history of medicine at that institution. She has published articles on individual nineteenth-century women physicians and is coeditor with Ruth Haimbach, Carol Fenichel, and Nina B. Woodside of *Women in Medicine: A Bibliography of the Literature on Women Physicians* (Metuchen, N.J.: Scarecrow Press, 1977).

RUTH DOELL is professor of biology at San Francisco State University. She is interested in the relationship of feminism to science and of gender to science. She has recently completed an article, with Helen Longino, on sex hormones and human behavior.

MARTHA R. FOWLKES is associate professor and associate dean of the School of Family Studies at the University of Connecticut at Storrs. She has published widely on topics including the interface of family and work, sex roles, career development, and the sociology of community and subcultures. She is the author of *Behind Every Successful Man: Wives of Medicine and Academe* (New York: Columbia University Press, 1980).

DONNA HARAWAY is professor of the history of consciousness at the University of California at Santa Cruz. She writes and teaches on the intersection of feminist theory, social and cultural studies of science and technology, and critical studies of colonial discourse. Her recent articles include "Manifesto for Cyborgs," *Socialist Review* 80 (1985): 65–108; "Teddy Bear Patriarchy," *Social Text* 11 (Winter 1984/85): 19–64; and "Primatology Is Politics by Other Means," in *Feminist Approaches to Science,* ed. Ruth Bleier (New York: Pergamon Press, 1986). She is currently coediting a book with Katie King on science fiction, the fictions of science, and feminist theory.

SANDRA HARDING is professor of philosophy and director of women's studies at the University of Delaware. She is the author of *The Science Question in Feminism* (Ithaca, N.Y.: Cornell University Press, 1986), coeditor with Merrill Hintikka of *Discovering Reality: Feminist Perspectives on Epistemology, Metaphysics, Methodology and Philosophy* (Dordrecht: D. Reidel Publishing Co., 1983), and the editor of *Feminism and Methodology* (Bloomington: Indiana University Press, 1987).

JOANNA HUBBS is associate professor of cultural history at Hampshire College. She is interested in the function of myth in the perception of a cultural identity. For the last decade she has worked primarily on a book about the feminine bias of Russian culture in history. Her publications also include articles in the journals *South Carolina Review* and *Fireweed* and in the collection *Mother Worship: Themes and Variations,* ed. James Preston (Chapel Hill: University of North Carolina Press, 1982). She is presently working on a novel.

EVELYN FOX KELLER is professor of mathematics and humanities at Northeastern University. She is the author of *A Feeling for the Organism: The Life and Work of Barbara McClintock* (San Francisco: W. H. Freeman, 1983) and *Reflections on Gender and Science* (New Haven, Conn.: Yale University Press, 1985). Her current research is on the language of individualism in evolutionary discourse.

HELEN H. LAMBERT is associate professor of biology at Northeastern University. Her laboratory interests are sexual differentiation, development, and behavior in nonhuman vertebrates. She is working on a book about human sex differences and their significance for feminist and antifeminist ideology.

HELEN LONGINO is associate professor of philosophy at Mills College. She has published articles on various aspects of scientific methodology, including science and feminist research, objectivity, and the nature of evidence. Her most recent publication is "Science Overrun" in *Governing Science and Technology in a Democracy,* ed. Malcolm Goggin (Knoxville: University of Tennessee Press, 1986). She is coeditor with Valerie Miner of *Competition among Women: A Feminist Forum* (New York: Feminist Press, in press). She is also completing a book on the role of ideology and community values in scientific inquiry; this work is prefigured in "Can There Be a Feminist Science?" Working Paper no. 163 (Wellesley College Center for Research on Women, Spring 1986).

JUDITH A. MC GAW is associate professor of the history of technology in the department of history and sociology of science at the University of Pennsylvania. She has published a number of articles on women and technological change in history, and she is the author of *Most Wonderful Machine: Mechanization and Social Change in Berkshire Paper Making, 1801–1885* (Princeton, N.J.: Princeton University Press, 1987). She is currently studying the domestic origins of the American Industrial Revolution in Philadelphia.

PATRICIA Y. MILLER is in the department of sociology at Smith College.

JEAN F. O'BARR is director of women's studies at Duke University and the editor of *Signs: Journal of Women in Culture and Society.* A political scientist by training, she writes in the area of women and education as well as women and politics. Her publications include *Language and Politics* (The Hague: Mouton, 1976), *Third World Women: Factors in Their Changing Status* (Durham, N.C.: Duke University Press, 1976), *Perspectives on Power: Women in Africa, Asia and Latin America* (Durham, N.C.: Duke University Press, 1982), and an introductory essay in Muthoni Kikimani's *Passbook Number F.47927: Women and the Mau Mau in Kenya* (London: Macmillan, 1985).

HILARY ROSE is professor of social policy and director of the West Yorkshire Centre for Research on Women at the University of Bradford, England. Her research in social policy and science is described in her contributions to *Feminist*

Approaches to Science, ed. Ruth Bleier (New York: Pergamon Press, 1986) and to *What Is Feminism?* ed. Juliet Mitchell and Ann Oakley (Oxford: Basil Blackwell, 1986). Her article "Victorian Values in the Test Tubes: The Politics of the Science and Technology of Reproduction" is forthcoming in *Reproduction Technologies,* ed. Michelle Stanworth (Oxford: Polity, 1987).

MARGARET W. ROSSITER is currently National Science Foundation visiting professor of the history of science at Cornell University. She is working on a sequel to her *Women Scientists in America: Struggles and Strategies to 1940* (Baltimore: Johns Hopkins University Press, 1982). A recent article, "Women and the History of Scientific Communication," appeared in the *Journal of Library History, Philosophy, and Comparative Librarianship* 21, no. 1 (Winter 1986): 39–59.

LONDA SCHIEBINGER is a National Endowment for the Humanities fellow and a visiting scholar in the department of history at New York University and at the Barnard College Women's Center at Columbia University. Her articles have been published in the journal *Isis* and in *A History of Women Philosophers,* vol. 3, ed. Mary Ellen Waithe (Dordrecht: Nijhoff, 1987). Her article "Skeletons in the Closet: The First Illustrations of the Female Skeleton in Eighteenth-Century Anatomy," which appeared in *Representations* 14 (Spring 1986): 42–82, will also be published in *Sexuality and the Social Body in the Nineteenth Century,* ed. Thomas Laqueur and Catherine Gallagher (Berkeley: University of California Press, 1987). She is also completing a book about women and the origins of modern science.

STEPHANIE A. SHIELDS is associate professor of psychology at the University of California at Davis. Her research is concerned with the impact of socialization on felt emotion. She has published articles on the history of the study of gender. Among her articles on emotion is "Women, Men, and the Dilemma of Emotionality," which will appear in *Review of Personality and Social Psychology,* vol. 7, ed. Phillip Shaver and C. Hendricks (Beverly Hills, Calif.: Sage Publications, 1987).

"Signs" Articles of Related Interest

Aldrich, Michele L., "Women in Science: Review Essay," 4, no. 1 (1978): 126–35.

Baker, Susan W., "Biological Influences on Human Sex and Gender: Review Essay," 6, no. 1 (1980): 80–96.

Bleier, Ruth, "Bias in Biological and Human Sciences: Some Comments," 4, no. 1 (1978): 159–62.

Bleier, Ruth, "Comment on Haraway's 'In the Beginning Was the Word: The Genesis of Biological Theory,' " 7, no. 3 (1982): 725–26.

Bremner, William J., and David M. de Kretser, "Contraceptives for Males," 1, no. 2 (1975): 387–96.

Briscoe, Anne M., "Phenomenon of the Seventies: The Women's Caucuses," 4, no. 1 (1978): 152–58.

Bullough, Vern L., "Merchandizing the Sanitary Napkin: Lillian Gilbreth's 1927 Survey: Archives," 10, no. 3 (1985): 615–27.

Deacon, Desley, "Political Arithmetic: The Nineteenth-Century Australian Census and the Construction of the Dependent Woman," 11, no. 1 (1985): 27–47.

Friedman, Richard C., et al., "Behavior and the Menstrual Cycle," 5, no. 4 (1980): 719–38.

Goodman, Madeleine, "Toward a Biology of Menopause," 5, no. 4 (1980): 739–53.

Gross, Harriet Engel, et al., "Considering 'A Biosocial Perspective on Parenting,' " 4, no. 4 (1979): 695–717.

Haraway, Donna, "Animal Sociology and a Natural Economy of the Body Politic, Part II: The Past Is the Contested Zone: Human Nature and Theories of Production and Reproduction in Primate Behavior Studies," 4, no. 1 (1978): 37–60.

Haraway, Donna J., "In the Beginning Was the Word: The Genesis of Biological Theory," 6, no. 3 (1981): 469–81.

Hoffmann, Joan C., "Biorhythms in Human Reproduction: The Not-So-Steady States," 7, no. 4 (1982): 829–44.

Kohlstedt, Sally Gregory, "In from the Periphery: American Women in Science, 1830–1880," 4, no. 1 (1978): 81–96.

Lantz, Alma, "Strategies to Increase the Number of Women in Science," 5, no. 1 (1979): 186–88.

Leavitt, Judith Walzer, "Birthing and Anesthesia: The Debate over Twilight Sleep," 6, no. 1 (1980): 147–64.

Lowe, Marian, "Sociobiology and Sex Differences: Viewpoint," 4, no. 1 (1978): 118–25.

MacCormack, Carol P., "Biological Events and Cultural Control," 3, no. 1 (1977): 93–100.

Magner, Lois N., "Women and the Scientific Idiom: Textual Episodes from Wollstonecraft, Fuller, Gilman, and Firestone," 4, no. 1 (1978): 61–80.

Mandelbaum, Dorothy Rosenthal, "Women in Medicine: Review Essay," 4, no. 1 (1978): 136–45.

Martin, Joan C., "Drugs of Abuse during Pregnancy: Effects upon Offspring Structure and Function," 2, no. 2 (1976): 357–68.

Monteiro, Lois A., "On Separate Roads: Florence Nightingale and Elizabeth Blackwell: Archives," 9, no. 3 (1984): 520–33.

Oakley, Ann, "A Case of Maternity: Paradigms of Women as Maternity Cases," 4, no. 4 (1979): 607–31.

Schwartz, Neena B., "Comment on Bremner and de Kretser's 'Contraceptives for Males,' " 2, no. 1 (1976): 247–48.

Sloan, Jan Butin, "The Founding of the Naples Table Association for Promoting Scientific Research by Women, 1897: Archives," 4, no. 1 (1978): 208–16.

Tanner, Nancy, and Adrienne Zihlman, "Women in Evolution, Part I: Innovation and Selection in Human Origins," 1, no. 3, pt. 1 (1976): 585–608.

Verbrugge, Martha H., "Women and Medicine in Nineteenth-Century America," 1, no. 4, (1976): 957–72.

Vetter, Betty M., "Women in the Natural Sciences: Review Essay," 1, no. 3, pt. 1 (1976): 713–20.

Zihlman, Adrienne L., "Women and Evolution, Part II: Subsistence and Social Organization among Early Hominids," 4, no. 1 (1978): 4–20.

Note.—The above articles appeared in various issues of *Signs*, from volume 1, number 1 through volume 12, number 2.

Index